C0-EDX-996

ADVANCES IN CHEMICAL ENGINEERING
Volume 32

Chemical Engineering Kinetics

ADVANCES IN
CHEMICAL ENGINEERING

Editor-in-Chief

GUY B. MARIN

Department of Chemical Engineering
Ghent University
Ghent, Belgium

Editorial Board

DAVID H. WEST

Research and Development
The Dow Chemical Company
Freeport, Texas, U.S.A.

PRATIM BISWAS

Department of Chemical and Civil Engineering
Washington University
St. Louis, Missouri, U.S.A.

ADVANCES IN CHEMICAL ENGINEERING

Volume 32

Chemical Engineering Kinetics

Edited by

GUY B. MARIN
*Department of Chemical Engineering,
Ghent University,
Ghent, Belgium*

Amsterdam – Boston – Heidelberg – London – New York – Oxford
Paris – San Diego – San Francisco – Singapore – Sydney – Tokyo
Academic Press is an imprint of Elsevier

Academic Press is an imprint of Elsevier
84 Theobald's Road, London WC1X 8RR, UK
Radarweg 29, PO Box 211, 1000 AE Amsterdam, The Netherlands
Linacre House, Jordan Hill, Oxford OX2 8DP, UK
30 Corporate Drive, Suite 400, Burlington, MA 01803, USA
525 B Street, Suite 1900, San Diego, CA 92101-4495, USA

First edition 2007

Copyright © 2007 Elsevier Inc. All rights reserved

No part of this publication may be reproduced, stored in a retrieval system
or transmitted in any form or by any means electronic, mechanical, photocopying,
recording or otherwise without the prior written permission of the publisher

Permissions may be sought directly from Elsevier's Science & Technology Rights
Department in Oxford, UK: phone (+44) (0) 1865 843830; fax (+44) (0) 1865 853333;
email: permissions@elsevier.com. Alternatively you can submit your request online by
visiting the Elsevier web site at http://elsevier.com/locate/permissions, and selecting
Obtaining permission to use Elsevier material

Notice
No responsibility is assumed by the publisher for any injury and/or damage to persons
or property as a matter of products liability, negligence or otherwise, or from any use
or operation of any methods, products, instructions or ideas contained in the material
herein. Because of rapid advances in the medical sciences, in particular, independent
verification of diagnoses and drug dosages should be made

ISBN: 978-0-12-373899-8
ISSN: 0065-2377

For information on all Academic Press publications
visit our website at books.elsevier.com

Printed and bound in USA

07 08 09 10 11 10 9 8 7 6 5 4 3 2 1

**Working together to grow
libraries in developing countries**

www.elsevier.com | www.bookaid.org | www.sabre.org

ELSEVIER   BOOK AID International   Sabre Foundation

CONTENTS

CONTRIBUTORS   ix
PREFACE   xi

Predictive Kinetics: A New Approach for the 21st Century
WILLIAM H. GREEN, JR.

I. Introduction 2
   A. The Historical Applications of Kinetic Modeling 2
   B. The Goal of Chemical Kinetic Modeling 5
II. Construction of Predictive Chemical Kinetic Models. 7
   A. The Challenge of Documenting Large Simulations 8
   B. Data Models for Chemical Kinetics 9
   C. Automated Construction of Lists-of-Reactions. 11
   D. A New Data Model for Chemical Reactions and Properties. 13
   E. The Reaction Mechanism Generator (RMG). 26
   F. Applications of RMG 27
III. Efficiently and Accurately Solving Large Kinetic Simulations 29
   A. Fast Solution of Large Systems of Chemistry Equations Using Sparsity 31
   B. Using Reduced Chemistry Models in Multidimensional Simulations without Introducing Error 32
   C. Constructing Reduced Chemistry Models Satisfying Error Bounds Over Ranges . 34
IV. Are Model Predictions Consistent with Experimental Data? 38
   A. How to Prove Inconsistency Even if Some Parameters Are Highly Uncertain . . . 39
   B. How Standard Operating Practice Must Change in 21st Century. . . . 41
   C. Estimating Error Bars on Model Predictions 42
V. Summary and Outlook 46
   Acknowledgments 47
   Nomenclature 47
   References 48

Kinetic Modelling of Pyrolysis Processes in Gas and Condensed Phase
MARIO DENTE, GIULIA BOZZANO, TIZIANO FARAVELLI, ALESSANDRO MARONGIU, SAURO PIERUCCI AND ELISEO RANZI

I. Introduction 52
II. Kinetic Modelling of Pyrolysis Reactions. 55
   A. Automatic Generation of Pyrolysis Mechanism 64
   B. Pyrolysis of Large Hydrocarbons 72

		C.	Detailed Characterization of Liquid Feeds	90
		D.	Extension of the Modelling to Liquid-Phase Pyrolysis	96
III.	Fouling Processes: Formation of Carbonaceous Deposits and Soot Particles			100
		A.	Fouling and Coking Mechanisms in Pyrolysis Coils and TLE's	101
		B.	Evolution of Structure and Properties of the Deposit	107
		C.	Soot Formation	114
IV.	Applications			124
		A.	Steam Cracking Process and Pyrolysis Coils	124
		B.	Visbreaking and Delayed Coking Processes	129
		C.	Thermal Degradation of Polymers	136
V.	Conclusions			150
	List of Symbols			151
	Appendix 1. MAMA Program: Automatic Generation of Primary Lumped Reactions			152
		A.1.	Evaluation of End-Product Composition	152
		A.2.	Lumping of Components	155
		A.3.	Database Structure	157
		A.4.	Molecule Homomorphism Algorithm	159
		A.5.	Schematic Functionality of the Kinetic Generator	160
	References			161

Kinetic Models of C_1–C_4 Alkane Oxidation as Applied to Processing of Hydrocarbon Gases: Principles, Approaches and Developments

MIKHAIL SINEV, VLADIMIR ARUTYUNOV AND ANDREY ROMANETS

I.	Introduction: Problem Statement		169
	A.	Modeling of Complex Reacting Systems: Purposes and Expectations	172
	B.	Circumscription of Subject and Area of Parameters	176
	C.	Accounting of Heterogeneous Processes	179
	D.	Accuracy of Modeling and Influence of Macro-Kinetic Parameters	183
II.	Alternative Approaches to Modeling		187
	A.	"Additive" Models	189
	B.	"Combinatorial" Models	193
	C.	Ruling Principles for Comprehensive Modeling	194
	D.	Reduction of Models	200
	E.	Modeling of Heterogeneous–Homogeneous Catalytic Reactions	201
III.	"Elemental Base"		203
	A.	Limitations of Species Taken into Account	203
	B.	Elementary Reactions	206
	C.	Selection of Rate Constants	210
	D.	Pressure-Dependent Reactions	210
	E.	Heterogeneous–Homogeneous Catalytic Reactions on Oxide Catalysts	213
	F.	Heterogeneous–Homogeneous Catalytic Reactions on Metal Catalysts	227
IV.	Modeling and Experimentation		231
	A.	Comparison with Experiment	231
	B.	Requirements of Experiment	233
V.	Some Examples and Comments		237
	A.	C_1–C_2 Joint Description	237
	B.	Expansion on Higher Hydrocarbons	239

	C. Reactions between Alkyl Radicals and Oxygen and Transformations of Alkylperoxy Radicals	243
	D. Capabilities of Process Influencing, Governing and Design	246
VI.	Concluding Remarks	250
	Acknowledgments	251
	List of Symbols	251
	References	253

Kinetic Methods in Petroleum Process Engineering

Pierre Galtier

I.	Introduction	260
II.	Kinetic Modelling by Single-Events	269
	A. Introduction	269
	B. Bifunctional Catalysis Mechanisms	270
III.	Generation of Reaction Networks	271
	A. Computer Representation of Species and Chemical Reactions	273
	B. Results Obtained for Computer Generation of Networks from nC8 to nC15	273
IV.	Kinetics by Single-Events	273
	A. Single-Events Microkinetic Concept	273
	B. Separation of Chemical and Structural Contributions	275
	C. Free Enthalpies of Reactants and Activated Complexes	276
	D. Thermochemical Restrictions and Constraints	278
V.	Late Lumping Kinetic Model	279
	A. Three-Phase Model	280
	B. Catalytic Act	280
	C. Composition of the Reaction Intermediates	281
	D. Lumping by Families	281
	E. Lumped Kinetics	282
	F. Summing Up	285
VI.	Extrapolation to Heavy Cuts	286
	A. Estimation of Kinetic Parameters	286
	B. Extension to Heavy Paraffins	289
	C. Extrapolation Capacities in Number of Carbon Atoms: Heavy Paraffinic Waxes	294
VII.	Perspectives	299
	References	302
	Further Reading	304

Index 305
Contents of Volumes in This Serial 313
Please See Color Plate Section in The Back of This Book

CONTRIBUTORS

Numbers in parentheses indicate the pages on which the authors' contribution begin.

V. ARUTYUNOV, *Laboratory of Hydrocarbon Oxidation, Semenov Institute of Chemical Physics, Moscow, Russia* (167)

G. BOZZANO, *Dipartimento di Chimica Materiali e ingegneria Chimica, Politecnico di Milano, Piazza L. da Vinci 32, 20133 Milano, Italy* (51)

M. DENTE, *Dipartimento di Chimica Materiali e ingegneria Chimica, Politecnico di Milano, Piazza L. da Vinci 32, 20133 Milano, Italy* (51)

T. FARAVELLI, *Dipartimento di Chimica Materiali e ingegneria Chimica, Politecnico di Milano, Piazza L. da Vinci 32, 20133 Milano, Italy* (51)

P. GALTIER, *Institut Français du Pétrole, IFP-Lyon, BP.3, F-69390 Vernaison, France* (259)

W. H. GREEN, JR., *Department of Chemical Engineering, Massachusetts Institute of Technology, Cambridge, MA 02139, USA* (1)

A. MARONGIU, *Dipartimento di Chimica Materiali e ingegneria Chimica, Politecnico di Milano, Piazza L. da Vinci 32, 20133 Milano, Italy* (51)

S. PIERUCCI, *Dipartimento di Chimica Materiali e ingegneria Chimica, Politecnico di Milano, Piazza L. da Vinci 32, 20133 Milano, Italy* (51)

E. RANZI, *Dipartimento di Chimica Materiali e ingegneria Chimica, Politecnico di Milano, Piazza L. da Vinci 32, 20133 Milano, Italy* (51)

A. ROMANETS, *Russian Research Center Kurchatov Institute, Moscow, Russia* (167)

M. SINEV, *Laboratory of Heterogeneous Catalysis, Semenov Institute of Chemical Physics, Moscow, Russia* (167)

PREFACE

Understanding and modeling the kinetics of chemical reactions certainly is crucial to any research and development effort aimed at process optimization and innovation. This issue provides four complementary points of view. It reflects the state-of-the-art as well as views on the way to proceed by reporting on the efforts of a, I hope representative, sample of research and development groups. I was in particular happy to find a group having close commercial ties with the process industry willing to communicate some of their results and methods. In order to stress as much as possible the generic nature of the messages that are conveyed, the focus is not on catalytic processes.

The blessing of the latter, i.e. the possibility to suppress specific undesired side reactions, renders any kinetic model strongly catalyst dependent which is a curse to the modeler. Accounting in a clever way for so-called catalyst descriptors deserves a separate issue.

A first contribution by W. H. Green Jr. from Massachusetts Institute of Technology sets the scene. The author advocates a paradigm shift in chemical kinetics from "postdictive" to predictive models. The latter would provide a sound basis for process and product design and, hence, not only accelerate the pace of innovation but also allow an *a priori* assessment of the societal, e.g. environmental, impact of new technologies. Predictive modeling requires further progress at three different levels. The construction of the models, the numerical solution of the conservation equations accounting also for the appropriate transport phenomena and finally the validation of the models are discussed.

A "data model", i.e. a standard way of expressing information different from the classical lists-of-reactions is proposed. The core of it is constituted by the concept of functional groups well known from organic synthesis. The latter is implemented in "hierarchical functional group tree structure(s) which can easily incorporate new chemistry information" because "it will be many decades before the functional group trees will be at all complete". Focusing on functional groups rather than elementary reactions allows to calculate rate coefficients based on "an extension of the thermochemical group-additivity to transition states". The elementary reactions are generated automatically by computer codes in which "reaction recipes" based on the reactivity of the functional groups are formalized. Criteria and computational techniques which allow to have the size of a reaction network large enough to account for the important features of a process but not larger than that are discussed. The importance of accounting for uncertainty is stressed throughout the paper. The propagation of uncertainties in both the reaction network and the rate coefficients via the solution of the model equations is presented in a very systematic way. This allows to give a statistically justified answer to the question of validation: "Are model predictions consistent with experimental data?"

At each of the levels addressed the author refers to available software and databases, which makes this chapter visionary as well as practical.

The contribution from the Politecnico di Milano reports on the tremendous experience accumulated over the years by the team of Mario Dente and Eliseo Ranzi in the field of steam cracking, one of the largest scale production processes of the petrochemical industry. The authors have been pioneering the substitution of global power law rates by what today is often called microkinetics, i.e. by kinetics accounting fully for the elementary reactions and the radicalar reaction intermediates. An overview is given of the challenges that have to be overcome in order to describe the kinetics of complex mixtures. The contribution is not limited to clearly explaining the developed methodology. It also explicitly indicates the introduction of simplifications. The latter are required not only to maintain the computational efforts within reasonable limits but also, and more fundamentally, because of the uncertainties concerning feed composition. It is fascinating to see how the introduced simplifications are based on insight in the reaction mechanism. Nevertheless, the authors illustrate the generic nature of their methodology by applying it to other free radicalar processes such as polyaromatic, i.e. coke or soot, formation and thermal degradation of polymers.

The Russian school of chemical kinetics is represented by a chapter on oxidation of alkanes by Mikhail Sinev and coworkers from Moscow. Their contribution addresses more "philosophical" issues. What should modeling be aimed at? According to the authors, obtaining insights in the important reaction paths rather than a "perfect" description of experimental data should be pursued. The latter are considered to be too often affected by phenomena such as transport limitations, certainly during the oxidation of alkanes. Most of the examples relate to the oxidation of methane to methanol/formaldehyde, to synthesis gas or, by oxidative coupling, to ethene. The importance of the interaction between homogeneous gas phase reactions and heterogeneously catalyzed steps is stressed. For the homogeneous reactions the use of kinetic parameters obtained from quantum chemical calculations is advocated. Although some examples are given, it is realized that this is much less within reach for the elementary reactions involving the catalytic active sites. The possible optimization of either the process conditions or the catalyst formulation for oxidative coupling of methane based on the insights provided by kinetic modeling is illustrated.

The last chapter gives an indication about the state-of-the-art in an industrial environment. The author Pierre Galtier reports on the activities at the Industrial Studies and Development Center of the French Petroleum Institute (IFP) by a "Kinetics and modeling" group of several researchers performed over a period of more than 10 years. The aim of the research work is to develop basic knowledge on the kinetics of the main reactions involved in refining and petrochemical processes. This methodological research is designed to reduce development costs by constructing detailed and predictive kinetic models. The results of kinetic studies conducted at laboratory and pilot scale allow to improve the economics of industrial processes and to minimize the risks inherent to industrialization of new processes. Galtier focuses on the so-called "single-event" methodology as applied to a bifunctionally, i.e. involving both a (de)hydrogenating and a carbon skeleton conversion function, catalyzed process: the hydrocracking of heavy crude oil cuts. The second function is provided by acid sites, which lead to reactive carbenium ion intermediates. As carbenium chemistry is well known, its formalization in reaction recipes is straightforward. It turns out that the number of elementary reaction *families* to be considered and, hence, the number of kinetic parameters is relatively small. Still, the number of elementary reactions becomes huge for heavy crude oil cuts.

The author shows how the single-event methodology can be extrapolated to heavy cuts without generating the corresponding reaction network.

Finally, let me point your attention to volume 28 of Advances in Chemical Engineering on "Molecular Modeling and Theory in Chemical Engineering" published five years ago. Several chapters elaborate on the use of quantum chemical calculations to obtain thermochemical and kinetic data. This is the only reason why I opted not to include similar material in this volume. Clearly, *ab initio* computational techniques, or at least the results of the latter, should be part of the toolkit to obtain chemical engineering kinetics.

Guy B. Marin
Ghent, Belgium
September 2006

PREDICTIVE KINETICS: A NEW APPROACH FOR THE 21ST CENTURY

William H. Green, Jr.[*]

Department of Chemical Engineering, Massachusetts Institute of Technology, Cambridge, MA 02139, USA

I. Introduction	2
A. The Historical Applications of Kinetic Modeling	2
B. The Goal of Chemical Kinetic Modeling	5
II. Construction of Predictive Chemical Kinetic Models	7
A. The Challenge of Documenting Large Simulations	8
B. Data Models for Chemical Kinetics	9
C. Automated Construction of Lists-of-Reactions	11
D. A New Data Model for Chemical Reactions and Properties	13
E. The Reaction Mechanism Generator (RMG)	26
F. Applications of RMG	27
III. Efficiently and Accurately Solving Large Kinetic Simulations	29
A. Fast Solution of Large Systems of Chemistry Equations Using Sparsity	31
B. Using Reduced Chemistry Models in Multidimensional Simulations without Introducing Error	32
C. Constructing Reduced Chemistry Models Satisfying Error Bounds Over Ranges	34
IV. Are Model Predictions Consistent with Experimental Data?	38
A. How to Prove Inconsistency Even if Some Parameters Are Highly Uncertain	39
B. How Standard Operating Practice Must Change in 21st Century	41
C. Estimating Error Bars on Model Predictions	42
V. Summary and Outlook	46
Acknowledgments	47
Nomenclature	47
References	48

[*]Corresponding author. E-mail: balkwill@MIT.EDU

Abstract

The capability to reliably predict the behavior of reactive chemical systems would allow rational *a priori* design of chemical reaction systems. Recent progress toward overcoming several technical obstacles to predictive kinetics for homogeneous gas-phase systems is reviewed. The focus is on (1) updates to the fundamental data model used in kinetic modeling, (2) methods for solving large kinetic simulations efficiently without relying on uncontrolled approximations, and (3) methods for determining whether or not the model predictions are consistent with experimental data. Appropriate handling and archiving of experimental data from different sources, and of the many uncertainties in the data embedded throughout the kinetic models, is a major challenge facing the kinetics community as kinetics becomes a predictive science.

I. Introduction

Chemical kinetic modeling has been one of the pillars of chemical engineering since the field was founded. In many respects, it is the aspect that distinguishes chemical engineering from other fields of engineering. However, despite a huge amount of work by thousands of researchers, we still have a very limited ability to predict what will happen in a newly proposed reactive system, even in well-characterized gas-phase systems with simple fluid mechanics. A number of technical problems significantly impeded progress in kinetics during the 20th century. Here we briefly review the history and the goals of chemical kinetic modeling, and highlight a few of the technical areas where significant advances have been made in the first years of the 21st century.

A. THE HISTORICAL APPLICATIONS OF KINETIC MODELING

1. Predictive vs. Postdictive Models

Chemical kinetic models play multiple roles. In the 20th century, these models were most often used to assist in the interpretation of experimental data, and for the interpolations used in process optimization and control. Kinetic models also played an essential role in the process of generalizing new chemical knowledge from experimental results. In both of these applications, a large amount of experimental data were measured first and then kinetic models were used to rationalize or interpolate between these experimental data. It would not be too unfair to characterize most of this work as "postdictive", i.e. the modelers were not able to say anything interesting until *after* a considerable amount of experimental work had been completed.

Of course, chemical kinetic models would be even more useful if they could accurately **predict** the behavior of reacting systems under conditions significantly different from those that have already been measured. If these extrapolative predictions were accurate enough, chemical kinetic models could become valuable tools in process and product design, and by reducing the need to do so many experiments in order to gain a small amount of information, the models could accelerate the pace of innovation. Reliable predictive kinetic models would be particularly helpful in situations where it is impractical to do the experiments, e.g. in the public policy arena, where a failed experiment could be prohibitively costly, or in situations where the experiment is impossible (e.g. predicting what happens in very slow or very fast processes).

In recent years, it has become possible to extrapolate accurately using detailed chemical kinetic models to predict quantitatively the behavior of some rather complicated chemical systems. The most famous examples of this success are the detailed atmospheric chemistry models whose predictions underlie the Montreal Protocol on ozone-depleting chemicals. However, these atmospheric chemistry models were developed through a huge international effort over several decades, based heavily on a large number of laboratory experiments. Much more rapid and efficient methods of model development are required for detailed predictive chemical kinetics to become a practical everyday design tool for chemical engineering.

2. Technical Hurdles in the 20th Century

Several technical issues significantly hampered progress in chemical kinetic modeling in the 20th century; among these were serious difficulties solving kinetic simulations numerically, difficulties constructing and checking (and peer-reviewing) large simulations, problems with estimating reaction rates when thermochemistry and experimental data were missing, and problems checking whether model predictions were consistent with experimental data.

First, significant numerical problems plagued the kinetic models. Almost all chemical kinetic systems are intrinsically "stiff", with a large separation of timescales between the most reactive and least reactive species in a mixture. Although a number of clever approximations were used to circumvent this problem, the situation was highly unsatisfactory (e.g. it was very difficult to prove that the approximations were valid) until William Gear invented the first algorithm capable of numerically solving stiff systems of differential equations with error control in the 1970s (Gear, 1971). Also, most chemical kinetic systems of interest to engineers involve a relatively large number of species, which made them hard to handle until computers with significant core memory became widely available ~1980. The development of the CHEMKIN® software in the 1980s, which dealt with the most common data-handling issues and also incorporated a reliable stiff differential equation solver that evolved from Gear's algorithm, was a major step-forward in chemical kinetic modeling (Kee et al., 1989). Toward the end of the last century, modelers benefited tremendously from the rapid improvements in computer hardware and from major advances in the capability to solve multidimensional, multi-species kinetic simulations. But there are still many important reactor types whose simulations are plagued by numerical issues and by hardware limitations; at present there are only a handful of well-tested widely distributed solvers with error control, and these solvers work only for specific idealized reactor types.

As the numerical solvers have become capable of solving rather complicated chemical kinetic simulations, modelers have had to accept that we do not really know the chemistry of most processes as well as one might hope, and in some cases the reacting mixtures are so complicated that it is doubtful anyone could write down all the reactions occurring, even if we did know all the chemistry. (For complicated reactor geometries, there is a similar problem associated with constructing a suitable mesh for solving the transport equations.) The problem of complex chemistry was recognized even before CHEMKIN became widely available, and has led to the development of dozens of automated mechanism-construction programs over the past two decades (Blurock, 2004a, b; Broadbelt et al., 1994; Chevalier et al., 1990; Chinnick et al., 1988; Dente et al., 1979; DiMaio and Lignola, 1992; Hillewaert et al., 1988; Kojima, 1994; Matheu et al., 2003a; Prickett and Mavrovouniotis, 1997a, b, c; Quann and Jaffe, 1992; Ranzi et al., 1995; Tomlin et al., 1997; Warth et al., 2000; Zarth et al., 2002). These programs have been successful in modeling some systems, but as discussed below there are many significant unresolved technical issues which have prevented these programs from having a broad impact on the field. A very important problem is that most of the model-construction software is not sufficiently flexible; in particular, it is rather hard to add additional chemistry and modify the rate parameters as knowledge improves. As a result, the software can easily become obsolete. Also, it is very difficult to check or peer-review large complicated simulations; this combined with the tendency for postdictive modeling makes many researchers skeptical about the usefulness of large simulations.

For most of the 20th century, the only solid basis for assigning a numerical value to a rate parameter was experimental data. However, in most experimental systems, the number of parameters determinable from the experimental rate data is much less than the number of rate parameters needed in the chemical kinetic model. Benson (1976) and others proposed methods to estimate some rate parameters, (primarily by analogy) to reactions whose rate constants had been measured, but there were still a large number of reactions that could not be estimated at all, and it was hard to know how confidently one could use the analogy-based estimates. Of necessity, researchers developing large kinetic models developed rules-of-thumb to estimate reaction rates for the reaction families of greatest importance in the systems they were simulating, but this was necessarily very *ad hoc* in the absence of any data with which to check the estimates. In the last decade of the 20th century, computation of reaction rates (and thermochemistry) using quantum chemistry became feasible, making it possible to check the accuracy of the analogy-based estimates. With the quantum chemistry techniques available at that time, this was most feasible for thermochemistry, and several researchers, notably Melius (Melius and Allendorf, 2000; Zachariah and Melius, 1998) and Bozzelli (Lay and Bozzelli, 1997a, b; Lay *et al.*, 1997) and their collaborators, used quantum chemical calculations to significantly improve the quality of thermochemical (and some rate) estimates used in combustion models. By the end of the 20th century, similar techniques were being widely used to improve barrier height and A-factor estimates, usually using transition state theory (Truhlar *et al.*, 1996). Sumathi and Green (2002) reviewed the state of the art in rate estimation for large-scale kinetic modeling around the turn of the century.

Finally, there are multiple hurdles that made it difficult to appropriately compare kinetic model predictions with experimental data in the 20th century, and which still make it very difficult today. As discussed in Section IV below, there are some numerical issues which make it hard to determine definitively whether or not a model is consistent with a data set. If the model is consistent, one should be able to refine the model parameters using the experimental data, but this is numerically challenging when one is working with large-scale non-linear kinetic simulations and large data sets. Response surface methods were used successfully for this parameter refinement step toward the end of the 20th century (Aghalayam *et al.*, 2000; Frenklach *et al.*, 1992).

A bigger problem than the numerical issues is the difficulty in coming up with the data and the error bars needed to make a meaningful comparison between the model and experiment. Very often one can no longer locate all the original experimental data in order to make a comprehensive comparison, and even the data one can find seldom comes with reliable error bars. Also, researchers rarely compute the error bars in a large chemical kinetic model's predictions. So, although thousands of kinetic model vs. data comparisons were published in the 20th century, in most cases these plots are insufficient to determine conclusively whether the model is really consistent with the data, and it would usually be difficult-to-impossible to determine the error bars on any of the model parameters based on the published comparison.

With this history of difficulties in mind, in the next section we reconsider what are the real goals of chemical kinetic modeling.

B. The Goal of Chemical Kinetic Modeling

The primary goal of chemical kinetic modeling is to make predictions: given our current understanding of chemistry, what do we expect to happen in a particular reacting mixture under specified reaction conditions? If desired, these

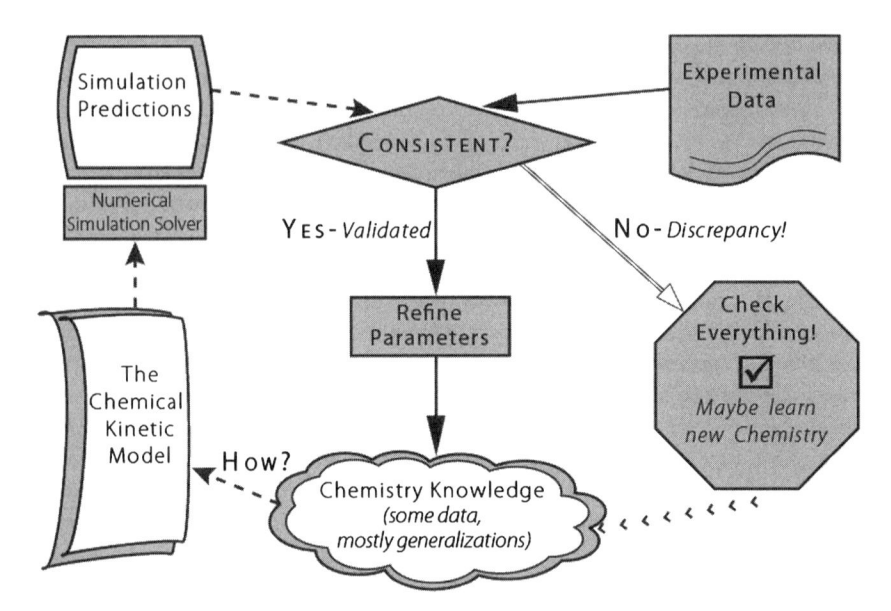

FIG. 1. The Model-Prediction-Data loop. This article is focused on the steps on the left side, from model-construction through consistency testing.

model predictions can then be quantitatively compared with experimental data, to determine if the two are consistent, Fig. 1. If the model and the data are inconsistent, at least one of them must be wrong, and we have identified an interesting opportunity to learn something new. If the model and the data are consistent within the error bars on each, then we say the model has been *validated* (never *proven*), and the experimental data can be used to refine the model parameters.

To the extent that we have confidence in the predictions of chemical kinetic models, we could then use these models to design new products, processes, and reactors to plan new experiments, to design model-based control systems and safety systems, and to inform critical business and public policy decisions. However, to develop the required level of confidence in the model predictions, the loop in Fig. 1 must be functioning effectively.

In the present paper we focus on a few of the steps involved in the left side of the loop shown in Fig. 1:

(1) Construction of predictive chemical kinetic models
(2) Efficient and accurate numerical solution of chemical kinetic simulations, and
(3) Quantitative tests whether the model and data are consistent.

Each of these three steps is addressed in a separate section below.

The right-hand half of the loop is, of course, equally important, and many of the difficult issues involved in moving from experimental kinetics data to chemical knowledge have been discussed in the literature by others (Feeley et al., 2004; Feng and Rabitz, 2004; Milford et al., 1992; Shenvi et al., 2002; Tomlin et al., 1997). The model should be tested for consistency against all relevant data, but there are significant difficulties assembling all the data in a form suitable for making this test, an issue being addressed by the PrIMe project (http://www.primekinetics.org). Combining data from different experiments can greatly increase the amount of information gained (Feeley et al., 2004). For some nice recent examples of the practical value of combining different data from different experiments, observe how the Active Tables approach has revolutionized determinations of high-accuracy molecular thermochemistry (Ruscic et al., 2004).

II. Construction of Predictive Chemical Kinetic Models

Constructing an accurate chemical kinetic model is quite a challenging task, both because "Chemistry Knowledge" is very large and somewhat amorphous, and because it often requires hundreds of differential equations and thousands of numerical parameters (rate constants, molecular thermochemistry, etc.) to accurately describe the details of the processes occurring in a reacting mixture. Constructing these models by hand is a very time-consuming and error-prone process, so over the past three decades many research groups have developed software to automate the process of constructing chemical kinetic simulations (Blurock, 2004a, b; Broadbelt et al., 1994; Chevalier et al., 1990; Dente et al., 1979; Kojima, 1994; Matheu, 2003; Prickett and Mavrovouniotis, 1997a, b, c; Quann and Jaffe, 1992; Tomlin et al., 1997; Warth et al., 2000; Zarth et al., 2002).

In brief, model-construction software works by applying reaction operators to a set of molecules, to generate new product molecules. The reaction operators break and make bonds following some pre-defined motif. Each time a new chemical reaction is constructed by this process, the corresponding rate parameters are estimated, drawing on some library of rate estimation parameters (e.g. Arrhenius parameters, or Evans–Polanyi parameters). The thermochemistry of each species is also estimated, drawing on another library of estimation parameters (e.g. Benson group values). Often there are some experimental data which can be used instead of estimates, and these can be drawn from another library. The reaction operators are then applied to the product species to generate byproducts and so on iteratively.

Computer memory limitations usually require that one terminate the iterative process before $\sim 10^6$ reactions have been generated. Several criteria have been used for terminating the model-construction process; this author prefers the rate-based species-selection method (Susnow et al., 1997). It should be noted that different model-construction software will construct different models for the same physical situation, both because different reactions/species will be included in the models and also because different estimates are made for the various rate and thermochemical parameters. Also model-construction software will often build different models for the same reactant(s) reacting under different conditions, since very different reactions might dominate under different conditions. Rather than having some gigantic model which includes everything that could take place, real-world models are quite finite, and so have limited ranges where they are accurate or valid.

Many of the technical issues involved in computer-aided model-construction have previously been reviewed by Tomlin et al. (1997). Several researchers, most notably Bozzelli, have extended Benson's method for estimating molecular thermochemistry using quantum chemistry (Lay and Bozzelli, 1997a, b; Lay et al., 1995). Sumathi and Green (2002) have discussed how quantum chemistry can supplement experiments in developing rate estimation. Matheu et al. have shown how to automate the computation of rates of chemically-activated (pressure-dependent) reactions (Matheu, 2003; Matheu et al., 2003a, b). Here we focus on a few issues which have not been so thoroughly discussed in the literature:

(1) The challenge of documenting large chemical kinetic simulations
(2) A data model for chemical reactions and molecular properties that facilitates extensibility.

A. THE CHALLENGE OF DOCUMENTING LARGE SIMULATIONS

Because the "Chemistry Knowledge" input, the procedure for developing the correct list of differential equations corresponding to this input, and the estimates of the numerical values of the rate and molecular parameters are all known imperfectly, it is quite important that each aspect be carefully documented to facilitate scientific progress. Ideally, all the assumptions and simplifications made along the way from "Chemistry Knowledge" through to quantitative predictions, Fig. 1, should be clearly and precisely documented, so each can be tested, and each number in both the inputs and the outputs should be assigned an uncertainty band. (Ideally, the correlations between all these uncertainties should also be documented, but this has seldom been done so far.) Clarity about the assumptions and simplifications would facilitate identification of the root causes why different models give different predictions, and why predictions differ from experimental results. Clear understanding and documentation of the origins of these discrepancies will dramatically improve the efficiency of the world-wide efforts to improve both chemical understanding and kinetic modeling methodology.

There is more than just scientific progress at stake. Predictions based on chemical kinetic models are increasingly used to inform major policy and business decisions, often with large impacts on society, so it is critical that the uncertainties and assumptions associated with these predictions be clearly enunciated and understood. Clarity about our current level of ignorance is essential both to avoid misleading decision makers and to facilitate future work to test the assumptions and reduce the uncertainties in the predictions.

B. Data Models for Chemical Kinetics

Clarity in expressing information (including information about assumptions and uncertainties) can be greatly facilitated by using a good "data model". A "data model" is a standard way of expressing information; it includes both the format and the relationships between various data objects.

Currently, the most popular way to document a chemical kinetic model is to list the reactions and corresponding rate parameters (typically in a CHEMKIN format (Kee *et al.*, 1989)), and separately to list the molecular parameters including enthalpies and heat capacities (typically in a NASA polynomial format).

This popular data model, Fig. 2, has many positive aspects: it unambiguously expresses all the information necessary to construct and solve the differential equations, and it is flexible enough that it can be used for a broad range of chemical systems.

However, the popular List-of-Reactions data model has several serious deficiencies. First of all, it is difficult in this data model to adequately document the origins of all the numerical parameters used in the simulation, and the associated (usually highly correlated) uncertainties in these parameters. The format is also not very user-friendly: it is almost impossible for a human to check that all the numbers and reactions in a large kinetic model expressed in this format are consistent and reasonable, much less correct. The conventional List format does not allow for graphical representation of chemical structures, leading to sometimes serious inconsistencies in naming conventions.

More fundamentally, the current data model does not provide any way to document why some chemical reactions were included in the simulation, while others were left out. Typically, a large number of assumptions about which reactions/species are likely to be important (under the reaction conditions of interest to the simulation's author) are made by the person (or computer program) who assembles the list of species and reactions, based on his or her (or its) chemistry knowledge. Because the current data model does not provide any convenient way to document these assumptions, most of these assumptions are never documented at all, and this contributes to the impression that this information is not worth recording. The fact that it is usually very difficult to uncover complete information about the assumptions behind a reaction list

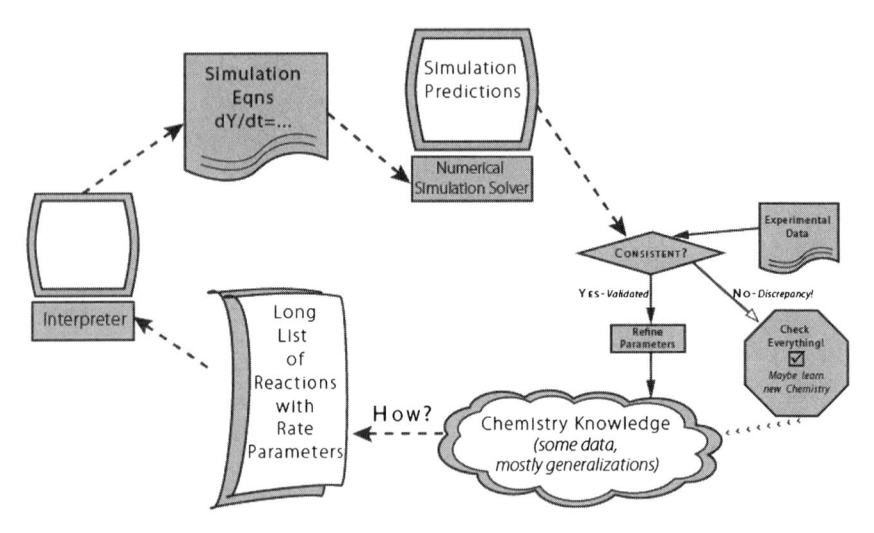

FIG. 2. Often the primary documentation of a kinetic model is the List-of-Reactions, and the steps from that List to the predictions are often well documented. However, the upstream steps that led to the List-of-Reactions are usually poorly documented.

discourages careful review of the critical steps "upstream" of the CHEMKIN-format file, Fig. 2.

It is important to realize that the CHEMKIN-format file really represents an intermediate step in the process of predicting chemical behavior. For large mechanisms, almost all of the parameter values in this data file are estimates that come from some simple (but not always documented, or consistently applied) rate estimation rule. Similarly, almost all of the numbers in the molecular property files are estimates and theoretical extrapolations. In large models, the thousands of NASA coefficients in the thermo file are typically derived from a much smaller set of Benson-type (Benson, 1976) group-additivity values. The molecular transport parameters are typically estimated even more approximately by rough analogy. So, although the typical CHEMKIN-type file contains thousands of numbers, the real information content is often much smaller: the numerical values in the simulation could usually be represented much more compactly in terms of group values and rate estimation parameters. Reducing the number of numerical parameters would greatly facilitate the important (but now rarely performed) task of comparing two competing kinetic models, to understand why they give different predictions.

Here we propose a fundamentally different data model, Fig. 3, where the crucial documentation lies upstream of the conventional List-of-Reactions. In this data model, the estimation rules for the rates and thermochemistry are the main data, supplemented by a set of numerical parameters for individual species and reactions (whose accurate values are known from experiments or

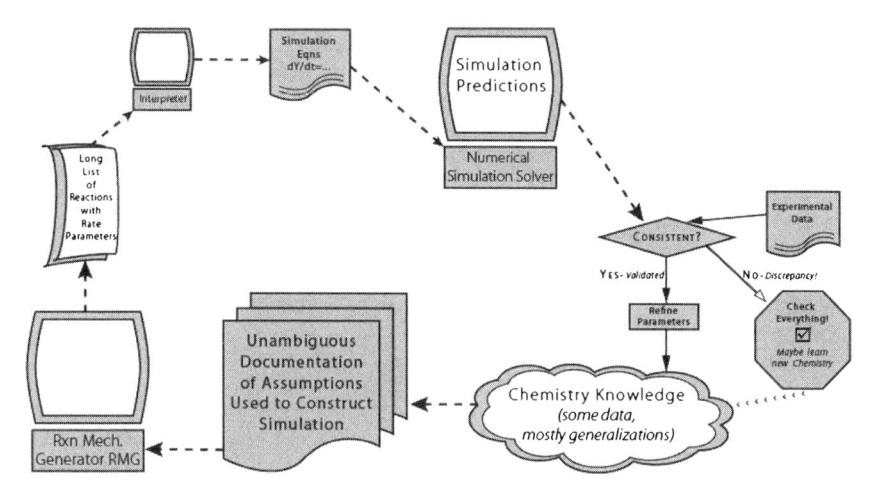

FIG. 3. In the proposed data model, the fundamental documentation is the information on how one estimates reaction rates (and molecular properties), and the other assumptions used to construct the simulation. All the subsequent steps in the process are automated, well-documented procedures.

high-quality quantum calculations). This new data model overcomes many of the objections to the List-of-Reactions data model it replaces. However, for this data model to be successful, it is necessary that unambiguous procedures, with clearly documented assumptions and tolerances, exist for converting the Estimation Rules into the corresponding List-of-Reactions, and thence (as is done currently) to Predictions. It is most convenient if these procedures can be performed automatically, using a computer.

C. AUTOMATED CONSTRUCTION OF LISTS-OF-REACTIONS

The awkwardness of the List-of-Reactions data model became apparent to many researchers modeling pyrolysis and combustion soon after CHEMKIN was developed, and since the 1980s many software packages have been developed to automate the process of constructing the long lists of reactions required, as reviewed by Tomlin *et al.* (1997). Many calculations have been published demonstrating that automated mechanism-construction can be successful. However, all of these software packages had significant flaws, many have been abandoned, and to our knowledge none of this software has yet been successfully distributed by its author to another research group.

To understand the problem, it is helpful to think about what the inputs are to these computer programs, i.e. what does one need to know in order to construct the appropriate List-of-Reactions chemistry model, and so correctly predict the behavior of a reacting system?

1. Fundamental Inputs for Chemical Kinetic Model-Construction

When one begins to construct a chemical kinetic model, there are several different types of required "input" information, Fig. 4. Obviously, one needs some specification of the initial concentrations of the reactants, and of the reaction conditions (e.g. T, P, timescale) of interest. Normally one wants to numerically solve the kinetic model to predict species and/or temperature profiles, so the inputs must also include some specification of numerical tolerances on these outputs, and options for the differential equation solver. The most complicated "input" information required to construct a kinetic model is the chemistry: what species, reactions, or reaction types will be considered? How will all the thermochemical and rate parameters be estimated?

A naïve answer is that one should just assemble the list of all the known species and all the known elementary-step reactions connecting them in the literature, and use all the literature values for the rate and thermochemical parameters. However, except in a very small number of very simple cases (e.g. H_2/O_2 combustion) this naïve approach is both inefficient and seriously inaccurate. First of all, unselectively constructing a kinetic model out of a very large set of reactions has the extremely undesirable effect of making it difficult or impossible for a human to understand/check/peer-review the model. Also, almost invariably, most of the reactions in any large compilation are unimportant under the specific reaction conditions of interest, so one is doing a lot more work both in checking and in solving the large model than is necessary. But the biggest problem is that in most branches of chemistry, only a very small fraction

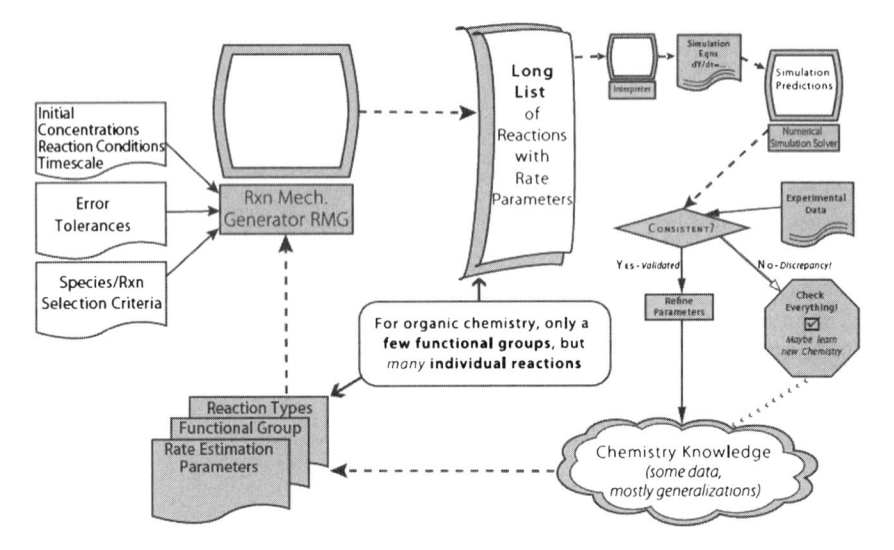

FIG. 4. The inputs needed to construct a chemical kinetic simulation. The Chemistry Knowledge underlying the model is embedded in the functional group database. This is the most important input to any large kinetic model, but also the most challenging to obtain and to document.

of the important elementary-step reactions have ever been studied at all. So models constructed by just assembling a list of previously studied reactions from the literature are almost always missing important reaction steps, and therefore give seriously erroneous predictions.

2. Estimations and Generalizations

One might ask: how can one make sensible predictions when no data on one or more of the important reaction steps can be found in the literature? The answer is that chemists have abstracted a tremendous number of generalizations from the limited number of experimental (and, recently, theoretical) studies that have been performed on particular reactions. So when it is said that a particular reaction is "well-understood", it is not meant that someone has actually measured that reaction's rate to high precision under every possible reaction condition; instead, it means that enough measurements and/or calculations have been done on that reaction or perhaps on some similar reaction so that one can reasonably generalize and confidently estimate any particular parameter that is needed. Using generalizations has a tremendous advantage compared to attempting to list every possible individual reaction: one generalization can be used to estimate the parameters of hundreds or thousands of individual reactions. In fact, most of the rate parameters in the existing large kinetic models are actually estimates from generalizations, not from individual reaction experiments or quantum calculations.

Most automated model-construction software uses only the conventional List-of-Reactions data model. Since there is no established data model for generalized rate estimates, nor for documenting the rules that determine which species and reactions are considered important, this information is typically not treated as "data" at all—instead, much of it is hard-coded into the software.

In our view, the main problem with existing software is that it is poorly designed to deal with the true complexity of the real chemistry and the corresponding level of detail needed to represent it accurately, nor was it designed to be easily modified to incorporate new or improved chemistry knowledge. Because the chemistry has many details, and is imperfectly known, it is imperative that the software be easily extensible, so that additional chemical detail can be added as desired (preferably by the wide community of chemists). In this way, the model predictions can be continuously improved, each time incorporating the latest Chemistry Knowledge. If a modeling package is not extensible in this way, it will soon become obsolete.

D. A New Data Model for Chemical Reactions and Properties

Here we propose replacing the conventional List-of-Reactions data model with a data model that accurately represents the chemistry knowledge that goes into constructing a large simulation: a relatively small number of data on individual reactions which have been extensively studied in the literature plus a relatively small number of generalized rate and thermo estimation procedures. These inputs are used to estimate most of the thousands of parameters in the kinetic differential equations. Despite early work along these lines by Blurock (1995), this approach of treating the reaction-construction procedures themselves as input data rather than as part of the software has not yet caught on with the kinetics community. However, very recently several papers have presented kinetics software based on this data model (Blurock, 2004a, b; Ratkiewicz and Truong, 2006). It appears that the reaction mechanism generator (RMG) program described below will be the first software package enabling use of the new data model to become widely disseminated throughout the kinetics community.

We have developed a convenient data format for storing the rate and thermochemistry estimation parameters, as well as a graphical user interface (GUI), which makes it much easier to improve the rate estimates, to add additional reaction types, and to handle complicated functional groups. As we have shown elsewhere (Song, 2004), one can devise transparent unambiguous procedures, with clearly documented assumptions, for converting these estimation rules into lists of reactions, and thence into differential equations and quantitative predictions as is done currently. The new database format, GUI, and algorithm for constructing the List-of-Reactions from our new tree-structured rate estimation database allows us to achieve the design shown in Fig. 2.

1. Hierarchical Tree Structure for Functional Group Parameters

The information needed to compute both the molecular thermochemistry and also the elementary-step rate parameters can typically be associated with functional groups. To use these estimation approaches during mechanism-construction, we need a reliable, efficient, and unambiguous method for rapidly identifying which functional group values should be used for any given molecule. To maximize scientific progress, we must display the functional groups and the numerical values we associate with each group in such a way that can be easily used, understood, scrutinized, criticized, amended, and extended by other researchers.

We have accomplished these goals by storing all the functional group definitions in a hierarchical tree database developed by R. Sumathi (Song, 2004). As chemistry knowledge improves, the functional groups can be specified more precisely, allowing distinctions between related functional groups.

For example, at the roughest level of approximation, all carbonyl carbons can be treated as identical: they all have similar thermochemistry and all can undergo certain types of reactions (e.g. they all can be formed by beta-scission reactions of the corresponding alkoxy radicals). However, if one looks at the situation more carefully, the carbonyl groups in ketones, ketenes, aldehydes,

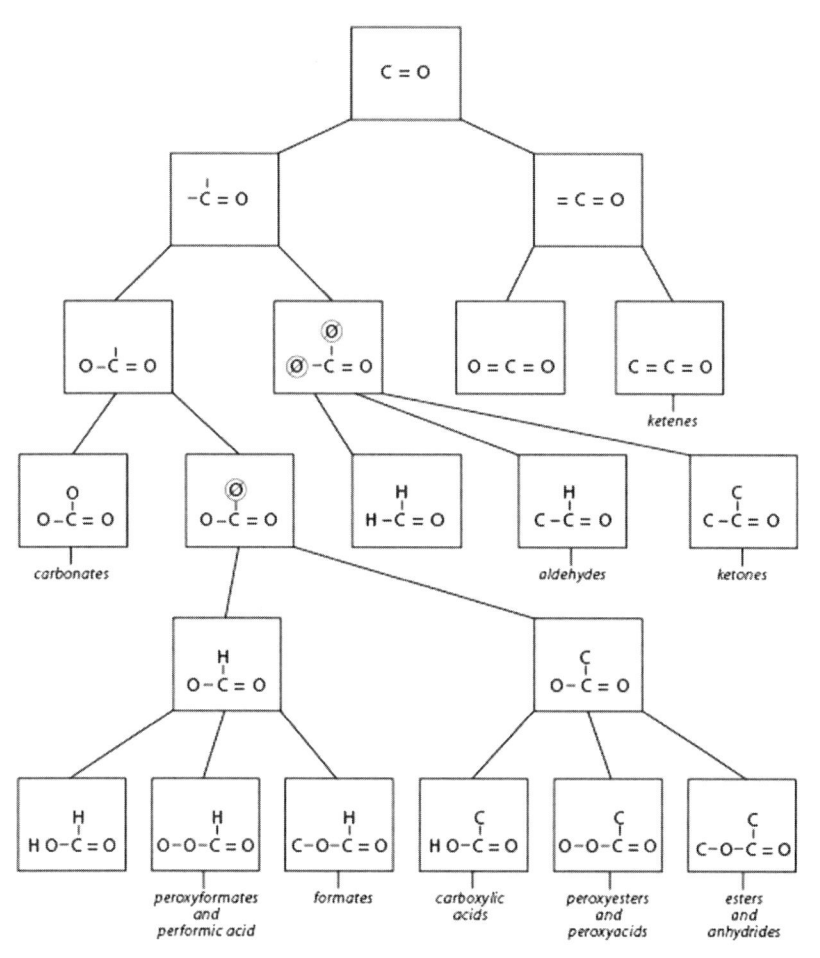

FIG. 5. Portion of a hierarchical functional group tree for classifying carbonyls. The slashed O means any atom except oxygen.

carboxylic acids, peroxyacids, and esters are all significantly different, Fig. 5. Still finer levels of distinction than are shown in Fig. 5 would certainly be justified chemically, e.g. unsaturated ketones certainly have different thermal properties than dialkyl ketones.

The functional group tree databases are read both by RMG and by a GUI developed by J. Robotham, Fig. 6. The RMG program performs substructure searches on each molecule in the model, to identify the most specific (i.e. most extended) functional group in the tree database that matches each portion of the molecule. The thermochemistry of each molecule and radical is estimated using Benson-type (Benson, 1976) group-additivity (one group value per non-terminal atom), but the functional group hierarchical tree allows one to define much

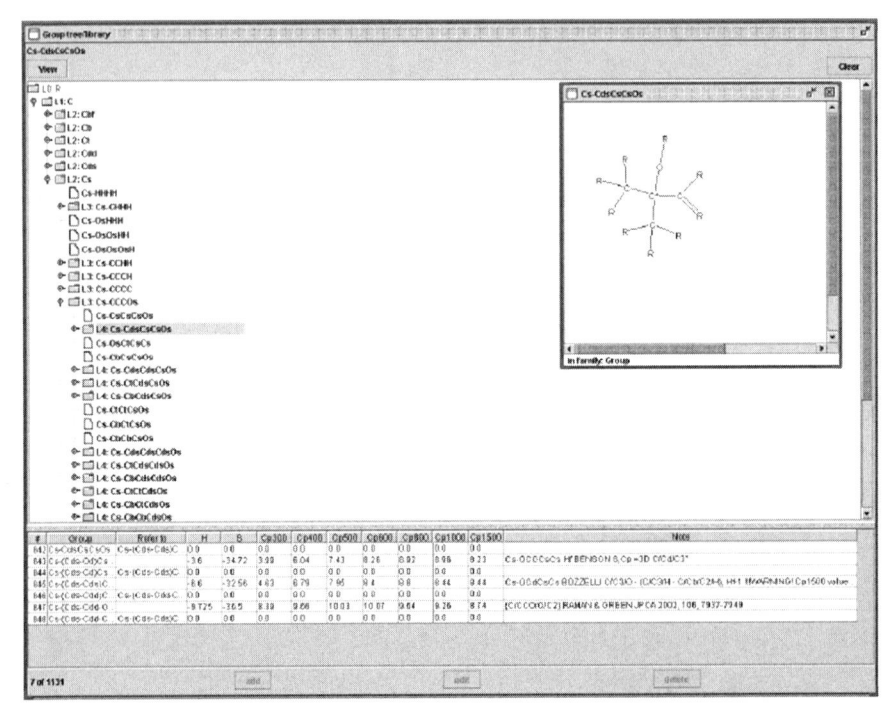

Fig. 6. The graphical user interface (GUI) for displaying and modifying the functional group trees used to estimate thermochemistry.

more specific, complicated, and extended functional groups than are typically used, for example the extended groups for computing the thermochemistry of polycyclic aromatic hydrocarbons (Yu et al., 2004). In each functional group the "central atom" of the group must be indicated; this is the non-terminal atom whose group thermochemistry is being estimated. In Fig. 5, the central atom is always the C double bonded to the oxygen.

At present, the functional group trees are certainly incomplete. For example, at present we only have reasonably comprehensive information on functional groups made of C, H, and O atoms. There is a limited amount of thermochemical (and detailed rate) information available in the literature on functional groups involving halogens, N, P, and S, but this is not yet reflected in the functional group trees. Very little generalized information is available for functional groups involving other elements. For certain classes of compounds, such as fused-ring compounds, there is some question about the correct method for generalizing from the limited available information to estimate the thermochemistry of never-studied species. It will be many decades before the functional group trees will be at all complete. Hence, it is essential to design a functional group tree structure which can easily incorporate new chemistry information.

Users can conveniently view and modify the thermochemical group values, or add new functional groups, using the GUI, Fig. 6. Most commonly, an expert will subdivide an existing functional group into several more detailed functional groups, each with distinct thermochemical parameters. The tree structure is intuitively obvious to chemists, since it groups similar chemical entities together. The tree structure of the database also facilitates efficient search for the best-matching functional group in each molecule encountered while the computer is building the reaction mechanism.

Following Bozzelli and Ritter (Lay et al., 1995; Ritter and Bozzelli, 1991), the thermochemistry of free radicals is estimated by adding Hydrogen Bond Increments (HBI) to the energy of the corresponding stable molecule where an H has capped the radical site. The HBI groups are also stored in a functional group tree.

It should be noted that there are some types of molecules where it is not clear if the functional group tree approach is viable. For example, some transition metal complexes have fluxional structures, so there can be some debate about which atoms are bonded to each other. In biochemistry, non-bonded interactions often have very large cumulative effects on the thermochemistry. For multi-ring compounds and molecules with chiral centers, there are often several distinct three-dimensional structures corresponding to the same connectivity diagram, and these might each have discernibly different thermochemistry. However, the functional group approach has been found to be quite effective for a very large fraction of all molecules studied to date.

So far, we have compiled and tree-classified more than 700 functional groups with estimates of their thermochemical group values. However, only about 110 of these group values have been carefully derived; the rest are mostly estimates based on analogy. More technical details on the tree classification of functional groups and the thermochemistry estimation procedure used in the RMG software package are given at http://web.mit.edu/greengp/RMG and elsewhere (Song, 2004).

2. Functional Group Trees for Reaction Rate Estimation

a. Functional group tree rate estimation: How it works. As with thermochemistry, functional groups can be used to classify chemical reactions in a way that is easily extensible and easy for a human to understand. The functional group concept also corresponds well to physical reality: at a microscopic level, the barrier height for a chemical reaction is controlled primarily by the local structure close to the bonds that are being made and broken. For many reactions, the Arrhenius A-factor is also strongly affected by the nearby functional groups.

However, this is considerably more complicated than for thermochemistry, because most chemical reactions involve two functional groups, not just one, and some types of reactions have a rather complex dependence on the extended molecular structure. When representing functional groups for chemical reactions, one usually needs to specify several special atoms rather than just one central atom: one must specify all the atoms whose bonding is changing in the course of the reaction.

For each type of reaction, we have first specified the "Reaction Recipe", which specifies which bonds are made and broken and which gives the numbering for the special "central" atoms. Then we have constructed functional group trees for each of the functional groups involved, typically two functional groups for a bimolecular reaction, one on each reactant.

In order to find the correct rate estimation parameters for a bimolecular reaction, one must first find the best-match functional group A in one reactant, using a substructure search algorithm very similar to that used to find thermochemical group values. Then one must find the best-match functional group B in the second reactant in a similar way. Then one can look up the rate estimation parameters for this reaction type using A and B as the keys.

As a simple example, consider the addition of ethyl radical to the O end of the carbonyl group in allyl acetate ($CH_3C(O)OCH_2CHCH_2$). The Reaction Recipe for radical-addition-to-double-bond is shown in Fig. 7.

This reaction will occur whenever one reactant has a radical center, and the other reactant has a double bond. In the example, the first reactant, ethyl radical, has a radical site, designated as atom 1 in the Reaction Recipe. The second reactant, allyl acetate, has two double bonds. The program would first select one of the double bonds, and then select one end of the double bond to be atom 2 in the Reaction Recipe, Fig. 7. (After completing this reaction, the computer will successively try all the other permutations.) Suppose the program selected the C=O double bond, and selected the O atom to be atom number 2. The carbonyl carbon would be designated atom number 3. Immediately, the computer would perform the reaction operation as shown in Fig. 7, generating a new product molecule, Reaction (1).

$$CH_3CH_2 + CH_3C(O)OCH_2CHCH_2 \rightarrow CH_3C(OEt)OCH_2CHCH_2 \quad (1)$$

If this product molecule did not yet exist in the reaction mechanism, the computer would immediately classify all of its functional groups using the thermochemical group value tree described above, to compute its thermochemistry. (The product thermochemistry will certainly be needed to compute the heat of reaction and the reverse reaction rate, and it might be needed to compute the forward reaction rate also.)

Now the computer must identify the best rate estimation parameters for the forward reaction. First the radical site in the first reactant, ethyl radical, will be classified as "primary alkyl" using the functional group tree shown in Fig. 8.

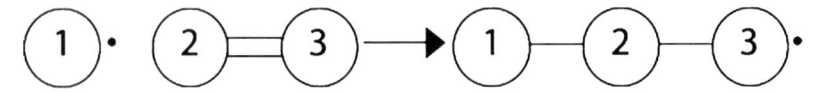

FIG. 7. Reaction Recipe for radical addition to a double bond.

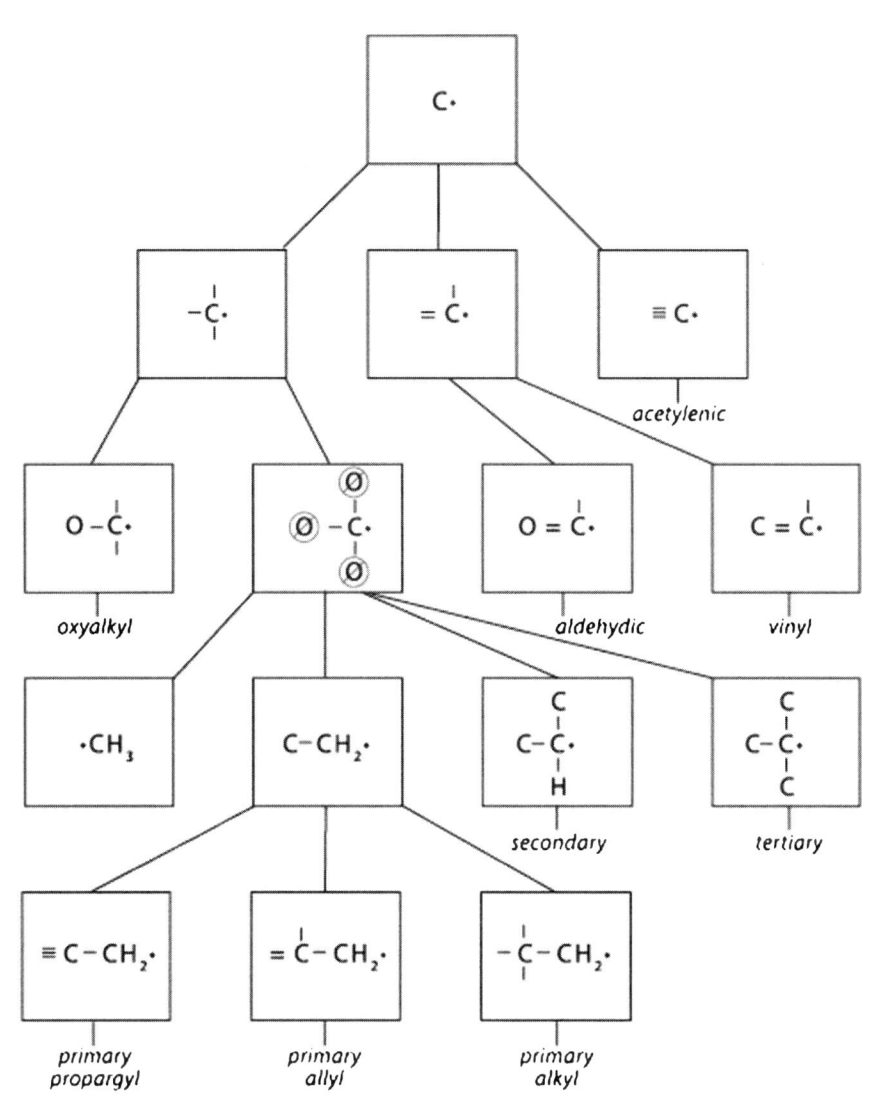

FIG. 8. Portion of one of the functional group trees for addition reactions: classification of site 1 (the radical center) in the Reaction Recipe shown in Fig. 7.

Then, from the tree in Fig. 9 the O-ended double bond in allyl acetate would be classified as an "ester O". Note the numbers 2 and 3 in this functional group tree. These numbers indicate which O and C atoms in the functional group participate directly in the reaction. (Of course the neighboring O and C atoms affect the rate to some extent, even though they do not change their valence bonding in this reaction.)

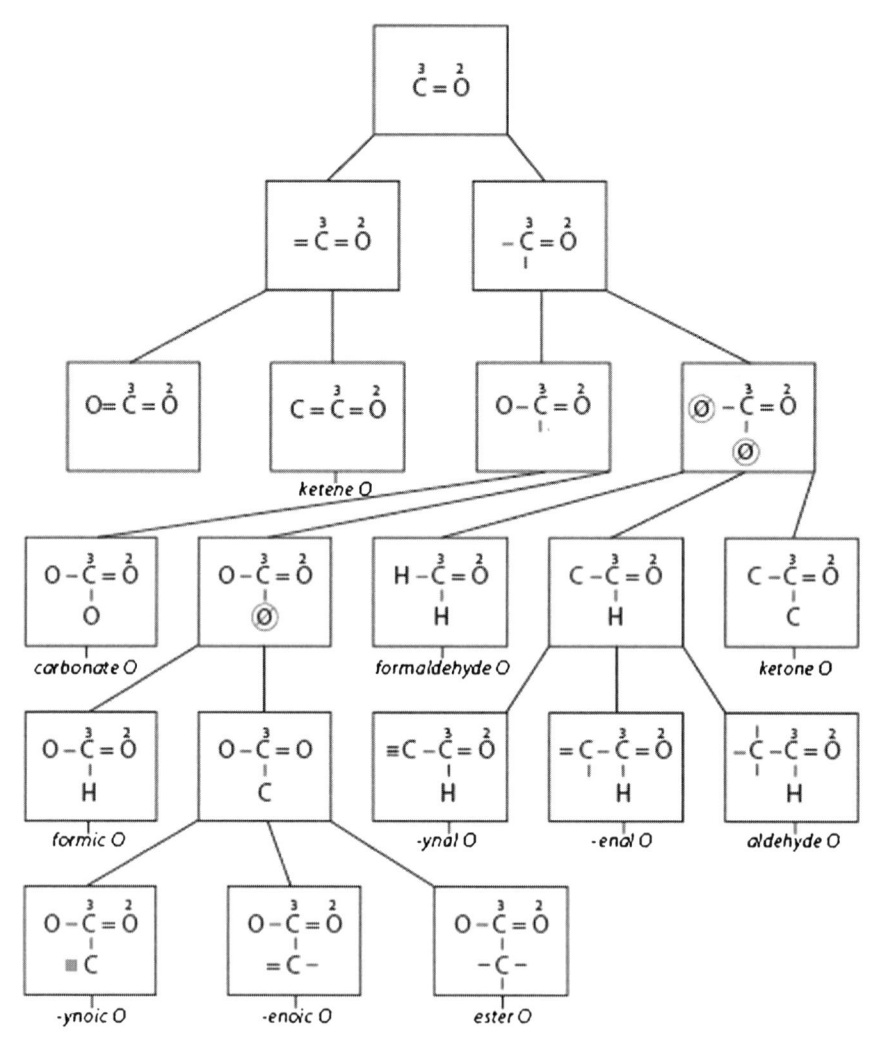

FIG. 9. Portion of the one of the functional group trees for addition reactions: classification of the 2–3 double bond in the Reaction Recipe shown in Fig. 7.

The computer would then look in the rate estimation library for Radical Addition reactions to find the rate parameters that correspond to "primary alkyl+ester O". There it would find a set of numerical parameters E_o, α, A, n with estimated uncertainties; one could then calculate the forward rate coefficient for the reaction of interest using this Evans–Polanyi modified Arrhenius form:

$$k_{\text{forward}} = AT^n \exp\left(\frac{-(E_o + \alpha \Delta H_{\text{reaction}})}{RT}\right) \qquad (1)$$

As discussed below, the k_{forward} computed this way is only accurate in the high-pressure-limit. The reverse rate would be computed from k_{forward} using the thermochemistry, for example if all the species are ideal gases:

$$k_{\text{reverse}} = k_{\text{forward}} \left(\frac{RT}{1\,\text{atm}}\right)^{\Delta\text{moles}} \exp\left(\frac{\Delta G_{\text{reaction}}}{RT}\right) \qquad (2)$$

where (Δmoles) = change in number of moles in the reaction. Note that both the forward and reverse rates depend exponentially on the enthalpy of the reaction, so it is critical that the thermochemistry be computed accurately.

The functional group tree classification of reaction rate parameters is conceptually related to an extension of the thermochemical group-additivity to transition states. In conventional transition state theory, for ideal gases and neglecting tunneling and recrossings, the reaction rate coefficient is given by:

$$k_{\text{forward}} = \left(\frac{RT}{1\,\text{atm}}\right)^{N_{\text{reactants}}-1} \left(\frac{k_B T}{h}\right) \exp\left(\frac{G_{\text{reactants}} - G_{\text{TS}}}{RT}\right) \qquad (3)$$

In automated model-construction programs, $G_{\text{reactants}}$ is normally estimated using group-additivity (as discussed above), i.e. $G_{\text{reactants}}(T) = \Sigma G_{\text{group},n}$ where the program looks up the value of $G_{\text{group}}(T)$ for each of the thermochemical groups in the molecule. It is possible to estimate G_{TS} in a similar way (Sumathi & Green, 2002), if one first divides the transition state structure (i.e. the geometry at the saddle point) into a "reacting" functional group which includes all the bonds that are being made and broken, and "inert" functional groups:

$$G_{\text{TS}}(T) = G_{\text{reactive}}(T) + \Sigma G_{\text{group},n} \qquad (4)$$

For the inert functional groups one can look up the same group values $G_{\text{group}}(T)$ as for any ordinary molecule; in Eq. (3) these will identically cancel the corresponding terms in $G_{\text{reactants}}(T)$. The values for the reacting functional groups are of course very different, but these group values can be computed using quantum chemistry and then stored in the functional group trees for reactions instead of the more conventional A, E_a, etc. In principle, it would be more accurate to store $G_{\text{reactive}}(T)$ rather than forcing all rates to conform to the modified Arrhenius form. However, with our current very imperfect knowledge of transition states and uncertainties in measured rates there aren't yet many cases where we can discriminate between the rate models in Eq. (1) and in Eqs. (3) and (4).

b. Difficulties with functional group rate estimates

1. Paucity of reliable data. A serious practical issue is that we do not yet have enough information to make accurate rate estimates for every possible combination of functional groups A + B. As noted above, we have identified about 700 individual functional groups for estimating thermochemistry of molecules $C_xH_yO_z$. We have identified about 30 distinct reaction types (with different reaction recipes) that are important in gas-phase thermal chemistry of C/H/O compounds; most of these involve more than one functional group. We would need $O(10^5)$ distinct sets of rate estimation parameters if we were to consider rate estimation at the same level of chemistry detail as we did thermochemistry (and surely if the thermochemistry varies with a change in the functional group, the reaction rate will be affected as well). But we have very few reliable data suitable for making these rate estimates. At present we are only able to use truncated functional group trees (or equivalently, use the same rate parameters for several similar functional groups) since we do not have estimates of the effects of subtler variations in the functional groups. Fortunately, the extensible tree structure makes it feasible to enlist the larger chemistry community in the task of converting "Chemistry Knowledge" into a more comprehensive listing of specific functional group-based rate estimation parameters that can be used in quantitative kinetic models.

One of the reasons for the paucity of data is that for many reaction families it is hard-to-impossible to set up an experiment that probes only the process we want to measure. For example, in the gas phase, it might be difficult to observe the reaction used as an example in Section II.D.2.a above, Reaction (1), for a couple of reasons. First, there are several competing reaction pathways (e.g. H-abstraction, addition to the C=C double bond instead of the C=O double bond) leading to different products. Second, as discussed further below, the product shown above may be so unstable that it cannot be detected, even if the reaction is actually running at a fast rate. Often researchers will report the rate for the sum of all the reactions, Reaction (2),

$$CH_3CH_2 + CH_3C(O)OCH_2CHCH_2 \rightarrow \text{All products} \qquad (2)$$

but it is difficult to use this information to derive rate estimation parameters for Reaction (1). To some extent this difficulty can be alleviated by quantum chemistry calculations, but if no experimental data are available to check quantum chemistry calculations on a new type of reaction, there is always some doubt about their reliability.

2. Chemical activation and fall-off. Gas-phase reactions that form an energized product cause particular difficulty in kinetic model-construction. For example, in Reaction (1) the unstable product will be formed with excess energy (due to the exothermicity of the bond-forming addition reaction), that it will rapidly dissociate to ethyl acetate + allyl radical, Reaction (3).

$$CH_3C(OEt)OCH_2CHCH_2 \rightarrow CH_3C(O)OCH_2CH_3 + CH_2CHCH_2 \qquad (3)$$

At the low pressures used in many laboratory kinetics experiments, the rate of formation of allyl radicals by this two-step pathway will be much higher than one would compute if one assumed that the energized intermediate $CH_3C(OEt)OCH_2CHCH_2$ was in thermal equilibrium with the bath gas; the rate for the apparently direct fast Reaction (4)

$$CH_3CH_2 + CH_3C(O)OCH_2CHCH_2 \rightarrow CH_3C(O)OCH_2CH_3 + CH_2CHCH_2 \quad (4)$$

will strongly depend on the bath gas temperature, pressure, and composition. This effect, where a reaction appears to go directly through several elementary steps in a single step, is called "chemical activation". (If the energized product does not fragment to form new species, but has a significant probability of falling apart back to the reactants before being thermalized, the effect is called "fall-off".)

Chemical activation causes many problems. First of all, rates are now a function of P (as well as T), adding another dimension to the rate estimation functions. Also, chemically-activated rates depend on properties of the complete adduct molecule, not just some of its functional groups. And, in order to compute the chemically-activated rates, one must enumerate all the important product channels, i.e. identify a large number (Matheu et al., 2003a, b) of different reactions and products, which could be of completely different types. As a result, each chemically-activated reaction has unique rate parameters, not amenable to easy generalization. It would not be practical to compute or to store the rate parameters for every conceivable chemically-activated reaction.

Our solution to these challenges is to have the computer compute the rate parameters for the chemically-activated reactions on-the-fly as needed during the model-construction process, using the algorithm of Matheu et al. (2003a). The density of states is estimated from the heat capacity using the three-frequency method devised by Bozzelli (Bozzelli et al., 1997) complemented by the Stein–Rabinovitch method for convolving rotors with vibrations (Stein and Rabinovitch, 1974, 1973). The micro-canonical rates $k(E)$ are estimated using the inverse Laplace transform method (Forst, 1973) or (often equivalently) Quantum Rice-Rampsberger-Kassel (QRRK) (Dean, 1985). The master equations are solved approximately using the modified strong collision approximation as implemented in CHEMDIS (Chang et al., 2000). For more details, see discussions by Matheu (2003).

In the database we only need to store high-pressure-limit (i.e. thermally-equilibrated intermediates) rate estimation parameters for elementary-step reactions; from this information, the molecular structure, and the thermochemical parameters (Section II.D.1) one can compute the pressure-dependent reaction rates. The elementary-step high-pressure-limit rate estimation parameters depend primarily on the local functional group structure, and so are very well-suited to functional group tree classification.

3. Intramolecular reactions. Intramolecular reactions, such as intramolecular addition reactions and intramolecular H-abstraction reactions, introduce serious complications. In these reactions, the reacting functional groups A and B are both contained in the same molecule, tied together by a third functional

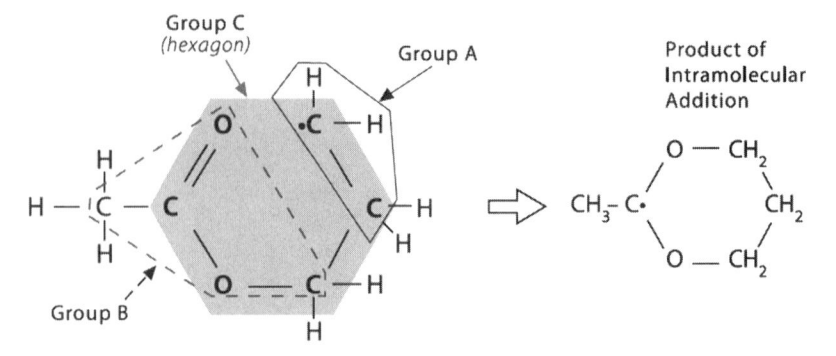

FIG. 10. A molecule undergoing an intramolecular addition (cyclization) reaction, indicating the definitions of functional groups A, B, and C.

group C, Fig. 10. In order to estimate the reaction rate, one must specify all three functional groups, each of which can be comparable in complexity to the functional groups used to describe thermochemical groups and bimolecular reactions. So one can imagine a very large number of distinct types of intramolecular reactions with unique rate constants specified by the functional groups involved (A, B, and C).

For the reaction illustrated in Fig. 10, groups A and B could be classified using the trees shown in Figs. 8 and 9, but the tether group C requires a new functional group tree, Fig. 11. In Fig. 11, the tether group is classified solely based on the bonding arrangement inside the ring; in the future this tree should be extended further to differentiate according to which atoms are at each node in the tether, and to account for substituent effects.

Unfortunately, only a very small fraction of the many conceivable distinct intramolecular reactions have been studied in the literature, due in part to technical problems. Experimentally, it is often difficult to prepare a run on high concentrations of thermalized molecules with a high-reactivity functional group A close to another functional group B that reacts with A. It is therefore very difficult to measure the (often rather high) rate at which this reaction occurs. To make the reaction slow enough to measure, one can run it in the endothermic direction—but then one needs a very efficient way to trap the product before it converts exothermically back to the reactants.

Also, in gas-phase experiments, many of these reactions are pressure-dependent, and special care must be taken in extrapolating the experimental measurements to the high-pressure-limit. Quantum chemical calculations are also rather difficult for most reactions of this type, since molecules that contain three distinct functional groups A, B, and C necessarily include a rather large number of heavy atoms. For this reason, so far relatively few high-level calculations have been performed on these sorts of reactions.

At present, these types of reactions are represented using three different extensible functional group trees, one for each of the groups involved. In principle,

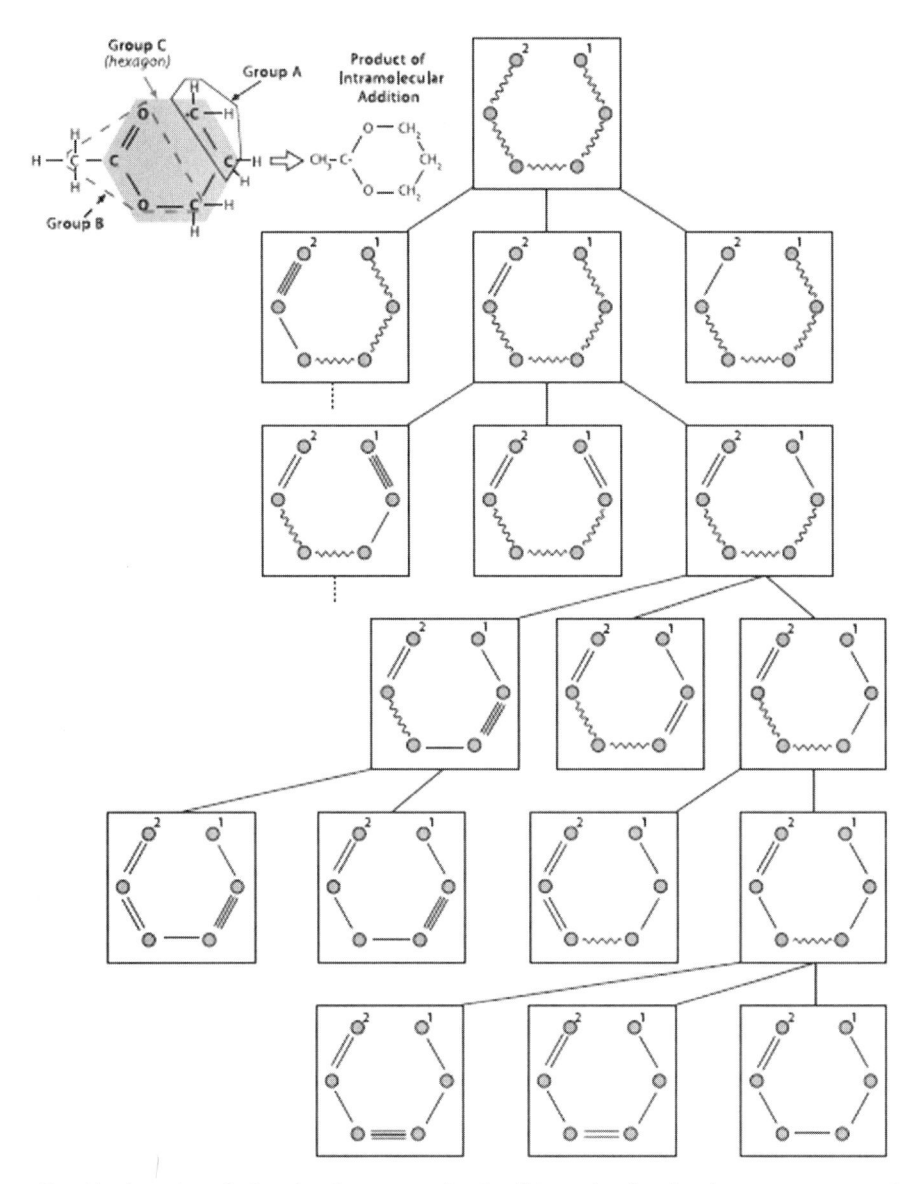

FIG. 11. A portion of a functional group tree for classifying tether functional groups, e.g. group C in Fig. 10.

this requires a much larger table of rate estimation parameters (keyed from A, B, and C). However, due to lack of information, we and many others working on these reactions often make untested separability assumptions, e.g. assuming that the high-pressure-limit rate parameters for an intramolecular reaction involving groups (A, B, and C) can be computed using Eq. (5):

$$k = k_{\text{bimolecular}}(A + B) \times D_C \exp\left(\frac{-E_C}{RT}\right) \qquad (5)$$

where $k_{\text{bimolecular}}(A+B)$ is the rate we estimate for the bimolecular reaction $A + B$ (i.e. if the tether C was cut somewhere in the middle), E_C is the estimated ring-strain involved in bringing reactive sites close to each other when they are connected by a tether C, and D_C is an estimate of the ratio between a bimolecular Arrhenius A factor and the corresponding intramolecular Arrhenius A factor; this depends significantly on the structure of the tether C (due to variations in the number loss of rotors in the transition state). By making the separability assumption in Eq. (5), one only needs to add a simple table that returns D_C and E_C given as a key a single tether group C. To move beyond this separability approximation and account for interactions between functional groups A, B, and C, one would need to extend to a much larger three-key table.

E. THE REACTION MECHANISM GENERATOR (RMG)

The new data model for chemical kinetics will work best if there is a reproducible, deterministic, automated procedure for going from the fundamental inputs (a library of the small number of known rate constants, and the hierarchical tree of rate estimation parameters for all the other reactions) to the desired outputs (e.g. predicted yield/selectivity profiles). Ideally, the software that performs this automated procedure will not need to be altered if someone changes an initial condition, a rate estimate, or even the structure of the hierarchical tree—reducing the need to touch the source code of the software dramatically reduces both the programming burden on the kineticist and the likelihood of introducing a bug into the software. This design also has the practical advantage that exactly the same software could be used by many kineticists for many different problems, making it much easier to reproduce work by another, while also reducing the difficulty of maintaining the software.

We have developed a software package, RMG, for exactly this purpose. The software is designed to take the hierarchical database trees as its main input, and to return a kinetic model as its main output. This kinetic model can be in the conventional List-of-Reactions format suitable for use in CHEMKIN or other integrators, or, if desired, the RMG program package can perform the integration itself and return product yield/selectivity profiles.

RMG, which was implemented by Jing Song (2004), uses a version of the rate-based model-construction algorithm described by Susnow *et al.* (1997), but it treats non-thermalized chemically-activated reaction paths on an equal basis with ordinary thermal reactions, as done by Matheu *et al.* (Matheu, 2003; Matheu *et al.*, 2003a, b). Byproducts and activated reaction intermediates are considered important species if their estimated instantaneous formation rates ever exceed $\varepsilon\|\omega(t)\|$, where $\|\omega(t)\|$ is the norm of the major species flux array (Song et al., 2002) and ε is the error tolerance. The concentrations needed to compute the rates of change $\omega(t)$ are estimated using the best mechanism available so far. The concentration estimates can be substantially in error early in the process when the mechanism may be missing important reactions. However, this procedure has been demonstrated to settle down and become quite stable and robust after a modest number of iterations even for rather complex systems (Matheu et al., 2003a). At convergence, all of the neglected byproducts are formed at rates at least a factor of ε lower than the rates of the important reactions in the system.

Unlike most model-generation software described in the literature, RMG correctly handles pressure and temperature variations; it does this by using $k(T,P)$ computed for the chemically-activated reactions at discrete (T,P) to determine coefficients in a Chebyshev form (Venkatesh et al., 1997) suitable for use in the differential equation solver.

At the heart of the RMG algorithm is the process of identifying all the possible reactions of some species of interest with itself and all the species included in the model so far, and estimating each reaction's rate. Unlike most previous mechanism generators, the RMG software handles this procedure in a very general way: all the specifics about functional groups and reaction types, as well as all the associated parameters, are external to the program, contained in the tree databases. This has the huge advantage that nothing inside the RMG program has to change when the chemistry information is updated: chemists only need to modify the tree databases using the GUI, they do not need to do any programming or re-compiling.

F. APPLICATIONS OF RMG

So far, RMG has been applied to several problems: the pyrolysis (steam-cracking) of n-hexane (Van Geem et al., 2006), the laser-initiated oxidation of neo-pentane (Petway, 2005), the supercritical water oxidation of methane [http://web.mit.edu/cfgold/www/RMG/testcases.html], and the pyrolysis and oxidation of butane (Song et al.). In each case, the software constructed models involving $O(100)$ species, and the model predictions were about as accurate as one would expect given the factor of 2–3 uncertainties in the rate parameters and equilibrium constants arising from the estimation techniques used to generate the model parameters.

As a simple example, consider the predicted yield vs. conversion data for butane oxidation at 715 K, Figs. 12 and 13, plotted on top of experimental data points measured by Cernansky's group at Drexel (Wilk et al., 1995). As is typical, most of the discrepancies between the model predictions and the experimental data can be safely attributed to the uncertainties in the rate

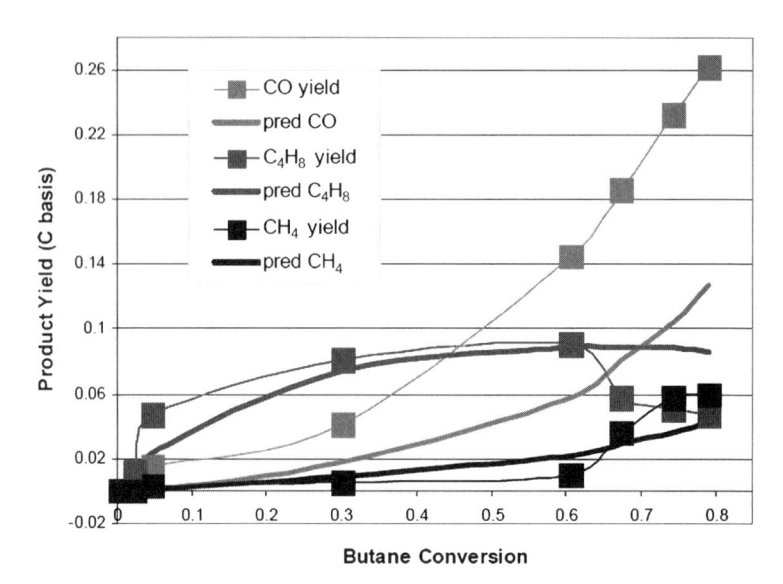

FIG. 12. Model predictions vs. experimental measurements of Wilk et al. (1995) from butane oxidation at 715 K. The predicted yields are generally within a factor of two of the experimental data, reflecting roughly factor of two uncertainties in rate constant estimates. As is typical, the discrepancies are largest at high conversions, both because the errors in the parameters cumulate, and because the model may be missing some reactions of the minor byproducts (Please see Color Plate Section in the back of this book).

constant estimates (and related uncertainties in the thermochemistry). However, the discrepancy between the predicted and observed CO_2 yields at very small conversion is likely due to the fact that the model does not accurately reproduce the true initial conditions (the model assumes instantaneous perfect mixing and a perfectly isothermal reactor).

Many detailed chemical kinetic models show discrepancies at high conversions similar to what is seen in Fig. 12, in large part due to the time-integrated effects of even relatively small errors in the estimated rate parameters, and in some cases also due to the omission of side reactions, which become important as products build up.

Several other mechanism-generation computer programs have been developed which have shown comparable ability to generate predictions in good accord with experimental data; for a couple of impressive recent examples see Matheu's models of methane (Matheu et al., 2003a) and ethane/ethene pyrolysis (Matheu and Grenda, 2005a, b) and the Nancy group's models for oxidation of fuels in the gasoline range (Buda et al., 2006; Glaude et al., 2002). The case of methane pyrolysis is particularly interesting as an example where the mechanism-generation program immediately identified critical reaction steps that the human experts had overlooked for more than 20 years (Matheu et al., 2003a).

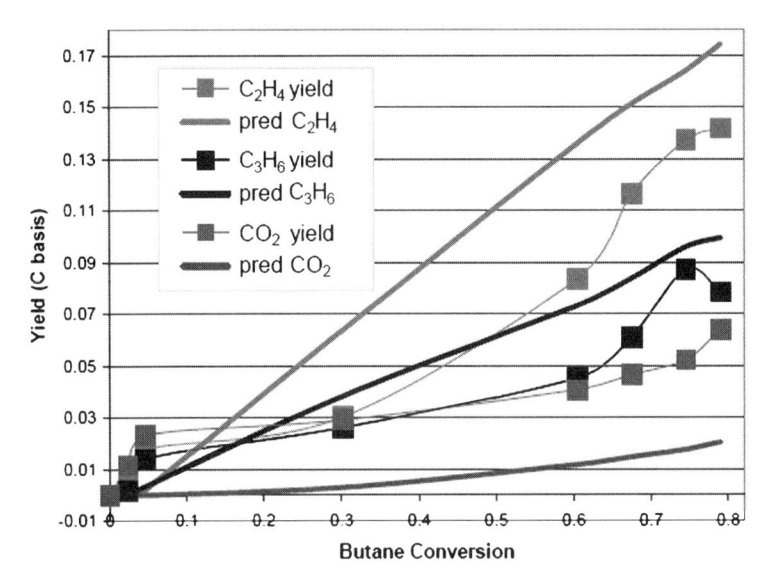

FIG. 13. Model predictions vs. experimental measurements of Wilk *et al.* (1995) for ethene, propene, and CO_2, same conditions as Fig. 12. The large discrepancy in the CO_2 predictions at the lowest conversions suggests that the model does not accurately represent the true boundary conditions at the inlet (Please see Color Plate Section in the back of this book).

The new RMG software combines high-accuracy chemistry estimation methods (including methods for automatically identifying chemically-activated reaction paths) with the extensible 21st century functional group tree data model described above, so it is now much easier for users to add, modify, and document the chemical assumptions that underlie the models than it was using the previous generation of model-construction software, and the software does not need to be modified or recompiled when new chemistry is added.

The RMG software, the rate and thermo estimation databases, and the GUI for reading and modifying the functional group tree databases are all available as open-source software at http://web.mit.edu/greengp/. It is hoped that experts in all branches of chemical kinetics will contribute to improving and extending the rate and molecular property estimates used by RMG.

III. Efficiently and Accurately Solving Large Kinetic Simulations

The methods described in Section II carry out the first step in Fig. 1, constructing a detailed chemical kinetic model from our current understanding of chemistry. However, this step is only useful if we can numerically solve the chemical kinetic simulation to obtain quantitative predictions. In many cases, solving the model is even more challenging than constructing it.

Most real reacting systems are accurately described by a large system of partial differential equations in (x, y, z, t); most commonly the spatial coordinates are discretized so that one is actually solving equations like this at each mesh point:

$$\frac{dY^p}{dt} = \Theta^p[Y] + \omega^p(Y^p) \qquad (6)$$

where Y is the long array of state variables, Θ are algebraic equations representing the transport between mesh point "p" and nearby mesh points, and ω^p the algebraic equations representing the rate of change of the state variables occurring at mesh point p due to chemical reactions, i.e. the chemical source terms. The differential equations Eq. (6) are usually supplemented by some algebraic equations (e.g. boundary conditions, equations of state). While a few of the state variables Y represent fluid dynamic variables such as momenta, for even relatively simple reacting chemical systems the great majority of the state variables are chemical species concentrations or mass fractions, and in most systems of interest to reaction engineers the (usually very stiff) chemistry terms consume most of the CPU time (Schwer et al., 2002).

Detailed chemical kinetic models are usually very stiff in time due to the large separation of timescales between reactive intermediates and the more stable major species. This is a serious problem for explicit methods and stochastic methods, but stiff ordinary differential equation (ODE) systems describing up to $O(10^4)$ variables can be handled routinely using modern implicit solvers like DASPK (Bernan et al., 1989) and VODE (Brown et al., 1989). However, spatially inhomogeneous reacting systems often lead to stiff systems with many more state variables, which cannot be efficiently solved with available off-the-shelf black-box software. Two-dimensional and three-dimensional simulations with large (>100 species) chemical kinetic models are almost impossible to solve with existing software. This is unfortunate, since mechanism-generation programs such as RMG described in Section II above regularly produce models containing more than 100 chemical species.

There are many ways one can try to reduce the computational burden. Ideally, one would find numerical methods which are guaranteed to retain accuracy while speeding the calculations, and it would be best if the procedure were completely automatic: i.e. it did not rely on the user to provide any special information to the numerical routine. Unfortunately, often one is driven to make physical approximations in order to make it feasible to reach a solution. Common approximations of this type are the quasi-steady-state approximation (QSSA), the use of reduced chemical kinetic models, and interpolation between tabulated solutions of the differential equations (Chen, 1988; Peters and Rogg, 1993; Pope, 1997; Tonse et al., 1999). All of these methods were used effectively in the 20th century for particular cases, but all of these approximated-chemistry methods share a serious problem: it is hard to know how much error is introduced by making these approximations (unless one is able to solve the simulation using the full chemistry model—in which case there may have been no need to make the approximation). Because there is no bound on the error introduced by reducing the chemistry this way, usually no one really knows how much accuracy one is sacrificing in order to make the calculation feasible.

Below we present 21st century numerical solution methods suitable for large chemical kinetic simulations. We present both a method that does not introduce any approximations, and an automated model-reduction method that allows rigorous error control in steady-state simulations.

A. Fast Solution of Large Systems of Chemistry Equations Using Sparsity

Before making approximations to speed up a calculation or reduce its memory requirements, it is worthwhile to consider if there are CPU time or RAM reducing alternatives that do not introduce any error. For large chemical kinetic simulations, one is typically solving rather large systems of differential-algebraic equations (DAEs), where the differential equation system is extremely stiff. The most popular solvers for these types of problems are DASSL (Petzold, 1982), DASAC (Caracotsios and Stewart, 1995), and DASSL's successor DASPK (Bernan et al., 1989). By time-profiling these types of solvers, one discovers that most of the CPU time is consumed constructing the Jacobians needed to solve stiff systems of chemistry equations.

The structure of these Jacobians depends on which variables one uses when setting up the equations; for most obvious choices the chemistry Jacobians are dense (i.e. more non-zero elements than zeroes in the Jacobian matrix). One can always rewrite the equations, however, to give unstructured sparse Jacobians (i.e. most of the elements in the matrix are identically zero) (Schwer et al., 2002). With standard Jacobian-construction and solvers, it makes little difference whether the Jacobian is dense or sparse. But if one uses software that takes advantage of the sparsity, one can reduce both the memory requirements and the CPU time required by a significant factor just by omitting all the entries known to be zero. Determining which Jacobian entries are always zero requires a lot of book-keeping in a large chemical kinetic system, but this book-keeping can be done automatically using programs like DAEPACK, which analyze FORTRAN programs symbolically (Schwer et al., 2002; Tolsma and Barton, 2000). Taking advantage of sparsity in homogeneous charge compression ignition (HCCI) engine simulations with very large chemistry models can speed up the calculations by a factor of 60, and reduce the memory requirements enough that rather large calculations can be run on ordinary laptop computers (Yelvington et al., 2004).

It is not hard to change the variables to make the Jacobian sparse, and it is simple to run a simulation using DAEPACK rather than DASPK or VODE, to see whether the sparsity-aware software is less demanding of computer resources than the conventional solvers. In our experience, if the chemistry model involves more than ~100 species, it is usually better to use the sparse solver approach. For more details on how to use DAEPACK to analyze a FORTRAN simulation and to handle the sparsity, see Schwer (Schwer et al., 2002) and http://yoric.mit.edu/

B. USING REDUCED CHEMISTRY MODELS IN MULTIDIMENSIONAL SIMULATIONS WITHOUT INTRODUCING ERROR

The CPU time required to solve large chemical kinetic simulations scales approximately linearly with the number of chemical reactions and with the number of finite volumes in the mesh, and approximately quadratically with the number of chemical species. (The LU decomposition of the Jacobian performed in most stiff solvers, which scales as N^3, has a much smaller prefactor than the N^2 terms and usually takes negligible time compared to other steps in chemical kinetic simulations (Schwer et al., 2002).) So for large chemical kinetic models ($10^2 \sim 10^3$ species, $10^3 \sim 10^4$ reactions) there is a significant potential to speed up the calculation by replacing the large chemistry model with a smaller model (fewer reactions, maybe also fewer species). This is particularly important in multidimensional reacting-flow simulations, where it is not unusual for calculations to take weeks to converge even on very powerful computers. As mentioned above, there are many ways to approximate the chemistry: the challenge is how to avoid losing accuracy when making these approximations.

It is relatively straightforward to bound the error in the transport and chemistry terms on the right-hand side of Eq. (6) caused by certain types of approximations to the Θ and ω operators, since Θ and ω are relatively simple algebraic operators. Methods for bounding/controlling the errors associated with the size of the mesh in the transport operator Θ are very well-understood, and form the basis of adaptive meshing technology (Berger and Oliger, 1984; MacNeice et al., 2000). Methods for bounding the error in ω made by using a reduced chemistry model are discussed below in Section III.C. For the moment, let us suppose we have constructed an approximation $\omega_{\text{approx}}(Y)$ to the true $\omega(Y)$, and that we can set bounds on the deviation between these two chemistry source terms.

We would like to solve the approximate model, Eq. (7) rather than the full model, Eq. (6) above, but we need to have some control over the error introduced by this approximation.

$$\frac{dY^p_{\text{approx}}}{dt} = \Theta^p(Y_{\text{approx}}) + \omega^p_{\text{approx}}(Y^p_{\text{approx}}) \qquad (7)$$

The difficulty is that $Y(t)$ and $Y_{\text{approx}}(t)$ can follow rather different trajectories; while we can bound the error $|\omega_{\text{approx}}(Y) - \omega(Y)|$ at a single point Y, it is not so easy to bound the error $|\omega_{\text{approx}}(Y_{\text{approx}}(t)) - \omega(Y(t))|$ along the whole trajectory. (Despite the difficulty, very recently Singer and Barton have developed a rigorous method to bound $|Y(t) - Y_{approx}(t)|$. So far this approach has only been applied to problems with only relatively small number of state variables; making it viable for large-scale chemical kinetic simulation is a challenge for the future.)

For cases where explicit integration methods are viable, one could adjust the time step to ensure that $Y_{approx}(t)$ very closely tracks $Y(t)$ (roughly speaking, to ensure that $|\omega - \omega_{approx}|^* \Delta t$ is small compared to other errors in the calculation). But for most large-scale chemical kinetic problems explicit integration methods are not viable due to the stiffness, so this idea is not very appealing.

There is, however, one important class of reacting-flow simulations where bounding the deviation between $\omega(Y)$ and $\omega_{approx}(Y)$ immediately allows one to control the approximation error: steady-state simulations. For steady-states, the left-hand side of Eq. (6) and Eq. (7) vanish, and one wants to ensure that the solution Y_{approx} to the approximate equation

$$0 = \Theta(Y_{approx}) + \omega_{approx}(Y_{approx}) \tag{8}$$

is very similar to the (unknown) solution Y to the original equation:

$$0 = \Theta(Y) + \omega(Y) \tag{9}$$

If one has already found a converged solution to Eq. (8) so that

$$-\text{tol}_{approx} < \Theta(Y_{approx}) + \omega_{approx}(Y_{approx}) < \text{tol}_{approx} \tag{10}$$

one can check whether Y_{approx} is an acceptable solution to the original equation by a single function evaluation

$$-\text{tol}_{total} < \Theta(Y_{approx}) + \omega(Y_{approx}) < \text{tol}_{total} \tag{11}$$

where tol_{total} is the user-specified convergence criterion for a solution to be considered acceptable. Note that Y_{approx} is probably not exactly the same solution as one would find if one had solved Eq. (9) directly, but that by the user's error specification Eq. (11), Y_{approx} is a perfectly acceptable solution.

This approach is great if one finds that the approximate solution satisfies Eq. (11). But one would prefer not to risk the possibility that after grinding a large-scale steady-state reacting-flow simulation to convergence, a process that could take months, one could discover that the solution obtained, Y_{approx}, does not satisfy the user-specified error tolerance Eq. (11); at that point it would not be clear how to proceed. Also, it would be helpful to know how to set the error tolerances for the reduced model ω_{approx} and the convergence criteria tol_{approx} in order to be certain of finding an acceptable solution. One does not want to set the tolerances too tight, otherwise one will waste CPU time, by using reduced chemistry models that are larger than necessary, or by running the calculation determining Y_{approx} to higher accuracy than is necessary.

If one were able to guarantee that the reduced chemistry model reproduced the full-chemistry source term within a tolerance ε for all Y's in some range $\mathbf{Y}$

$$-\varepsilon < \omega(Y) - \omega_{approx}(Y) < \varepsilon \quad \forall Y \in \mathbf{Y} \tag{12}$$

and if the Y_{approx} found by solving the reduced-model computational fluid dynamics (CFD) simulation in Eq. (8) numerically fell in the range, i.e. $Y_{approx} \in \mathbf{Y}$, then by adding Eqs. (10) and (12)

$$-(tol_{approx} + \varepsilon) < \theta(Y_{approx}) + \omega(Y_{approx}) < tol_{approx} + \varepsilon \tag{13}$$

Comparison of Eqs. (13) and (11) reveals that if one ran the CFD calculation to a computational to a convergence $tol_{approx} = tol_{total} - \varepsilon$, one could be certain that the resulting Y_{approx} would be a satisfactory solution of the full chemistry model, i.e. Eq. (11) would be satisfied. Methods for constructing reduced models guaranteed to satisfy Eq. (13) over a known range of conditions are discussed in detail in Section III.C below. The practical effectiveness of this overall error control approach, once one has an approximate source term ω_{approx} that satisfies Eq. (13) has recently been demonstrated for 1–d and 2–d steady laminar flame simulations.

The challenge is to construct reduced chemistry models which are fast to evaluate, yet which still satisfy the error tolerance Eq. (13). One effective approach to this is the Adaptive Chemistry method (Schwer *et al.*, 2003a, b), where different reduced chemistry models are used under different local reaction conditions. For example, in the 1–d steady premixed flame studied by Oluwole *et al.* (Oluwole *et al.*, 2006), six different reduced chemistry models were used, and the full chemistry model only had to be used at about 20% of the grid points, Fig. 14.

This approach makes it easier to satisfy Eq. (13) using small chemistry models, but it requires a method to easily obtain a set of reduced models appropriate for different conditions. In the next section, we present a method for constructing the various reduced models, and for identifying the range of reaction conditions where each should be used.

C. Constructing Reduced Chemistry Models Satisfying Error Bounds Over Ranges

The algebraic equations in the chemistry operator ω can be much more complicated than the equations usually used for the transport operator Θ, but it is still possible to bound the error introduced by approximating ω, by using interval analysis. Interval analysis is a branch of mathematics that considers how mathematical operations affect intervals (ranges) $[Y_{low}, Y_{high}]$ rather than single points Y. For error control, one wants to rigorously bound the error

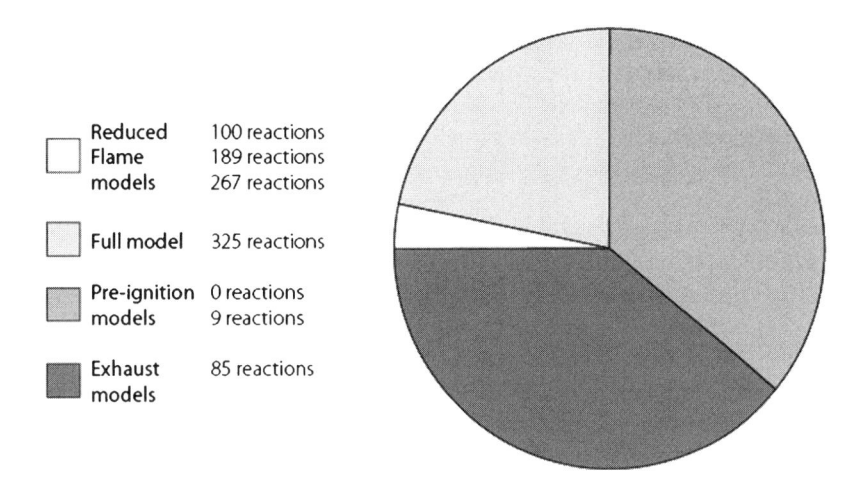

FIG. 14. Fraction of the grid points used by various models during a 1-d Adaptive Chemistry simulation of a premixed stoichiometric methane–air flame. The full chemistry model is GRI-Mech 3.0. Reduced models were accurate at about 80% of the grid points.

Reduced Flame models	100 reactions
	189 reactions
	267 reactions
Full model	325 reactions
Pre-ignition models	0 reactions
	9 reactions
Exhaust models	85 reactions

made by approximating ω not just at a single point Y, but over entire ranges of Y values that might be encountered by the solver in the process of solving Eq. (8). This is particularly important since ω is usually a highly nonconvex function of Y, i.e. just because an approximation to $\omega(Y)$ works well at certain points Y_1, Y_2, etc., does not guarantee that the approximation will be accurate at other points lying between Y_1, Y_2, etc. (Oluwole et al., 2006; Sirdeshpande et al., 2001). Hence understanding the behavior of ω and its approximations over intervals $[Y_{\text{low}}, Y_{\text{high}}]$ is critical to achieve effective error control.

Interval analysis is trickier than ordinary mathematics, since in interval analysis some arithmetic operations give outputs which are overestimates of the true range. In other words, if a function $f(x)$ takes on values between $f_{\min}$ and $f_{\max}$ for input values of x in the interval $[x_{\text{low}}, x_{\text{high}}]$, interval analysis may give you an output range $[f^*_{\min}, f^*_{\max}]$, where $f^*_{\max} > f_{\max}$ and/or $f^*_{\min} < f_{\min}$.

As a result, interval analysis tends to overestimate error bounds. However, there are clever ways to reduce this overestimation, for example by appropriately grouping terms. One of the best and most computationally efficient ways to minimize overestimation of the bounds on the approximation error is to replace terms $f(Y)$ in $\omega(Y)$ by their Taylor models, e.g. the first-order Taylor model is given by Eq. (14):

$$f(Y) = f(Y^{(o)}) + \sum_n \left(\frac{\partial f}{\partial Y_n}\right)_{Y^{(o)}} (Y_n - Y_n^{(o)})$$
$$+ \frac{1}{2} \sum_{m,n} (Y_m - Y_m^{(o)}) \left(\frac{\partial^2 f}{\partial Y_m \partial Y_n}\right)_\zeta (Y_n - Y_n^{(o)}) \quad (14)$$

In ordinary arithmetic one normally neglects the final term, making this an approximation rather than a rigorous equality, since all one knows is the interval in which ζ lies:

$$\zeta \in [Y^{(o)}, Y] \quad (15a)$$

But with interval analysis, it is no problem to rigorously compute a range that is guaranteed to bound that final term in Eq. (14) over any specified range of Y. The first terms are simple polynomials which can be bounded easily and accurately using interval analysis. So, although we do not know the value of ζ, we can still rigorously bound any function $f(Y)$ over this range of Y as long as we know its value and its first derivative evaluated at $Y^{(o)}$, and the algebraic form of its second derivatives. The Taylor models themselves can be constructed automatically by the computer using automatic differentiation (AD) technology, for example using the DAEPACK software to do all the book-keeping and to compute all the derivatives (Tolsma and Barton, 2000). So with this approach one can easily construct a fairly tight bound on the error made by straightforward approximations to $\omega(Y)$ over any specified range $[Y_{min}, Y_{max}]$.

1. Reaction Elimination via Optimization

A particularly simple method of approximating the chemical source term $\omega(Y)$ is "reaction elimination": all one does is delete reactions which are numerically insignificant under the current reaction conditions. As shown by Bhattacharjee *et al.* (2003), at any set of single point reaction condition(s), one can rigorously identify the smallest possible set of reactions which reproduces the rate at which species are made or consumed chemically, and the rate at which heat is released due to chemical reactions, cumulatively $\omega(Y)$, to within a user-specified tolerance vector "tols", Eq. (16). Mathematically, the process of finding the smallest possible model is a constrained interval optimization

$$\min \Sigma z_m \quad (15b)$$

subject to the constraint

$$-\text{tols} < \omega_{true} - \omega_{approx}(Z) < \text{tols} \quad (16)$$

where Z is a vector of ones and zeros; if $z_m = 1$ reaction m is included in the approximate model. Because both the function to be optimized and the constraint depend linearly on the optimization variables z_m, it can be solved easily to global optimality.

A free computer service that automatically performs this type of chemistry model-reduction is available through the CMCS Web portal (http://www.cmcs.org). This web software can also provide an interval where the reduced model is guaranteed to replicate the full model to within user-specified tolerances, i.e. a range $[Y_{low}, Y_{high}]$ where the error constraint Eq. (16) is guaranteed to be satisfied (Oluwole et al., 2006). However, sometimes the intervals returned by this software are inconvenient for use in reacting-flow simulations (e.g. the computed range where a reduced model is guaranteed to be accurate might be too narrow to be practically useful), and the range-computing software is limited to models containing about 100 species.

2. Constructing Reduced Models with User-Specified Valid Ranges

For CFD reacting-flow simulation applications, instead of having the computer it is often preferable to pre-specify a desired range of reaction conditions $[Y_{low}, Y_{high}]$, and then have the computer find a reduced model which is guaranteed to be accurate over this range. To accomplish this, we replace the error constraints Eq. (16) with Eqs. (17)

$$\sum \max \text{TaylorModel}_m([Y_{low}, Y_{high}]) \times (1 - z_m) < \text{tols} \quad m = 1 \ldots N_{\text{reactions}}$$
$$\sum \min \text{TaylorModel}_m([Y_{low}, Y_{high}]) \times (1 - z_m) > -\text{tols} \quad (17)$$

where maxTaylorModel$_m$ is a guaranteed upper bound on $\omega_m(Y)$ over the range $[Y_{low}, Y_{high}]$ and minTaylorModel$_m([Y_{low}, Y_{high}])$ is a guaranteed lower bound on $\omega_m(Y)$. The two expressions in Eq. (17) thus represent guaranteed upper and lower bounds, respectively, on each element of $\omega_{\text{true}} - \omega_{\text{approx}}(Z)$ over the range $[Y_{low}, Y_{high}]$. These bounds are computed automatically by interval analysis using DAEPACK (Tolsma and Barton, 2000).

The constraints Eqs. (17) are still linear in z_m, so one can still find the global optimum, i.e. the reduced model with the smallest number of reactions that satisfies error constraints Eqs. (17). However, because applying interval analysis to the Taylor Model of $\omega_{\text{true}}(Y) - \omega_{\text{approx}}(Y)$ overestimates the deviation between ω_{true} and ω_{approx}, Eqs. (17) are a tighter constraint than Eq. (16), and the reduced model obtained will generally contain more reactions than that obtained using the first formulation. However, despite the loss in speed up due to the overestimate, this second approach is often much more convenient in practice, and our Taylor Model software can handle very large reaction mechanisms easily. For more technical details, see Oluwole et al. (Oluwole et al., 2007).

An example: The temperature field computed for a partially-premixed radially-symmetric methane/air flame is shown in Fig. 15. This is the same $\phi = 2.464$ laminar flame simulated by Bennett et al. (2000). We used the same 217 reaction full chemistry model used by Bennett et al. (2000) to compute the temperature field shown on the left-hand side of Fig. 15. On the left-hand side is shown the temperature field computed using the full chemistry model everywhere. On the right-hand side is shown the temperature field computed by the Adaptive Chemistry method using 13 different reduced models ranging in size from zero reactions to 156 reactions. As guaranteed by the error control

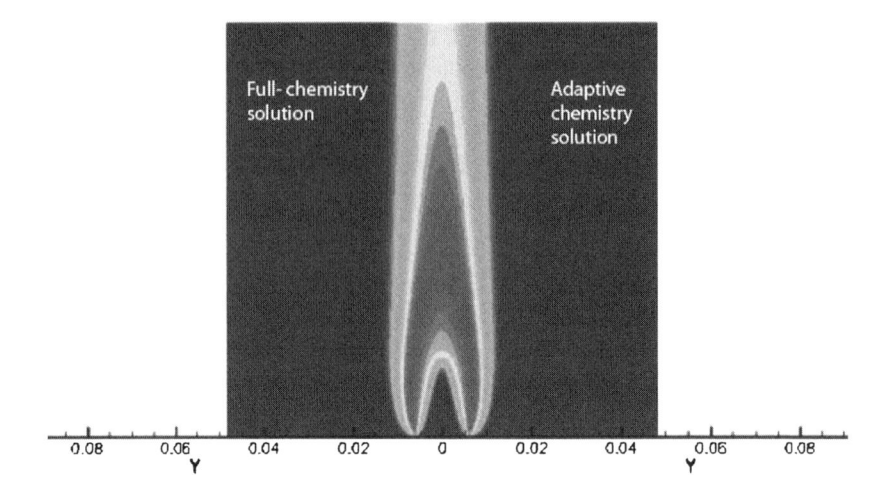

FIG. 15. Computed temperature field for a radially-symmetric partially-premixed methane–air laminar jet flame. The left-hand side shows the temperature field computed using the full chemistry simulation, and the right-hand side shows the temperature field computed using a set of reduced chemistry models that satisfy the error control constraints.

equations above, the two sets of model predictions agree to within the specified error tolerance.

IV. Are Model Predictions Consistent with Experimental Data?

Frequently, the next step after numerically solving a chemical kinetic simulation is to compare the model predictions with some experimental data, to check whether it is consistent with reality at least in one case. This is called "validating" the model. In the 20th century, it was a common practice to plot chemical kinetic model predictions with some experimental data, without any attempt to indicate the uncertainties in either. The reader then had to make his or her own judgment about whether the model and the data were close enough to be considered "consistent", or whether the data had disproved the model.

If the uncertainty information is available, then one can make a much more satisfactory quantitative comparison between the model predictions M_i (uncertainty range u_i) and the corresponding data D_i (uncertainty range σ_i), to determine whether the two are truly consistent. The extent of deviation between the model and the data scaled by the uncertainties is measured by χ^2, Eq. (18).

$$\chi^2 = \Sigma \left[\frac{(D_i - M_i)}{(\sigma_i + u_i)} \right]^2 \qquad (18)$$

From statistics, assuming normally distributed errors, the probability that the data and the model would differ so much that $\chi^2 > \Omega$ is given by Eq. (19).

$$P(\Omega) = \int_{\Omega}^{\infty} \frac{t^{m-1} e^{-t/2}}{\Gamma(m) 2^m} \, dt \tag{19}$$

where $m = (N_{\text{data}} - N_{\text{adjustable}})/2$. If we have a pure prediction (no parameters adjusted to improve the fit with the data) then $m = N_{\text{data}}/2$. If $P(\chi^2) < 0.1$ that means that there is less than 10% chance that the model and the data are consistent. Or in other words, we can say with 90% confidence that the model and the data are inconsistent, i.e. that the data disproved the model.

A. How to Prove Inconsistency Even if Some Parameters Are Highly Uncertain

The presumption of the approach above is that all the model parameters that significantly affect the model predictions M_i are known with reasonable accuracy, so it is not completely unreasonable to assume that the errors in the M_i are normally distributed (although a normal distribution is only reasonable if the model prediction could vary over the full range $-\infty$ to $+\infty$; for bounded quantities appropriately bounded distributions should be used). However, very often some of the important model parameters are not well established from prior work, so we do not know if their uncertainties are normally distributed. In fact, often it would be most reasonable to use the experimental data to try to determine these highly uncertain parameters **x**.

When experimental data is unavailable the critical question becomes: is there any physically reasonable choice of these very uncertain parameters **x** that makes the model and the data consistent? If not, we can be confident that the model and the data are inconsistent, and if we trust the data and the estimated error bars we can reasonably conclude that the data have disproved the model. Mathematically, we are confident the model and the data are inconsistent if and only if

$$\max_{\mathbf{x}} P(\chi^2(\mathbf{x})) < 0.1 \tag{20}$$

where

$$\chi^2(\mathbf{x}) = \Sigma \frac{(D_i - M_i(\mathbf{x}))^2}{(\sigma_i + u_i(\mathbf{x}))^2} \tag{21}$$

$$M_i(\mathbf{x}) = M_i[Y(t; \mathbf{x})] \tag{22}$$

Where $Y(t; \mathbf{x})$ is not known explicitly, but only implicitly as the solution of a system of stiff differential or DAEs such as Eq. (7) which depend parametrically on $\mathbf{x}$, and $\mathbf{x}$ is confined to lie in its physically reasonable range. The functional M_i that relates the recorded experimental measurement M_i to the underlying species concentrations and other state variables could in principle be quite complicated, but often it is linearly dependent on just a few of the species concentrations at some time $t_{\text{measurement}}$. As discussed in a later section, our estimate of the error bar in the model prediction for a fixed $\mathbf{x}$ might depend on $\mathbf{x}$, that is why $u_i(\mathbf{x})$ is written in Eq. (21).

While this is conceptually simple, numerically it is very difficult to find the global maximum of P, because of the fact that the dependence of Y on $\mathbf{x}$ is only known implicitly through a complicated numerical procedure (solving the whole simulation).

Note that if at any point we find any physically acceptable $\mathbf{x}$ which causes $P(\chi^2(\mathbf{x})) > 0.1$, then we know immediately that the model could be consistent with the data, and there is then no need to continue to try to find the global maximum of P. Once one has found any reasonable choice of model parameters that makes the model consistent with the data, then one should move onto questions about what ranges of parameters give predictions consistent with the data, and how the experimental data can be used to help refine and tighten the uncertainty ranges on the parameter values. This parameter-refinement step is indicated in Fig. 1 and has been discussed at length in the literature. It is complicated for nonlinear models with many significantly uncertain correlated parameters, since the range of parameter values consistent with the experiment will generally have a rather complex, often nonconvex, shape in a high-dimensional space. For a recent discussion of how one can deal with the parameter correlations in practice, see Feeley et al. (2004).

In the present work, we focus on the opposite case: if we cannot find any choice of $\mathbf{x}$, which makes the model consistent with the data, how do we prove that the model and data are inconsistent? To prove this, we must establish that all physically acceptable $\mathbf{x}$'s give $P(\chi^2(\mathbf{x})) < 0.1$, and for this we need to find or at least set an upper bound on the global maximum of P over the entire range of $\mathbf{x}$. It is in general much more difficult to find guaranteed-globally-optimal maxima than it is to find local maxima. Many of the most efficient methods for finding global optima rely on knowing the explicit function you are trying to optimize. But here the function we are trying to optimize is an integral involving the solution of a complicated stiff system of nonlinear differential equations that depend on the value of $\mathbf{x}$.

In the 20th century there was no way known to solve this type of optimization problem. However, very recently Singer and Barton (2006) have developed a branch-and-bound algorithm that is guaranteed to find the global optimum of $P(\chi^2(\mathbf{x}))$ for the important case where $Y(t;\mathbf{x})$ is specified by a system of nonlinear ODEs (as an initial value problem).

$$\frac{dY}{dt} = F(Y; \mathbf{x}) \quad (23a)$$

$$Y(t_0) = Y_0(\mathbf{x}) \quad (23b)$$

$$M = M[Y(t; \mathbf{x}); \mathbf{x}] \quad (23c)$$

The numerical procedure used in Singer's global dynamic optimization code (GDOC) software involving establishing bounds on the algebraic equations on the right-hand sides of Eqs. 23a, 23b and 23c over the physically reasonable range of the parameters $\mathbf{x}$ using interval analysis techniques similar to those used in Section III above. Then one constructs linear systems of differential equations that bound Y and P. Finally, one can systematically remove regions of $\mathbf{x}$ space which probably cannot contain the maximum of P. As the $\mathbf{x}$ space is subdivided, all the bounds tighten, until the procedure converges to the true global minimum of $\chi^2(\mathbf{x})$ (which is the global maximum of P). The details with proofs are presented elsewhere. A more concise explanation aimed at kineticists, with some practical examples, has been published recently. Practical kineticists interested in using the GDOC software to convincingly disprove chemical kinetic models are welcome to download it from http://yoric.mit.edu/

At present, the GDOC software can only handle problems with <6 highly uncertain parameters $\mathbf{x}$. It is hoped and expected that in the near future, more efficient algorithms will be found for handling systems with many more uncertain parameters, so that one could more easily determine, even in messy real-world systems, whether the discrepancies one observes between model predictions and data really indicate that the model must be wrong, or just indicate that one has not yet found the best values for the uncertain parameters.

B. How Standard Operating Practice Must Change in 21st Century

As discussed above, numerical tools now exist that make it possible to rigorously test whether chemical kinetic models and data are quantitatively consistent for many important cases. There is good reason to expect that the numerical methods will continue to improve rapidly, so that these quantitative models vs. data tests will become possible for ever more complex experiments. As discussed in Sections II and III, our capability to construct and solve complex simulations is also improving rapidly.

However, it is far from certain that we will be able to access the experimental data we need for these consistency checks (and for subsequent refinement of the models and model parameters) in a useful format. Historically, the field of kinetics has not been consistent in requiring that complete sets of experimental data be archived in the open literature, and in many journals error bars are optional. Instead, researchers often report only fitting parameters that they extracted from their experimental data using a model. The raw data is often lost in a few years, when the researcher or student who measured it moves on to a different position, and the media on which it is stored becomes obsolete. This makes it very difficult to try to reconstruct what was actually observed; in the common case where the model originally used to interpret the data had some flaw this can make the literature report about the experimental measurement almost useless.

The PrIMe project (http://www.primekinetics.org) is addressing these serious problems by developing software tools to make it easier for experimentalists to archive their measurements in an electronic format, so that it will be easier for future researchers to use the data for consistency checks and to refine model parameters.

In many respects, the situation is much worse in kinetic modeling than in experimental work. One seldom sees any kind of uncertainty estimate on a chemical kinetic model prediction, and as discussed extensively in Section II, many models have been incompletely documented, in part due to their reliance on a 20th century data model which made it inconvenient to document uncertainties and information on the origin of rate/thermo/transport parameter estimates. Very commonly, estimated rate constants and thermochemical parameters have been copied from prior reports (without error bars) without any serious attempt to check their veracity. After this cycle repeats several times, rough estimates made 50 years ago eventually become "firmly established" because they have been repeated so often in the literature, not because the numbers were ever really determined precisely. As a consequence, it seems likely that most chemical kinetic models in the literature will be discarded and replaced by new models as efficient reaction-mechanism-generation software based on well-documented rate and thermochemical estimates become widely deployed over the next few years: why bother to track down all the poorly documented assumptions in prior modeling work, when one can construct a brand new, possibly more complete, model-based on the latest rate and thermochemical data and estimates in just a few hours? This paradigm shift in kinetics will be most useful if it is accompanied by much more consistent treatment of uncertainty and uncertainty propagation.

C. ESTIMATING ERROR BARS ON MODEL PREDICTIONS

In order to estimate the uncertainties in a model's predictions (the u_i in Eq. (21)) for consistency checking, one must have estimates of the uncertainties in the model's input parameters. In the 20th century, model input uncertainties were usually converted into model prediction uncertainties using normalized first-order sensitivity coefficients S_{ij}, Eq. (24a). For ODEs and DAEs, first-order sensitivities can be computed very efficiently at the same time the simulation is solved (Maly and Petzold, 1996).

$$S_{ij} = \frac{\partial(\ln M_i)}{\partial(\ln k_j)} \quad i = 1\ldots N_{\text{predictions}}, \quad j = 1\ldots 2N_{\text{reactions}} \quad (24a)$$

Knowing which of the S_{ij} are large in magnitude can be very helpful to a modeler, but Eq. (24a) requires some care. First, in systems where T and P are not constant, one usually means the sensitivity with respect to the pre-exponential factor in the rate constant, i.e. a uniform scaling of $k_j(T,P)$ at all T,P (Kee et al., 1989).

More importantly, as written, the partial derivative implies that a single rate constant is to be varied, holding all the others constant, and indeed this is the way it is implemented in many sensitivity analysis routines. The index "j" runs out to $2N_{\text{reactions}}$ because each reaction has two rate constants, one for the forward direction, and one for the reverse. However, in order for the model to remain consistent with the laws of thermodynamics, the rate constant for the reverse of reaction "j" must vary simultaneously with the forward reaction, since the two rate constants must maintain a detailed-balance ratio related to the $\Delta G_{\text{reaction}}$, Eq. (2). This can be assured by specifying that the partial derivative is taken only for the forward reaction, while holding the thermochemistry fixed. Note that this also cuts the number of partial derivatives to be computed in half. These sensitivities should then be supplemented with sensitivities to the individual species' thermochemistry as in Eq. (25); overall the number of partial derivatives to be computed per model prediction M_i is $(N_{\text{reactions}} + N_{\text{species}})$, not $2N_{\text{reactions}}$.

$$S_{ij} = \frac{\partial(\ln M_i)}{\partial(\ln k_j)} \quad i = 1\ldots N_{\text{predictions}}, \quad j = 1\ldots N_{\text{reactions}} \text{ (forward directions only)}$$
$$(24b)$$

$$S_{im} = \frac{\partial(\ln M_i)}{\partial(\Delta H_f(\text{species } m))} \quad i = 1\ldots N_{\text{predictions}}, \quad m = 1\ldots N_{\text{species}} \quad (25)$$

If one (optimistically) believes the boundary conditions, initial conditions, and other molecular properties (e.g. transport properties, heat capacities) have all been specified very precisely, and one thinks the uncertainties in the rate constants and the enthalpies are not highly correlated, one might then estimate the uncertainty in prediction M_i by Eq. (26).

$$\delta M_i = u_i \sim [\Sigma|S_{ij}^*(\delta \ln k_j)|^2 + \Sigma|S_{im}^*\delta\Delta H_f(\text{species } m)|^2]^{1/2} \quad (26)$$

If one believes that the uncertainties in rate constants and thermo are highly correlated, a more conservative (pessimistic) estimate would be

$$\delta M_i = u_i \sim \Sigma |S_{ij}^*(\delta \ln k_j)| + \Sigma |S_{im}^* \delta \Delta H_f \text{ (species } m)| \qquad (27)$$

1. Moving Beyond First-Order Sensitivity Analysis

Of course, Eq. (24b) and Eq. (25) are only the first derivatives, and Eqs. (26) and (27) are just low-order approximations to u_i; for cases where the first derivative S_{ij} is small or where the uncertainties $\delta \ln(k_j)$ and $\delta \Delta H_f$ are large, one must consider higher-order effects (Li et al., 2002; Phenix et al., 1998; Vuilleumier et al., 1997). Many of the methods of including the higher-order effects, including Monte Carlo simulations, can be computationally expensive if the original kinetic simulation is hard to solve, but these calculations are certainly not impossible with modern computers, particularly if one only wants to consider the effects of large variations in a few key parameters. Certain quantities involving second-order sensitivities can be computed efficiently even for very large systems using adjoint methods (Oezyurt and Barton, 2005). But in the near future, for most cases involving large simulations, we expect that the routine first approximation to the error will be something like Eq. (26) or Eq. (27). In a limited number of important cases, e.g. when an apparent discrepancy has been identified, a more accurate method for uncertainty propagation, and a careful scrutiny of the most important estimated input uncertainties $\delta \ln(k_j)$ and $\delta \Delta H_f$(species m), will be required.

2. Correlated Uncertainties in Model Input Parameters

Often the uncertainties in the rate constants are highly correlated. For example, in large kinetic models, most of the rate constants are derived from a small number of functional group-based estimates, such as Evans–Polanyi parameters (Eq. (1)), some of which are significantly uncertain. In these cases, it would make sense to vary the Evans–Polanyi parameters, and so adjust the whole group of related rate constants, rather than allowing each rate constant to vary separately from all of the rest. For example, the high-pressure-limit rate for pent-2-yl radical addition to butadiene is not known very precisely, it could be uncertain by a factor of 10 in certain temperature ranges, but it almost certainly must be within a factor of two of the high-pressure-limit rate for but-2-yl + butadiene, because of the similarity between the functional groups involved. Similarly, many of the enthalpy values used in large kinetic models are very highly correlated, many of them coming from group-additivity estimates using the group values published more than 30 years ago in Benson's book (Benson, 1976). It would be very reasonable as a first approximation to vary the group values rather than varying the thermochemistry of all the individual species independently. However, to our knowledge, no one has yet produced software that makes it convenient to compute how these correlated uncertainties in the inputs to large kinetic models affect the model predictions.

3. Model Truncation Error

Model predictions can also be inaccurate due to the incompleteness of the chemical model, e.g. if some reactions or species were incorrectly omitted from the mechanism. If the missing species or reactions are completely missing from the database used by the model-construction software, there is no easy way to detect them (though perhaps a human expert might notice the omission in the tree databases described in Section II). There is certainly chemistry which is not well understood, even in the well-studied thermal gas-phase chemistry of small organic molecules; for example some of the important reactions of peroxyl radicals are still unclear (Taatjes, 2006), the true reaction path for $CH + N_2$ was only recently identified (Moskaleva and Lin, 2000), and recently some reactions that occur over ridges rather than saddle points have been identified (Townsend et al., 2004). It will be some time before there is a community consensus on how to correctly generalize from some of these observations.

It is also possible that reactions or species might be omitted from the model even if the chemistry is well-understood and was correctly included in the databases, for example if the species-selection tolerances were not set tightly enough during model-construction. Song et al. (2002) have discussed methods for trying to detect and avoid this type of omission. It should be possible to improve this approach using interval analysis techniques (Oluwole et al., 2007; Oluwole et al., 2006); perhaps it is possible to rigorously bound the error in the model predictions due to the finite error tolerance in species-selection, at least for certain types of simulations.

4. Other Sources of Error in Model Predictions

Finally, the model predictions always include some error because the simulation never exactly models the physical situation in the experiment, and because of numerical errors in solving the simulations. Both types of errors can be very important, even dominant, though in many systems these errors are smaller than the uncertainties in the chemistry model. It is very difficult to deal with the distressingly common problem that the model does not match the physical experimental situation; at present only human experts can address this problem (e.g. by numerically tests which demonstrate that including more experimental details does not materially change the simulation predictions). On the other hand, most of the numerical errors that arise when solving the simulations can be rigorously addressed through correctly designed algorithms and software.

At present, black-box programs for solving initial value problems typically maintain fairly rigorous automatic control of the numerical errors. However, for multidimensional reacting-flow simulations (boundary value problems), error control is usually not automated, relying on the user performing the correct checks to try to ensure that the calculation has actually converged to something close to the true solution. Schwer *et al.* (2003b) show how a relatively straightforward calculation can converge to an unphysical "solution". Many algorithms use low-order upwinding, but this introduces "numerical diffusion" which can lead to significantly erroneous conclusions in some cases; this issue is often difficult to deal with because removing the upwinding approximation can make the simulations numerically unstable. A major effort is needed to upgrade the numerical solvers for multidimensional reacting-flows and to incorporate error control, to make it easier to ensure that numerical errors are not significantly perturbing the simulation results, and to increase our confidence that two different researchers will arrive at the same converged solution if they start with the same reacting-flow equations. After this is achieved, we would then be in a better state to rationally decide which algorithms are really the most efficient and robust for solving different types of reacting-flow simulations.

V. Summary and Outlook

In this article, we have presented an overview of predictive chemical kinetics, a very broad field which is rapidly advancing in many ways. This manuscript also discusses some recent technical progress addressing a few of the sub-problems that must be overcome before predictive chemical kinetics can completely fulfill its potential:

(1) making the automated chemical mechanism-construction more extensible
(2) speeding multidimensional reacting-flow simulations involving complex chemistry, while maintaining rigorous error control
(3) conclusively determining whether or not model predictions are consistent with experimental data, when both are uncertain.

While there has been very significant recent progress on all three sub-problems, as discussed in Sections II, III, and IV, clearly much remains to be done before the iterative predict-measure-learn loop shown in Fig. 1 can function smoothly for most systems of interest to chemical engineers. For example, Section II describes a good method for storing and updating chemical information, but these data trees will need to be significantly expanded before predictive chemical kinetics will be usefully accurate for most systems. Similarly, in Section III we present a promising method for faster, error-controlled reacting-flow simulations, but this method is limited to steady-state processes, and these computations are still extremely difficult. Section IV discusses a method for performing rigorous consistency checks, but this method presumes that one has access to a lot of data and error bars on both data and model predictions that are usually not available at present. The overall message is that predictive chemical kinetics looks very promising in the early 21st century, but very significant efforts, including changes in the day-to-day functioning of the kinetics community, will be required in order to achieve this promise.

This manuscript certainly does not address all of the challenges to the advance of predictive chemical kinetics; for example it focuses only on homogeneous systems, and it is not even a complete review of efforts on any of the three sub-problems discussed in detail. However, this overview will hopefully be helpful to researchers trying to comprehend this broad and somewhat fragmented area of chemical engineering research, a field that will grow rapidly in the 21st century, both in technical capability and in practical importance.

ACKNOWLEDGMENTS

The work reported here was made possible through financial support from the US Department of Energy, the US National Science Foundation, and Alstom Power. The RMG program was written by Jing Song, with contributions from David Matheu, Sarah Petway, Sandeep Sharma, Paul Yelvington, and Joanna Yu. The functional group trees were designed by Sumathy Raman, who also computed many of the parameters, and the GUI was written by John Robotham. The numerical methods presented in Sections III and IV were developed in Prof. Paul Barton's group at MIT by John Tolsma and Adam Singer, and were adapted to chemical kinetic problems by Binita Bhattacharjee, O. Oluwole, Douglas Schwer, and James Taylor. Assistance from Greg Beran and Susan Lanza in the preparation of this manuscript is gratefully acknowledged. The figures were drawn by Tammy Keithley.

NOMENCLATURE

χ^2	weighted sum of squares of deviations between model and experiment
α	Evans–Polanyi slope
n	temperature exponent
P	probability that one would measure data giving such a low χ^2 value if the model were true
Ω	cut-off in χ^2 for a model to acceptably fit the data
Z	integer array indicating which reactions are kept or deleted
$T\,[=]$Kelvin	temperature
$P\,[=]$atm	pressure
$t\,[=]$sec	time

E_o [=]kJ/mole	Evans–Polanyi barrier for hypothetical isothermic reaction
k [=]sec^{-1} or [=]liter/mole-second	Rate constant (unimolecular or bimolecular)
A [=](units of k)/(Kelvin)^n	modified-Arrhenius prefactor
E_a [=]kJ/mole	activation energy
R [=]J/mole-Kelvin	gas constant
k_B [=]J/Kelvin	Boltzmann's constant
h [=]Js	Planck's constant
G [=]kJ/mole	Gibb's free energy
H [=]kJ/mole	enthalpy
Θ [=]mole/s	rate of change in species due to transport
ω [=]mole/s	rate of change in species due to chemical reaction Θ, ω are generalized to include change in other state variables as well as species.
Y	the state variables in a CFD simulation
D	the array of measured data values
M	the array of model predictions for these data
σ	uncertainties in D
u	uncertainties in M
x	adjustable parameters in a model
F	function returning rate of change of Y
M	functional for computing M from Y
S	normalized sensitivity of model predictions M

REFERENCES

Aghalayam, P., Park, Y. K., and Vlachos, D. G. *AIChE J.* **46**, 2017–2029 (2000).

Bennett, B. A. V., McEnally, C. S., Pfefferle, L. D., and Smooke, M. D. *Combust. Flame* **123**, 522–546 (2000).

Benson, S. W., "Thermochemical Kinetics". 2d ed. Wiley, New York (1976).

Berger, M. J., and Oliger, J. *J. Comput. Phys.* **53**, 484–512 (1984).

Bernan, K. E., Campbell, S. L., and Petzold, L. R., "Numerical Solution of Initial-Value Problems in Differential-Algebraic Equations". 2nd ed. Elsevier, New York (1989).

Bhattacharjee, B., Schwer, D. A., Barton, P. I., and Green, W. H. *Combust. Flame* **135**, 191–208 (2003).

Blurock, E. S. *J. Chem. Inf. Comput. Sci.* **35**, 67 (1995).

Blurock, E. S. *J. Chem. Inf. Comput. Sci.* **44**, 1336–1347 (2004a).

Blurock, E. S. *J. Chem. Inf. Comput. Sci.* **44**, 1348–1357 (2004b).

Bozzelli, J. W., Chang, A. Y., and Dean, A. M. *Int. J. Chem. Kinet.* **29**, 161–170 (1997).

Broadbelt, L. J., Stark, S. M., and Klein, M. T. *Ind. Eng. Chem. Res.* **33**, 790–799 (1994).

Brown, P. N., Byrne, G. D., and Hindmarsh, A. C. *SIAM J. Sci. Stat. Comput.* 1038–1064 (1989).

Buda, F., Heyberger, B., Fournet, R., Glaude, P.-A., Warth, V., and Battin-Leclerc, F. *Energy Fuels* **20**, 1450–1459 (2006).

Caracotsios, M., and Stewart, W. E. *Comput. Chem. Eng.* **19**, 1019–1030 (1995).
Chang, A. Y., Bozzelli, J. W., and Dean, A. M. *Zeitschrift fuer Physikalische Chemie (Muenchen)* **214**, 1533–1568 (2000).
Chen, J. Y. *Combust. Sci. Technol.* **57**, 89–94 (1988).
Chevalier, C., Warnatz, J., and Melenk, H. *Berichte der Bunsen-Gesellschaft* **94**, 1362–1367 (1990).
Chinnick, S. J., Baulch, D. L., and Ayscough, P. B. *Chemom. Intell. Lab. Syst.* **5**, 39 (1988).
Dean, A. M. *J. Phys. Chem.* **89**, 4600–4608 (1985).
Dente, M., Ranzi, E., and Goossens, A. G. *Comput. Chem. Eng.* **3**, 61–75 (1979).
DiMaio, F. P., and Lignola, P. G. *Chem. Eng. Sci.* **47**, 2713 (1992).
Feeley, R., Seiler, P., Packard, A., and Frenklach, M. *J. Phys. Chem. A* **108**, 9573–9583 (2004).
Feng, X.-J., and Rabitz, H. *Biophys. J.* **86**, 1270–1281 (2004).
Forst, W., "Theory of Unimolecular Reactions". Academic Press, New York (1973).
Frenklach, M., Wang, H., and Rabinowitz, M. J. *Prog. Energy Combust. Sci.* **18**, 47 (1992).
Gear, C. W. *Commun. ACM* **14**, 176–179 (1971).
Glaude, P. A., Conraud, V., Fournet, R., Battin-Leclerc, F., Come, G. M., Scacchi, G., Dagaut, P., and Cathonnet, M. *Energy Fuels* **16**, 1186–1195 (2002).
Hillewaert, L. P., Diericks, J. L., and Froment, G. F. *AIChE J.* **34**, 17 (1988).
Kee, R. J., Rupley, F. M., and Miller, J. A. Sandia National Laboratories, Livermore, CA, USA (1989).
Kojima, S. *Combust. Flame* **99**, 87–136 (1994).
Lay, T. H., and Bozzelli, J. W. *Chem. Phys. Lett.* **268**, 175–179 (1997a).
Lay, T. H., and Bozzelli, J. W. *J. Phys. Chem. A* **101**, 9505–9510 (1997b).
Lay, T. H., Bozzelli, J. W., Dean, A. M., and Ritter, E. R. *J. Phys. Chem.* **99**, 14514–14527 (1995).
Lay, T. H., Yamada, T., Tsai, P.-L., and Bozzelli, J. W. *J. Phys. Chem. A* **101**, 2471–2477 (1997).
Li, G., Wang, S.-W., Rabitz, H., Wang, S., and Jaffe, P. *Chem. Eng. Sci.* **57**, 4445–4460 (2002).
MacNeice, P., Olson, K. M., Mobarry, C., de Fainchtein, R., and Packer, C. *Comput. Phys. Commun.* **126**, 330–354 (2000).
Maly, T., and Petzold, L. R. *Appl. Numer. Math.* **20**, 57 (1996).
Matheu, D. M. Thesis, PhD. Dissertation Massachusetts Institute of Technology (2003).
Matheu, D. M., Dean, A. M., Grenda, J. M., and Green, W. H. *J. Phys. Chem. A* **107**, 8552–8565 (2003a).
Matheu, D. M., Green, W. H., and Grenda, J. M. *Int. J. Chem. Kinet.* **35**, 95–119 (2003b).
Matheu, D. M., and Grenda, J. M. *J. Phys. Chem. A* **109**, 5332–5342 (2005a).
Matheu, D. M., and Grenda, J. M. *J. Phys. Chem. A* **109**, 5343–5351 (2005b).
Melius, C. F., and Allendorf, M. D. *J. Phys. Chem. A* **104**, 2168–2177 (2000).
Milford, J. B., Gao, D., Russell, A. G., and McRae, G. J. *Environ. Sci. Technol.* **26**, 1179–1189 (1992).
Moskaleva, L. V., and Lin, M. C. *Proc. Combust. Inst.* **28**, 2393–2401 (2000).
Oezyurt, D. B., and Barton, P. I. *Ind. Eng. Chem. Res.* **44**, 1804–1811 (2005).
Oluwole, O. O., Bhattacharjee, B., Barton, P. I., and Green, W. H. *Combust. Flame* **146**, 348–365 (2006).
Oluwole, O. O., Bhattacharjee, B., Tolsma, J., Barton, P. I., and Green, W. H. *Combust. Theor. Model.* **11**(1), 127–146 (2007).
Peters, N., and Rogg, B., "Reduced Kinetic Mechanisms for Applications in Combustion Systems". Springer-Verlag, Berlin and New York (1993).
Petway, S. V. Application of Automatic Reaction Model Generation to Chlorine-Initiated Neopentane Oxidation, (Poster Presentation) 6th International Conference on Chemical Kinetics, NIST (Gaithersburg, MD) (2005).
Petzold, L. R. Report SAND82-8637. Sandia National Laboratories (1982).
Phenix, B. D., Dinaro, J. L., Tatang, M. A., Tester, J. W., Howard, J. B., and McRae, G. J. *Combust. Flame* **112**, 132–146 (1998).
Pope, S. B. *Combust. Theor. Model.* **1**, 41–63 (1997).

Prickett, S. E., and Mavrovouniotis, M. L. *Comput. Chem. Eng.* **21**, 1219–1235 (1997a).
Prickett, S. E., and Mavrovouniotis, M. L. *Comput. Chem. Eng.* **21**, 1237–1254 (1997b).
Prickett, S. E., and Mavrovouniotis, M. L. *Comput. Chem. Eng.* **21**, 1325–1337 (1997c).
Quann, R. J., and Jaffe, S. B. *Ind. Eng. Chem. Res.* **31**, 2483–2497 (1992).
Ranzi, E., Faravelli, T., Gaffuri, P., and Sogaro, A. *Combust. Flame* **102**, 179–192 (1995).
Ratkiewicz, A. T., and Truong, T. N. *Int. J. Quantum Chem.* **106**, 244–255 (2006).
Ritter, E. R., and Bozzelli, J. W. *Int. J. Chem. Kinet.* **23**, 767–778 (1991).
Ruscic, B., Pinzon, R. E., Morton, M. L., von Laszevski, G., Bittner, S. J., Nijsure, S. G., Amin, K. A., Minkoff, M., and Wagner, A. F. *J. Phys. Chem. A* **108**, 9979–9997 (2004).
Schwer, D. A., Lu, P., and Green, W. H. *Combust. Flame* **133**, 451–465 (2003a).
Schwer, D. A., Lu, P., Green, W. H.Jr., and Semiao, V. *Combust. Theor. Model* **7**, 383–399 (2003b).
Schwer, D. A., Tolsma, J. E., Green, W. H., and Barton, P. I. *Combust. Flame* **128**, 270–291 (2002).
Shenvi, N., Geremia, J. M., and Rabitz, H. *J. Phys. Chem. A* **106**, 12315–12323 (2002).
Singer, A. B., and Barton, P. I. *J. Global Optim.* **34**, 159–190 (2006).
Sirdeshpande, A. R., Ierapetritou, M. G., and Androulakis, I. P. *AIChE J.* **47**, 2461–2473 (2001).
Song, J. Thesis, PhD. Dissertation Massachusetts Institute of Technology (2004).
Song, J., Stephanopoulos, G., and Green, W. H. *Chem. Eng. Sci.* **57**, 4475–4491 (2002).
Song, J., Sumathi, R., Green, W. H., and Van Geem, K. *Int. J. Chem. Kinet.* (submitted, under revision).
Stein, S. E., and Rabinovitch, B. S. *J. Chem. Phys.* **58**, 2438–2445 (1973).
Stein, S. E., and Rabinovitch, B. S. *J. Chem. Phys.* **60**, 908–917 (1974).
Sumathi, R., and Green, W. H. Jr. Theor. Chem. Acc. **108**, 187–213 (2002).
Susnow, R. G., Dean, A. M., Green, W. H., Peczak, P., and Broadbelt, L. J. *J. Phys. Chem. A* **101**, 3731–3740 (1997).
Taatjes, C. A. *J. Phys. Chem. A* **110**, 4299–4312 (2006).
Tolsma, J., and Barton, P. I. *Ind. Eng. Chem. Res.* **39**, 1826–1839 (2000).
Tomlin, A. S., Turanyi, T., and Pilling, M. J. *Compr. Chem. Kinet.* **35**, 293–437 (1997).
Tonse, S. R., Moriarty, N. W., Brown, N. J., and Frenklach, M. *Isr. J. Chem.* **39**, 97–106 (1999).
Townsend, D., Lahankar, S. A., Lee, S. K., Chambreau, S. D., Suits, A. G., Zhang, X., Rheinecker, J., Harding, L. B., and Bowman, J. M. *Science* **306**, 1158–1161 (2004).
Truhlar, D. G., Garrett, B. C., and Klippenstein, S. J. *J. Phys. Chem.* **100**, 12771–12800 (1996).
Van Geem, K., Reyniersa, M. F., Marina, G., Song, J., Matheu, D. M., and Green, W. H. *AIChE J.* **52**(2), 718–730 (2006).
Venkatesh, P. K., Chang, A. Y., Dean, A. M., Cohen, M. H., and Carr, R. W. *AIChE J.* **43**, 1331–1340 (1997).
Vuilleumier, L., Harley, R. A., and Brown, N. J. *Environ. Sci. Technol.* **31**, 1206–1217 (1997).
Warth, V., Battin-Leclerc, F., Fournet, R., Glaude, P. A., Come, G. M., and Scacchi, G. *Comput. Chem.* **24**, 541–560 (2000).
Wilk, R. D., Cohen, R. S., and Cernansky, N. P. *Ind. Eng. Chem. Res.* **34**, 2285–2297 (1995).
Yelvington, P. E., Rallo, M. B. I., Liput, S., Tester, J. W., Green, W. H., and Yang, J. *Combust. Sci. Technol.* **176**, 1243–1282 (2004).
Yu, J., Sumathi, R., and Green, W. H. *J. Am. Chem. Soc.* **126**, 12685–12700 (2004).
Zachariah, M. R., and Melius, C. F., Bond-additivity correction of ab initio computations for accurate prediction of thermochemistry, *Comput. Thermochem.* **677**, 162–175 (1998).
Zarth, C., Tirtowidjojo, M., and West, D. Annual Meeting Archive – American Institute of Chemical Engineers, Indianapolis, IN, United States (November 3–8, 2002), pp. 2627–2646 (2002).

KINETIC MODELLING OF PYROLYSIS PROCESSES IN GAS AND CONDENSED PHASE

Mario Dente, Giulia Bozzano, Tiziano Faravelli, Alessandro Marongiu, Sauro Pierucci and Eliseo Ranzi[*]

Dipartimento di Chimica, Materiali e ingegneria Chimica, Politecnico di Milano, Piazza L. da Vinci 32, 20133 Milano, Italy

I. Introduction	52
II. Kinetic Modelling of Pyrolysis Reactions	55
A. Automatic Generation of Pyrolysis Mechanism	64
B. Pyrolysis of Large Hydrocarbons	72
C. Detailed Characterization of Liquid Feeds	90
D. Extension of the Modelling to Liquid-Phase Pyrolysis	96
III. Fouling Processes: Formation of Carbonaceous Deposits and Soot Particles	100
A. Fouling and Coking Mechanisms in Pyrolysis Coils and TLE's	101
B. Evolution of Structure and Properties of the Deposit	107
C. Soot Formation	114
IV. Applications	124
A. Steam Cracking Process and Pyrolysis Coils	124
B. Visbreaking and Delayed Coking Processes	129
C. Thermal Degradation of Polymers	136
V. Conclusions	150
List of Symbols	151
Appendix 1. MAMA Program: Automatic Generation of Primary Lumped Reactions	152
A.1. Evaluation of End-Product Composition	152
A.2. Lumping of Components	155
A.3. Database Structure	157
A.4. Molecule Homomorphism Algorithm	159
A.5. Schematic Functionality of the Kinetic Generator	160
References	161

Abstract

The main goal of this paper is to discuss the common features in the development and validation phases of the reaction schemes able to describe pyrolysis processes in both gas and condensed phases. The

---

[*]Corresponding author. E-mail: eliseo.ranzi@polimi.it

complexity of these systems is due not only to the large number of elementary reactions involved but also to the difficulty of properly characterizing the reacting mixtures. This is typical of the pyrolysis process when liquid feedstocks, such as naphtha and gasoil, are vaporized and cracked to produce ethylene and alkenes. However, it is also true for the refinery processes involved in upgrading heavy feeds and in the combustion of liquid feeds, such as gasoline, diesel or jet propulsion fuels. Consequently, apart from the analysis of the chemistry involved in the reacting systems, another important step is the characterization of these complex hydrocarbon mixtures. High-temperature radical reactions, typical of pyrolysis processes, are characterized by their modular and hierarchical structure. This feature means that pyrolysis reactions can be studied by starting from the simpler systems and then progressively extending the simulation capability of the model to more complex situations. High-temperature pyrolysis of large hydrocarbon species rapidly gives rise to small radicals and species, and their interactions are common ground shared by all pyrolysis systems. The interactions of small hydrocarbon species, such as hydrogen, methane, ethane and ethylene, together with their parent radicals are the true core of the kinetic scheme and they constitute the first hierarchical step in all the pyrolysis models. The same mechanisms and similar radical reactions are also extended to the kinetic modelling study of the pyrolysis of liquid and condensed phases. Using a similar approach, it is possible to deal with the kinetic modelling of carbon residue and carbonaceous deposit formation on pyrolysis coils, and, more generally, on the metallic walls of different process units. These pyrolysis and condensation reactions help explain the soot and carbon particle formation in combustion processes. Always a similar kinetic approach and the same lumping techniques are conveniently applied moving from the simpler system of ethane dehydrogenation to produce ethylene, up to the soot formation in combustion environments. A brief discussion on the mathematical modelling of steam cracking, visbreaking and delayed coking processes shows the practical and direct interest of these kinetic models. The thermal degradation of the plastics, of very great environmental interest, is a further and final application example of pyrolysis reactions in the condensed phase and concludes this kinetic analysis.

I. Introduction

For several years now, detailed and complex radical reaction schemes have been widely used by the scientific and technical community to describe pyrolysis and combustion processes. At the beginning of 1970s, the first detailed kinetic models for methane and methanol oxidation (Bowman, 1975; Seery and Bowman, 1970; Smoot *et al.*, 1976) were developed by taking advantage of the early stiff kinetic solvers (Gear, 1968), as Westbrook neatly summarized recently (Westbrook *et al.*, 2004). Using a detailed kinetic scheme of several thousand elementary steps to describe pyrolysis reaction chemistry was considered enormously innovative at the time and largely unknown to the scientific community at large (Dente *et al.*, 1970; Sundaram and Froment, 1978). The numerical solution of the very large ordinary differential equations (ODE) system of chemical component mass balances was obtained using a specifically conceived original and efficient numerical method (Dente *et al.*, 1979). More recently, theoretical concepts and methods, derived from vapour phase modelling experience and various modifications, were extended to describe pyrolysis reactions in the condensed phase. These applications first focused on refinery processes for heavy distillation residues, which essentially meant visbreaking and delayed coking. The second most important extension was to the thermal degradation of polymers. This kinetic analysis and approach also provided practical support for the kinetic studies on fouling phenomena in different process units.

Nowadays, improved computing facilities and, more importantly, the availability of the Chemkin package (Kee and Rupley, 1990) and similar kinetic compilers and processors have made these complex kinetic schemes more user-friendly and allows the study of process alternatives as well as the design and optimization of pyrolysis coils and furnaces. In spite of their rigorous, theoretical approach, these kinetic models of pyrolysis have always been designed and used for practical applications, such as process simulation, feedstock evaluations, process alternative analysis, reactor design and optimization, process control and so on. Despite criticisms raised recently by Miller *et al.* (2005), these detailed chemical kinetic models constitute an excellent tool for the analysis and understanding of the chemistry of such systems.

The main focus of this paper is to review and discuss certain features that are crucial to the various development and validation phases of these complex reaction schemes. In combustion practice as well as pyrolysis and refinery processes, the complexity of the reacting system is due not only to the large number of elementary reactions involved but also to the difficulty of properly characterizing the reacting mixture. This is typical of the pyrolysis process when liquid feedstocks, such as naphtha and gasoil, are vaporized and cracked to produce ethylene and alkenes. However, it is also true for the refinery processes involved in upgrading heavy feeds and in the combustion of liquid feeds, such as gasoline, kerosene, diesel or jet propulsion fuels. Consequently, another important step, apart from the analysis of the chemistry involved in the reacting systems, is the reliable characterization of these complex hydrocarbon mixtures. It is mostly for this reason that there is such great interest in the combustion community in studying properly designed and defined mixtures of reference components (surrogate fuels) with the aim of analysing the decomposition and combustion behaviour of real transportation fuels in a reproducible way (Edwards and Maurice, 2001; Hudgens, 2003; Ranzi, 2006).

Gas-phase pyrolysis processes are characterized by high-temperature radical reactions. One relevant aspect of these complex and detailed reaction schemes is their modular and hierarchical structure (Westbrook and Dryer, 1984). This feature means that combustion or pyrolysis reactions can be studied by starting from the simpler systems and then progressively extending the simulation capability of the model to new and more complex situations. High-temperature pyrolysis of large hydrocarbons rapidly gives rise to small radicals and species, and their interactions are common ground shared by all pyrolysis systems. The interactions of small hydrocarbon species, such as hydrogen, methane, ethane and ethylene, together with their parent radicals are the true core of the kinetic scheme. In fact, these reactions constitute the first hierarchical step in the pyrolysis model. For example, successive dehydrogenation and condensation reactions of small alkenes, vinyl and allyl radicals are of vital importance in the characterization of ethylene formation from the degradation of all the different feeds, gases and liquids. Similarly, the high-temperature combustion of heavy fuels proceeds with the initial breaking down of the fuel to smaller radicals before the formation of the ultimate oxidation products. With the exception of certain resonance stabilized radicals, their decomposition rates are so fast that they form smaller unsaturated compounds. The smaller compounds, in rich conditions, can follow pyrolytic paths too and can form poly-aromatic hydrocarbons (PAH) and carbon particles. Pyrolysis reactions are thus critical not only when it comes to describing different refinery and petrochemical processes but also in extending the available databases to cover the decomposition behaviour of liquid fuels as well. The relative weight of these radical reactions is dependent on the operating conditions of the reacting system. The competitive nature of the pyrolysis and combustion process is of key importance. Thus, the gas-phase pyrolysis reactions of light hydrocarbon species hierarchically precede the kinetic models of heavier species.

As previously mentioned the same mechanisms and similar radical reactions can be extended to the kinetic modelling study of the pyrolysis of liquid and condensed phases. Using a similar approach, it is also possible to deal with the kinetic modelling of carbon residue and carbonaceous deposit formation on pyrolysis coils, and, more generally, on the metallic walls of different process units. These pyrolysis and condensation reactions help explain the soot and carbon particle formation in combustion processes. Finally, the last section of this chapter features brief discussion on the kinetic modelling of steam cracking, visbreaking and delayed coking processes. The thermal degradation of the plastics, of very great environmental interest, is a further example of pyrolysis reactions in the condensed phase and concludes this kinetic analysis.

II. Kinetic Modelling of Pyrolysis Reactions

It is now firmly established that free radical reactions dominate the thermal degradation of hydrocarbon species (Benson, 1976; Dente *et al.*, 1983; Poutsma, 2000; Savage, 2000). Only mechanistic radical kinetic schemes can provide reliable descriptions of the pyrolysis process.

In summary, the main reaction classes during pyrolysis are:

1. Initiation and termination (or recombination) reactions

$$\text{Unimolecular} \quad CH_3-CH_2-CH_2-CH_3 \rightleftharpoons C_2H_5\bullet + C_2H_5\bullet$$

$$\text{Bimolecular} \quad CH_3-CH_3 + CH_2=CH_2 \rightleftharpoons C_2H_5\bullet + C_2H_5\bullet$$

2. Propagation reactions
   H-abstraction (or H-metathesis) on molecules

$$CH_3\bullet + CH_3-CH_3 \rightleftharpoons CH_4 + C_2H_5\bullet$$

Alkyl radical isomerization via (1,4), (1,5) and (1,6) H-transfer

$$CH_2\bullet-CH_2-CH_2-CH_2-CH_3 \rightleftharpoons CH_3-CH_2-CH_2-CH\bullet-CH_3$$

$$CH_2\bullet-(CH_2)_3-CH_2-CH_3 \rightleftharpoons CH_3-(CH_2)_3-CH\bullet-CH_3$$

$$CH_2\bullet-(CH_2)_4-CH_2-CH_3 \rightleftharpoons CH_3-(CH_2)_4-CH\bullet-CH_3$$

Addition of radicals to unsaturated molecules and alkyl radical decomposition

$$CH_3\bullet + CH_2=CH-CH_3 \rightleftharpoons CH_3-CH_2-CH\bullet-CH_3$$

It is worthwhile pointing out that certain concerted path molecular reactions can play a significant role too. Typical examples are *cyclo*-alkane and olefin isomerization as well as dehydrogenation and "ene" decomposition via four- and six-centre reactions

$$CyC_6H_{12} \rightarrow CH_2=CH-CH_2-CH_2-CH_2-CH_3$$
$$CH_3-CH=CH-CH_3 \rightarrow CH_2=CH-CH_2-CH_3$$
$$CH_2=CH-CH_2-CH_3 \rightarrow CH_2=CH-CH=CH_2 + H_2$$
$$CH_2=CH-CH_2-CH_2-CH_3 \rightarrow CH_3-CH=CH_2 + CH_2=CH_2$$

There is an abundance of groundwork in the scientific literature regarding both the fundamental and applied chemical kinetics of pyrolysis processes.

The n-butane pyrolysis is analysed here as an initial, simple example of a pyrolysis reaction mechanism. It is important to note that the pyrolysis reactions of small hydrocarbons are fundamental to the proper understanding of the whole process. In fact, the pyrolysis mechanism displays a typical hierarchical structure and the small hydrocarbons must be analysed first. Fig. 1 shows the main and minor products from n-butane decomposition, under isothermal conditions, at 1,093 K and 1 atm. Ethylene, propylene and methane are the main products, while only trace amount of butenes, ethane, benzene and cyclopentadiene are observed. These model predictions have been confirmed and validated by several experimental measurements (Dente and Ranzi, 1983).

The most common initiation or homolysis reaction is the breaking of a covalent C–C bond with the formation of two radicals. This initiation process is highly sensitive to the stability of the formed radicals. Its activation energy is equal to the bond dissociation enthalpy because the reverse, radical–radical recombination reaction is so exothermic that it does not require activation energy. C–C bonds are usually weaker than the C–H bonds. Thus, the initial formation of H radicals can be ignored. The total radical concentration in the reacting system is controlled both by these radical initiation reactions and by the termination or radical recombination reactions. In accordance with Benson (1960), the rate constant expressions of these unimolecular decompositions are calculated from the reverse reaction, the recombination of two radical species to form the stable parent compound, and microscopic reversibility (Curran et al., 1998). The reference kinetic parameters for the unimolecular decomposition reactions of n-alkanes for each single fission of a C–C bond between secondary

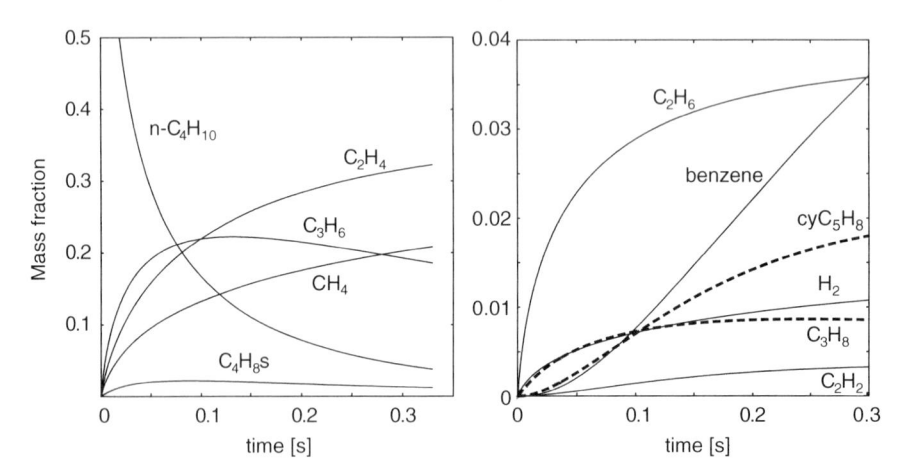

FIG. 1. Mass fractions of the main products from n-butane decomposition, at 1,093 K and 1 atm (Model predictions).

C-atoms, are

$$k_{\text{ref}} = 5.0 \times 10^{16} \exp\left(\frac{-81,000}{RT}\right) \quad [\text{s}^{-1}]$$

Thus, the initiation reaction of n-butane is

$$n\text{C}_4\text{H}_{10} \rightleftharpoons \text{C}_2\text{H}_5\bullet + \text{C}_2\text{H}_5\bullet$$

where the kinetic parameters suggested by Tsang (1978c), $k = 2.51 \times 10^{16} \exp(-82,000/RT)$, by Warnatz (1984), $k = 2 \times 10^{16}\exp(-81,200/RT)$, and finally by Dean (1985), $7.94 \times 10^{16}\exp(-80,150/RT)$, agree well with the reference rate parameters. About 2,000 kcal/kmol of extra energy is required to split off a terminal methyl group. Taking into account the hydrocarbon symmetry, it is therefore possible to assume

$$k = 10^{17} \exp\left(\frac{-83,500}{RT}\right) \quad [\text{s}^{-1}]$$

for the second initiation reaction of n-butane

$$n\text{C}_4\text{H}_{10} \rightarrow 1\text{-}\text{C}_3\text{H}_7\bullet + \text{CH}_3\bullet$$

Once initiation reactions generate the propagating radicals, the primary and main products are explained simply by the H-abstraction reactions and the subsequent, fast decomposition of 1- and 2-butyl radicals. All the different H-abstracting radicals (R•) can produce the two n-butyl radicals.

$$\text{H}\bullet + n\text{C}_4\text{H}_{10} \rightarrow \text{H}_2 + 1\text{-}\text{C}_4\text{H}_9\bullet$$
$$k = 1.29 \times 10^4 T^2 \exp\left(\frac{-6,500}{RT}\right) \quad [\text{m}^3/\text{kmol s}]$$

$$\text{H}\bullet + n\text{C}_4\text{H}_{10} \rightarrow \text{H}_2 + 2\text{-}\text{C}_4\text{H}_9\bullet$$
$$k = 0.86 \times 10^4 T^2 \exp\left(\frac{-4,000}{RT}\right) \quad [\text{m}^3/\text{kmol s}]$$

$$\text{CH}_3\bullet + n\text{C}_4\text{H}_{10} \rightarrow \text{CH}_4 + 1\text{-}\text{C}_4\text{H}_9\bullet$$
$$k = 2.34 \times 10^2 T^2 \exp\left(\frac{-7,500}{RT}\right) \quad [\text{m}^3/\text{kmol s}]$$

$$\text{CH}_3\bullet + n\text{C}_4\text{H}_{10} \rightarrow \text{CH}_4 + 2\text{-}\text{C}_4\text{H}_9\bullet$$
$$k = 1.56 \times 10^2 T^2 \exp\left(\frac{-5,000}{RT}\right) \quad [\text{m}^3/\text{kmol s}]$$

$$C_2H_5\bullet + nC_4H_{10} \rightarrow C_2H_6 + 1\text{-}C_4H_9\bullet$$
$$k = 1.40 \times 10^2 T^2 \exp\left(\frac{-10,500}{RT}\right) \quad [\text{m}^3/\text{kmol s}]$$

$$C_2H_5\bullet + nC_4H_{10} \rightarrow C_2H_6 + 2\text{-}C_{10}H_{21}\bullet$$
$$k = 0.92 \times 10^2 \, T^2 \exp\left(\frac{-7,700}{RT}\right) \quad [\text{m}^3/\text{kmol s}]$$

$$1\text{-}C_3H_7\bullet + nC_4H_{10} \rightarrow C_3H_8 + 1\text{-}C_4H_9\bullet$$
$$k = 0.81 \times 10^2 T^2 \exp\left(\frac{-9,400}{RT}\right) \quad [\text{m}^3/\text{kmol s}]$$

$$1\text{-}C_3H_7\bullet + nC_4H_{10} \rightarrow C_3H_8 + 2\text{-}C_4H_9\bullet$$
$$k = 0.54 \times 10^2 T^2 \exp\left(\frac{-6,600}{RT}\right) \quad [\text{m}^3/\text{kmol s}]$$

Butyl radicals decompose quickly to form ethylene and propylene. At high temperatures, alkyl radical decomposition reactions constitute an important reaction class and the prevailing fate of alkyl radicals. Take 1-propyl and 1-butyl radicals, for example. These primary alkyl radicals give rise to the following β-decomposition reactions:

$$1\text{-}C_3H_7\bullet \rightarrow C_2H_4 + CH_3\bullet \quad k = 3 \times 10^{13} \exp\left(\frac{-32,000}{RT}\right) \quad [\text{s}^{-1}]$$

$$1\text{-}C_4H_9\bullet \rightarrow C_2H_4 + C_2H_5\bullet \quad k = 3 \times 10^{13} \exp\left(\frac{-30,000}{RT}\right) \quad [\text{s}^{-1}]$$

These values agree closely with the ones proposed by Warnatz (1984), Dean (1985) and Tsang (1988). At the usual pyrolysis temperatures, these rate constants are $10^{6.5}$ and $10^7 \, \text{s}^{-1}$, respectively, i.e. the lifetime of alkyl radicals is shorter than 10^{-6} s. Similarly, 2-butyl radical can form propylene and methyl radical,

$$2\text{-}C_4H_9\bullet \rightarrow C_3H_6 + CH_3\bullet \quad k = 3 \times 10^{13} \exp\left(\frac{-32,000}{RT}\right) \quad [\text{s}^{-1}]$$

Via this mechanism, n-butyl radicals decompose directly to form ethylene and propylene, with ethyl and methyl radicals, respectively. The successive dehydrogenation reaction of ethyl radical forms ethylene and H radicals,

$$C_2H_5\bullet \rightarrow C_2H_4 + H\bullet \quad k = 10^{14} \exp\left(\frac{-40,000}{RT}\right) \quad [\text{s}^{-1}]$$

The high-pressure limit of the kinetic constant of this reaction clearly indicates that the dehydrogenation reactions are less favoured than the dealkylation ones.

Only about 3 wt% of ethane is observed in the steam cracking products, indicating that the formation of ethylene is the preferential fate of ethyl radicals. Note that most of ethane is formed via ethyl radical H-abstraction reactions, while less than 10% is due to the recombination reaction of methyl radicals. Similarly, propane formation is mostly due to the H-abstraction reactions of propyl radicals and only marginally to the recombination of methyl and ethyl radicals.

It is important to stress that all the alkyl radicals can also undergo dehydrogenation reactions with rate constants, which depend on the type of radical and the number of hydrogen atoms involved

$$1\text{-}C_3H_7\bullet \to C_3H_6 + H\bullet \quad k = 1.0 \times 10^{14} \exp\left(\frac{-37,300}{RT}\right) \quad [s^{-1}]$$

$$1\text{-}C_4H_9\bullet \to 1\text{-}C_4H_8 + H\bullet \quad k = 1.0 \times 10^{14} \exp\left(\frac{-39,940}{RT}\right) \quad [s^{-1}]$$

$$2\text{-}C_4H_9\bullet \to 1\text{-}C_4H_8 + H\bullet \quad k = 1.58 \times 10^{13} \exp\left(\frac{-39,550}{RT}\right) \quad [s^{-1}]$$

$$2\text{-}C_4H_9\bullet \to 2\text{-}C_4H_8 + H\bullet \quad k = 3.16 \times 10^{12} \exp\left(\frac{-36,960}{RT}\right) \quad [s^{-1}]$$

Of course, this high activation energy means that dehydrogenation reactions are less favoured than the side β-decomposition reactions (Dean, 1985; Weissman and Benson, 1984). This is why butene formation is limited to less than 2–3 wt%.

A complete analysis of the products reported in Fig. 1 requires some more comments on cyclopentadiene and benzene. Both are typical secondary products, and are mainly the result of successive addition and condensation reactions of alkenes and unsaturated radicals. Once a significant amount of ethylene and propylene is formed, vinyl and allyl radicals are present in the reacting system and form butadiene, via butenyl radicals. Successive addition reactions of vinyl and allyl-like radicals on alkenes and dialkenes sequentially explain the formation of cyclopentadiene and benzene. These reactions are discussed in-depth in the literature and will be also analysed in the coming paragraphs (Dente et al., 1979). It seems worthwhile mentioning that these successive reactions and interactions of small unsaturated radicals and species constitute the critical sub-mechanism for the correct evaluation of ethylene selectivity. In fact, once the primary decomposition of the hydrocarbon feed has largely completed, the primary products and mainly small alkenes can be involved in successive reactions with detrimental effect on the overall selectivity of the process.

The simple pyrolysis of *n*-butane gives only a partial idea of how complex pyrolysis mechanisms are. The pyrolysis of *n*-decane provides a further example of the complexity of pyrolysis reactions.

There are five different C–C bonds in the *n*-decane chain, and therefore there are five different initiation reactions. For symmetry reasons, it is possible to assume $k = 10^{17} \exp(-81{,}000/RT)$ [s^{-1}] for the following initiation reactions:

$$nC_{10}H_{22} \rightarrow 1\text{-}C_8H_{17}\bullet + C_2H_5\bullet$$

$$nC_{10}H_{22} \rightarrow 1\text{-}C_7H_{15}\bullet + 1\text{-}C_3H_7\bullet$$

$$nC_{10}H_{22} \rightarrow 1\text{-}C_6H_{13}\bullet + 1\text{-}C_4H_9\bullet$$

while for the reaction:

$$nC_{10}H_{22} \rightarrow 1\text{-}C_5H_{11}\bullet + 1\text{-}C_5H_{11}\bullet$$

only $5 \times 10^{16} \exp(-81{,}000/RT)$ [s^{-1}] is assumed. Finally, $10^{17} \exp(-83{,}500/RT)$ [s^{-1}] is the kinetic constant of the last initiation reaction:

$$nC_{10}H_{22} \rightarrow 1\text{-}C_9H_{19}\bullet + CH_3\bullet$$

Once radicals are formed in the system, propagation reactions are the main basis of the initial feed decomposition. Figure 2 shows the complete reaction path of H-abstraction reactions on *n*-decane and itself provides proof of the complexity of the mechanism. All the different H-abstracting radicals (R•) can produce the five different isomers of *n*-decyl radical. These radicals can then isomerize and/or decompose.

The H-abstraction or metathesis reactions can be systematically described in their general form:

$$R\bullet + R'H \rightleftharpoons R'\bullet + RH$$

R and RH stand for all the possible H-abstracting radicals and for the corresponding saturated species. It is possible to assume that the rate constant for the reaction is only function of the H-abstracting radical (R•) and of the molecule from which H-atom is removed (R'H). As a first approximation, it is possible to decompose this kinetic constant as the product of two separate terms:

$$k_{\text{Habstr,ref}} \approx k_{\text{ref},R}^0 C_{R'H}$$

where $k_{\text{ref},R}^0$ represents the intrinsic reactivity of the radical and $C_{R'H}$ is the relative reactivity of the removed H-atom. This assumption simply means that the contributions for evaluating the rate constant only come from properties

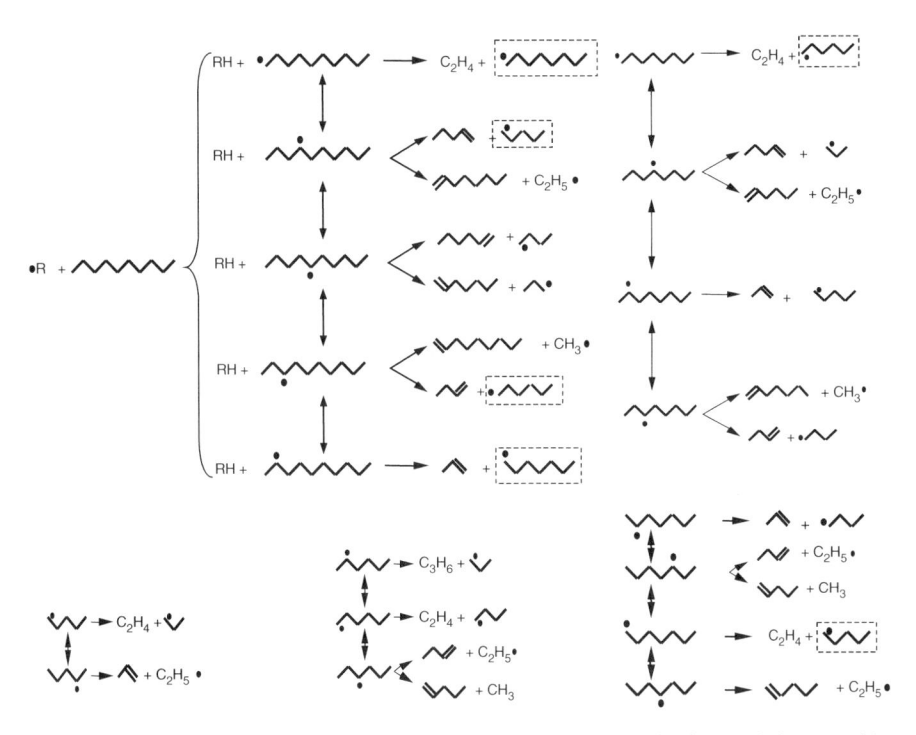

FIG. 2. H-abstraction reactions of *n*-decane and successive isomerization and decomposition reactions of alkyl radicals larger than C4.

related to the abstracting radical and to the type of the hydrogen atom to be abstracted. The theoretical basis for this simplification lies partially in the assumption that the forces between atoms are very short range (Benson, 1976), i.e. in the order of magnitude of the bond lengths: each atom contributes constant amounts to the molecule properties. Details of this approach can be found elsewhere (Ranzi *et al.*, 1994). On this basis, the H-abstraction reactions on *n*-decane are estimated taking into consideration the presence of 16 secondary H-atoms, with the corresponding formation of the four different secondary radicals (s–$C_{10}H_{21}$, $s = 2,5$) and the presence of six primary H-atoms, with the formation of the primary *n*-decyl radical. Depending on the attacking radical, the estimated rate constants for each *n*-decyl isomer are

$$H\bullet + nC_{10}H_{22} \rightarrow H_2 + 1\text{-}C_{10}H_{21}\bullet$$
$$k = 6.43 \times 10^3 T^2 \exp\left(\frac{-6,500}{RT}\right) \quad [\text{m}^3/\text{kmol s}]$$

$$H\bullet + nC_{10}H_{22} \rightarrow H_2 + s\text{-}C_{10}H_{21}\bullet$$
$$k = 8.60 \times 10^3 T^2 \exp\left(\frac{-4,000}{RT}\right) \quad [\text{m}^3/\text{kmol s}]$$

$$CH_3\bullet + nC_{10}H_{22} \rightarrow CH_4 + 1\text{-}C_{10}H_{21}\bullet$$
$$k = 1.17 \times 10^2 T^2 \exp\left(\frac{-7,500}{RT}\right) \quad [m^3/\text{kmol s}]$$

$$CH_3\bullet + nC_{10}H_{22} \rightarrow CH_4 + s\text{-}C_{10}H_{21}\bullet$$
$$k = 1.56 \times 10^2 T^2 \exp\left(\frac{-4,900}{RT}\right) \quad [m^3/\text{kmol s}]$$

$$C_2H_5\bullet + nC_{10}H_{22} \rightarrow C_2H_6 + 1\text{-}C_{10}H_{21}\bullet$$
$$k = 0.69 \times 10^2 T^2 \exp\left(\frac{-10,500}{RT}\right) \quad [m^3/\text{kmol s}]$$

$$C_2H_5\bullet + nC_{10}H_{22} \rightarrow C_2H_6 + s\text{-}C_{10}H_{21}\bullet$$
$$k = 0.92 \times 10^2 T^2 \exp\left(\frac{-7,700}{RT}\right) \quad [m^3/\text{kmol s}]$$

Similar H-abstraction reactions are easily evaluated for all the H-abstracting radicals.

Internal isomerization reactions also play an important role during pyrolysis. In fact, 1–4, 1–5 and 1–6 H-transfer reactions are easily explained on the basis of internal H-abstraction reactions, via five-, six- and seven-membered ring intermediates. The rate constants of these isomerization reactions are estimated in terms of the number of atoms in the transition-state ring structure (including the H-atom) and the type of sites involved in the H-transfer (Benson, 1976).

For instance, 1-decyl radicals can undergo the following isomerization reactions:

$$1\text{-}C_{10}H_{21}\bullet \rightarrow 4\text{-}C_{10}H_{21}\bullet \quad k_{1,4} = 1.9 \times 10^{11} \exp\left(\frac{-18,300}{RT}\right) \quad [s^{-1}]$$

$$1\text{-}C_{10}H_{21}\bullet \rightarrow 5\text{-}C_{10}H_{21}\bullet \quad k_{1,5a} = 3.0 \times 10^{10} \exp\left(\frac{-12,200}{RT}\right) \quad [s^{-1}]$$
$$k_{1,5b} = 1.2 \times 10^{10} \exp\left(\frac{-17,200}{RT}\right) \quad [s^{-1}]$$

The (1,5) isomerization reaction can be explained on the basis of both a six- and a seven-membered ring intermediate due to the symmetry of position 5 and 6 of n-decane. The 1–6 H-transfer (seven-membered ring intermediate) is less important than other isomerization reactions. This is due to the extra activation energy for the ring strain and to the decrease in the A-factor for the tie-up of the additional rotor (Benson, 1976; Curran et al., 1998; Matheu et al., 2003).

KINETIC MODELLING OF PYROLYSIS PROCESSES 63

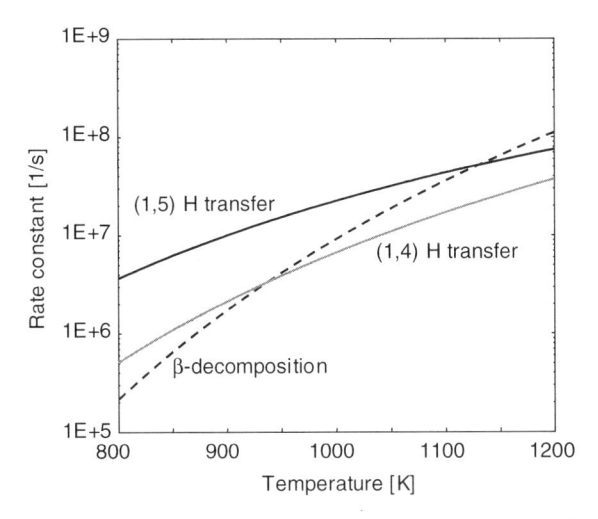

FIG. 3. Isomerization and decomposition reactions of alkyl radicals.

Figure 3 compares the rate values of these isomerization reactions with the 1-decyl radical β-decomposition:

$$1\text{-}C_{10}H_{21}\bullet \to C_2H_4 + 1\text{-}C_8H_{17}\bullet \quad k = 3 \times 10^{13} \exp\left(\frac{-30,000}{RT}\right) \quad [s^{-1}]$$

In the temperature range of interest for steam cracking, isomerization and decomposition reactions compete and decomposition prevails only at temperatures higher than 1,100 K, while the 1–5 isomerization reaction dominates at temperatures lower than 900 K. Moreover, alkyl radicals can undergo competitive dehydrogenation reactions.

Besides these primary reactions, the successive decompositions of decyl radicals must also be taken into account as must the pyrolysis of the resulting alkenes. It is quite evident that a manual compilation of the whole set of reactions would be unmanageable, particularly when the molecular weight of the hydrocarbon components is increased.

As already observed above, the pyrolysis of naphthas and gasoils to produce light alkenes is a process whose chemical complexity is dictated both by the characterization of the hydrocarbon mixture and by the complete definition of the kinetic mechanism. Due to the large number of species involved and the need to take into account all the relevant interactions between the different species, the number of elementary reactions becomes very large. For this reason, it is useful to classify the reactions on the one hand and very convenient to apply automatic procedures in order to generate the kinetic scheme, on the other. Likewise, as the molecular weight of the molecules rises, it is not only useful but sometimes necessary to adopt carefully evaluated simplifying rules.

A. Automatic Generation of Pyrolysis Mechanism

The use of computer generation systems in modelling the pyrolysis of large hydrocarbons is no longer considered simply an alternative to manual mechanism construction. It has become a necessity. The quantity of species and reactions becomes enormous, increasing molecular weight. This is particularly true if the focus is not merely on linear alkanes but also on other typical components of naphthas and gasoils, such as *iso*-alkanes or *cyclo*-alkanes, where the number of possible isomers increases exponentially with the number of carbon atoms in the molecule.

The generation of kinetic mechanisms is performed by computer programs which, at their best, decompose components via a defined set of rules. The procedures included in the programs can then be readily modified if changes in the understanding of the process require modifications to the available rules. It is important to stress, however, that this automatic generation is normally applied to the extension of existing kinetic schemes. Despite the fact that these programs are so easy to use and the relevance of the generated results, unresolved critical points in mechanism development still remain. These include, for instance, the proper definition of the species and reactions involved in the core of the kinetic mechanism where the interactions of small and stable species require greater direct in-depth analysis. The automatic generation of detailed mechanisms is not a feature peculiar to pyrolysis processes but has also been used extensively in different processes such as combustion, catalytic cracking and coking. Thus, many basic characteristics are worth determining not only in the pyrolysis area but also in other ones, as will be revealed in the description that follows.

Two different approaches are commonly referred to in the literature when it comes to the automatic generation of reaction mechanisms. The first approach involves combinatorial algorithms based mainly on the pioneering work of Yoneda (1979). These generate the whole set of possible reactions by only taking into account the congruence of the electronic configuration of reactants and products. Bond electron matrices are used to represent the chemical species and matrix operators describe all the possible reactions.

The second approach is simpler and generates the mechanism on the basis of specific reaction classes. Only a limited set of reaction rules is applied to all those molecules undergoing the same specific rearrangement. Reaction classes are founded on the basic principle that the reactivity of a molecule is based solely on the structural features around the bonds and atoms that change in the course of a reaction. These structural features are typically related to the influence of primary, secondary and tertiary H-atoms, to the stability of the formed radicals, and to the influence of neighbouring functional groups. Although this method easily produces reaction mechanisms, it does require a complete knowledge of the relevant reaction classes. As Blurock (Blurock, 2004) correctly observed, the goal of this mechanism generation system is to simulate the procedures used for producing mechanisms by hand. It is a formalization of the procedures used to generate and manipulate all objects associated with mechanism development. This formalization leads to the automation of the steps and higher level operations as well as allowing more complex problems be dealt with. The tedious details, which could be error-prone if done by hand, are left to the system. Initial and partial applications to the steam cracking process date back to 1980s (Clymans and Froment, 1984; Dente and Ranzi, 1983; Froment, 1992; Hillewaert et al., 1988). The approach was later extended to the catalytic cracking and hydropyrolysis processes of petroleum mixtures (Liguras and Allen, 1992). Quann and Jaffe (1992, 1996), in particular, proposed a very comprehensive example of a computer-generated reaction mechanism. Combustion processes also take this approach although they exhibit a more complex variety of phenomena and require a much higher number of species (intermediate and products) and reactions. This is as a result of the presence of oxygen as a further element together with carbon and hydrogen. Chevalier et al. (1988) proposed a detailed mechanism for the oxidation of *n*-hexadecane consisting of 1,200 species and 7,000 elementary reactions. Their computational technique, based on the LISP programming language, was also applied to the branched hydrocarbons such as *iso*-octane or different heptane isomers. Ranzi et al. (1997b) extended the automatic generation of pyrolysis reactions to combustion processes by including the interactions between alkyl radicals and oxygen with the aim of highlighting a few classes of primary propagation reactions responsible for the low-temperature phenomena during the hydrocarbon oxidation. Analogous to this, the research group in chemical kinetics in Nancy developed software (EXGAS) for the oxidation of alkanes, alkenes, ethers and *cyclo*-alkanes and mixtures of hydrocarbons. The programming of this system is based mainly on a referenced canonical tree-like description of molecules and free radicals, and can handle both acyclic and cyclic compounds (Warth et al., 2000). Applications to normal and *iso*-alkanes up to *n*-hexadecane are referred to (Fournet et al., 2001) and the extension to *n*-hexene was also discussed (Touchard et al., 2004) recently.

Other characteristics besides the approaches described above, which may distinguish the automatic generators, are also adopted for the molecule and reaction paths descriptions.

Graph theory, substitution matrices and Boolean algebra have been extensively used for the automatic generation of the primary product distribution from the pyrolysis of hydrocarbon species. On this basis, Broadbelt and co-workers more recently proposed the NetGen program for automatic mechanism generation (Broadbelt et al., 1994, 1995, 1996). A matrix representation of valence-bond connectivities is used to generate the chemical reactions. The unambiguous identification of different species is based on a general planar algorithm for the determination of homomorphism. The XMG (Exxon Mobil Mechanism Generation) generation code is the next extension of this work and was developed in cooperation with MIT (Green et al., 2001; Grenda *et al.*, 1998, 2003). Despite the claims regarding the potential of these programs, they have mostly been applied to relatively small hydrocarbons. An interesting exception is the kinetic model of tetradecane pyrolysis, presented by De Witt *et al.* (2000).

A further key feature of each automatic generation of kinetic models is the rapid estimation of the thermodynamics properties and the rate constants. Thermochemistry generally refers to Benson's group additivity (Benson, 1976). Specific programs were developed by some of the research groups involved in this activity. In Nancy, specific heats, standard enthalpies of formation and entropies of molecules or free radicals were calculated using the software THERGAS (Muller *et al.*, 1995). The GAPP program and interface is a modified and improved version of the THERM program (Bozzelli and Dean, 1990; Ritter and Bozelli, 1991).

As reaction rates are often expressed in a modified Arrhenius form, simple approaches like those based on linear free energy relationships, such as Evans–Polanyi, are adopted (Susnow *et al.*, 1997). Automatic generators usually refer to thermochemical kinetics methods (Benson, 1976) and the kinetic parameters rely on a limited number of reference rate constants and are extended to all the reactions of specific classes adopting analogy rules (Battin-LeClerc *et al.*, 2000; Ranzi *et al.*, 1995). Recently, extensive adoption of *ab initio* calculations of activation energies and reaction rates are adopted (Saeys *et al.*, 2003, 2004, 2006).

Finally, it is worth pointing out that one intrinsic limitation of these programs in terms of the automatic generation of pyrolysis mechanisms is the explosive number of possible reactions and intermediates products. The figures involved increase dramatically with the size of the initial radicals. While the number of end products can be controlled by adopting different solutions, such as lumping techniques (i.e. the grouping of species, generally isomers or homologous species, with the same functional groups and the same reactivity), the reduction of intermediate products as well as the grouping of the reactions involved depends on the hypothesis assumed about the interactions between the propagation paths of the different initial radicals. Specific hypothesis may lead to several reactions being expressed as a single equivalent one.

As an example of application of automatic generation, Table I gives the complete set of the primary propagation reactions of *n*-decyl radicals: isomerization, β-decomposition and dehydrogenation reactions. These reactions are produced directly by the MAMA program which was specifically developed for pyrolysis mechanism generation (Dente and Ranzi, 1983; Dente *et al.*, 2005; Pierucci *et al.*, 2005).

The main characteristics and peculiarities of this model are reported in Appendix 1. This automatic generation is performed quite simply on the basis of the definition of the different classes of primary reactions with the related small set of reference kinetic parameters, as reported in Table II.

TABLE I
Primary Propagation Reactions of n-Decyl Radicals

Isomerization reactions

Reaction	Rate
•C–C–C–C–C–C–C–C–C–C → C–C–C–•C–C–C–C–C–C–C	$1.89 \times 10^{11} \exp(-18,300)$
•C–C–C–C–C–C–C–C–C–C → C–C–C–C–•C–C–C–C–C–C	$3.00 \times 10^{10} \exp(-12,200)$
•C–C–C–C–C–C–C–C–C–C → C–C–C–C–•C–C–C–C–C–C	$1.20 \times 10^{10} \exp(-17,200)$
C–•C–C–C–C–C–C–C–C–C → C–C–C–C–•C–C–C–C–C–C	$1.89 \times 10^{11} \exp(-19,300)$
C–•C–C–C–C–C–C–C–C–C → C–C–C–C–•C–C–C–C–C–C	$3.00 \times 10^{10} \exp(-13,200)$
C–•C–C–C–C–C–C–C–C–C → C–C–C–•C–C–C–C–C–C–C	$1.20 \times 10^{10} \exp(-18,200)$
C–C–•C–C–C–C–C–C–C–C → C–C–C–C–•C–C–C–C–C–C	$1.89 \times 10^{11} \exp(-19300)$
C–C–•C–C–C–C–C–C–C–C → C–C–C–C–•C–C–C–C–C–C	$3.00 \times 10^{10} \exp(-13,200)$
C–C–C–•C–C–C–C–C–C–C → •C–C–C–C–C–C–C–C–C–C	$2.84 \times 10^{11} \exp(-21,600)$
C–C–C–•C–C–C–C–C–C–C → C–C–•C–C–C–C–C–C–C–C	$3.00 \times 10^{10} \exp(-13,200)$
C–C–C–•C–C–C–C–C–C–C → C–•C–C–C–C–C–C–C–C–C	$1.20 \times 10^{10} \exp(-18,200)$
C–C–C–C–•C–C–C–C–C–C → C–•C–C–C–C–C–C–C–C–C	$1.89 \times 10^{11} \exp(-19,300)$
C–C–C–C–•C–C–C–C–C–C → C–C–•C–C–C–C–C–C–C–C	$1.89 \times 10^{11} \exp(-19,300)$
C–C–C–C–•C–C–C–C–C–C → •C–C–C–C–C–C–C–C–C–C	$4.50 \times 10^{10} \exp(-15,500)$
C–C–C–C–•C–C–C–C–C–C → C–•C–C–C–C–C–C–C–C–C	$3.00 \times 10^{10} \exp(-13,200)$
C–C–C–C–•C–C–C–C–C–C → •C–C–C–C–C–C–C–C–C–C	$1.80 \times 10^{10} \exp(-20,500)$

β-decomposition reactions

Reaction	Rate
•C–C–C–C–C–C–C–C–C–C → C=C + •C–C–C–C–C–C–C–C	$1.00 \times 10^{14} \exp(-30,000)$
C–•C–C–C–C–C–C–C–C–C → C–C=C + •C–C–C–C–C–C–C	$1.00 \times 10^{14} \exp(-31,000)$
C–C–•C–C–C–C–C–C–C–C → C=C–C–C–C–C–C–C–C + •C	$1.00 \times 10^{14} \exp(-33,000)$
C–C–•C–C–C–C–C–C–C–C → C–C–C=C + •C–C–C–C–C–C	$1.00 \times 10^{14} \exp(-31,000)$
C–C–C–•C–C–C–C–C–C–C → C=C–C–C–C–C–C–C–C + C–•C	$1.00 \times 10^{14} \exp(-31,000)$
C–C–C–•C–C–C–C–C–C–C → C–C–C–C=C + •C–C–C–C–C	$1.00 \times 10^{14} \exp(-31,000)$
C–C–C–C–•C–C–C–C–C–C → C=C–C–C–C–C–C–C + C–C–•C	$1.00 \times 10^{14} \exp(-31,000)$
C–C–C–C–•C–C–C–C–C–C → C–C–C–C–C=C + •C–C–C–C	$1.00 \times 10^{14} \exp(-31,000)$

Dehydrogenation reactions

Reaction	Rate
•C–C–C–C–C–C–C–C–C–C → H + C=C–C–C–C–C–C–C–C–C	$1.00 \times 10^{14} \exp(-39,500)$
C–•C–C–C–C–C–C–C–C–C → H + C=C–C–C–C–C–C–C–C–C	$1.00 \times 10^{14} \exp(-41,000)$
C–•C–C–C–C–C–C–C–C–C → H + C–C=C–C–C–C–C–C–C–C	$1.00 \times 10^{14} \exp(-41,000)$
C–C–•C–C–C–C–C–C–C–C → H + C–C=C–C–C–C–C–C–C–C	$1.00 \times 10^{14} \exp(-41,000)$
C–C–•C–C–C–C–C–C–C–C → H + C–C–C=C–C–C–C–C–C–C	$1.00 \times 10^{14} \exp(-41,000)$
C–C–C–•C–C–C–C–C–C–C → H + C–C–C=C–C–C–C–C–C–C	$1.00 \times 10^{14} \exp(-41,000)$
C–C–C–•C–C–C–C–C–C–C → H + C–C–C–C=C–C–C–C–C–C	$1.00 \times 10^{14} \exp(-41,000)$
C–C–C–C–•C–C–C–C–C–C → H + C–C–C–C=C–C–C–C–C–C	$1.00 \times 10^{14} \exp(-41,000)$
C–C–C–C–•C–C–C–C–C–C → H + C–C–C–C–C=C–C–C–C–C	$1.00 \times 10^{14} \exp(-41,000)$

The resulting mechanism includes 16 isomerization, eight β-decomposition and nine dehydrogenation reactions, which are the primary propagation reactions of five different n-decyl radicals. As already discussed elsewhere, it is very easy to extend this generation to heavier species (Ranzi et al., 2005). Of course, the complexity, or rather the number of elementary reactions, and the number of intermediate radicals and molecules in particular, rapidly increase with the number of C-atoms. The need to introduce the primary reactions of the primary products, such as alkenes, is also taken into account.

TABLE II
Reference Kinetic Parameters of Pyrolysis Reactions (Units are kmol, m, s and kcal)

Initiation reactions: Unimolecular decomposition of C–C bonds

CH_3–$Csec$	$Csec$–$Csec$	$Csec$–$Cter$	$Csec$–$Cquat$
$5 \times 10^{16} \exp(-83{,}500/RT)$	$5 \times 10^{16} \exp(-81{,}000/RT)$	$5 \times 10^{16} \exp(-80{,}000/RT)$	$5 \times 10^{16} \exp(-78{,}000/RT)$

H-abstraction reactions of alkyl radicals

	Primary H-atom	Secondary H-atom	Tertiary H-atom
Primary radical	$10^8 \exp(-13{,}500/RT)$	$10^8 \exp(-11{,}200/RT)$	$10^8 \exp(-9{,}000/RT)$
Secondary radical	$10^8 \exp(-14{,}500/RT)$	$10^8 \exp(-12{,}200/RT)$	$10^8 \exp(-10{,}000/RT)$
Tertiary radical	$10^8 \exp(-15{,}000/RT)$	$10^8 \exp(-12{,}700/RT)$	$10^8 \exp(-10{,}500/RT)$

Isomerization reactions (Transfer of a primary H-atom)[a]

	1–4 H-transfer	1–5 H-transfer	1–6 H-transfer
Primary radical	$10^{11} \exp(-20{,}600/RT)$	$1.58 \times 10^{10} \exp(-14{,}500/RT)$	$3.16 \times 10^9 \exp(-19{,}500/RT)$
Secondary radical	$10^{11} \exp(-21{,}600/RT)$	$1.58 \times 10^{10} \exp(-15{,}500/RT)$	$3.16 \times 10^9 \exp(-20{,}500/RT)$
Tertiary radical	$10^{11} \exp(-22{,}100/RT)$	$1.58 \times 10^{10} \exp(-16{,}000/RT)$	$3.16 \times 10^9 \exp(-21{,}000/RT)$

Alkyl radical decomposition reactions (to form primary radicals)

Primary radical	Secondary radical	Tertiary radical
$10^{14} \exp(-30{,}000/RT)$	$10^{14} \exp(-31{,}000/RT)$	$10^{14} \exp(-31{,}500/RT)$

Corrections of decomposition rates to form

Methyl radical	Secondary radical	Tertiary radical	Allyl radical
$\exp(-2{,}500/RT)$	$\exp(1{,}500/RT)$	$\exp(2{,}500/RT)$	$0.316 \times \exp(8{,}000/RT)$

[a] Corrections for secondary and tertiary H-atoms are the same as for H-abstractions.

1. Lumping of Reactions

The previous paragraph highlights the rapid increase in the number of species and reactions as the molecular weight of the hydrocarbons rises. Appropriate reduction techniques are thus very useful in taking a more viable approach to the problem.

As previously discussed, alkyl radicals decomposition reactions constitute an important fate and reaction path of alkyl radicals. Due to the very short lifetimes of alkyl radicals, Rice and Herzfeld (1933, 1934) suggested a complete decomposition mechanism where all the radicals larger than methyl were considered instantaneously decomposed into alkenes and H and CH_3 radicals. In this mechanism, all the intermediate alkyl radicals decompose to directly form alkenes and smaller alkyl radicals. This would mean that the final ethylene production from a steam cracking process would be significantly overestimated when compared with the experimental measurements. For instance, the net and final result of the successive decomposition mechanism of 1-decyl radical would be 5 moles of ethylene and one H radical.

This typical feature of pyrolysis systems is very useful in improving the handling of the decomposition paths of large hydrocarbon species. Within the pressure and temperature range of the pyrolysis processes, the absence of interactions between large radicals and other species in the cracking mixture allows the direct substitution of n-decyl radicals with their primary isomerization and decomposition products, as shown in Fig. 2. Table III reports the product distribution from the isomerization and β-decomposition reaction of n-decyl radicals, as a net result of the H-abstraction reactions on n-decane. These distributions, which were evaluated at three different temperatures, show a limited temperature effect. In fact, the largest deviations in product distribution relate to the increasing importance of the primary radical and its successive decomposition to form ethylene and n-octyl radicals. The greater stability of the 2-decyl radical is due to the single β-decomposition of this

TABLE III
TEMPERATURE EFFECT ON PRIMARY PRODUCT DISTRIBUTION FROM β-DECOMPOSITION REACTION OF n-DECYL RADICAL

	800 K	1,000 K	1,200 K
$C_9H_{18} + CH_3\bullet$	0.0424	0.0516	0.0588
$C_8H_{16} + C_2H_5\bullet$	0.1478	0.1332	0.1193
$C_7H_{14} + C_3H_7\bullet$	0.1519	0.1475	0.1346
$C_6H_{12} + C_4H_9\bullet$	0.1519	0.1475	0.1346
$C_5H_{10} + C_5H_{11}\bullet$	0.1479	0.1332	0.1193
$C_4H_8 + C_6H_{13}\bullet$	0.1492	0.1412	0.1359
$C_3H_6 + C_7H_{15}\bullet$	0.1526	0.1569	0.1677
$C_2H_4 + C_8H_{17}\bullet$	0.0563	0.0889	0.1300

secondary radical. Moreover, it is important to stress that at temperatures higher than 1,200 K the lifetime of alkyl radicals falls to less than 10^{-8} s and their internal distribution becomes of limited importance.

In principle, heavy radicals could undergo also H-abstraction, addition on unsaturated bonds and recombination reactions. It is quite easy to demonstrate how little relevance these reactions have compared with the isomerization and decomposition ones. This helps drastically reduce the total number of radicals and reactions to be considered. All of the intermediate alkyl radicals, higher than C4, are supposed to be instantaneously transformed into their final products. With reference to the primary products of Table III, the heavy radicals from pentyl up to octyl undergo direct isomerization and decomposition reactions to form smaller radicals and alkenes. Therefore, large sections of the kinetic scheme can be reduced to a few equivalent or lumped reactions whilst still maintaining a high level of accuracy. The complete kinetic scheme shown in Fig. 2 can be then simply reduced to this single, equivalent or lumped reaction:

$$R\bullet + nC_{10}H_{22} \rightarrow RH + 0.0205\ H\bullet + 0.0803\ CH_3\bullet + 0.2593\ C_2H_5\bullet$$
$$+ 0.4061\ nC_3H_7\bullet + 0.2339\ 1\text{-}C_4H_9\bullet + 0.3785\ C_2H_4$$
$$+ 0.3127\ C_3H_6 + 0.2114\ 1\text{-}C_4H_8 + 0.1870\ 1\text{-}C_5H_{10}$$
$$+ 0.1815\ 1\text{-}C_6H_{12} + 0.1461\ 1\text{-}C_7H_{14} + 0.1284\ 1\text{-}C_8H_{16}$$
$$+ 0.0540\ 1\text{-}C_9H_{18} + 0.0025\ 1\text{-}C_{10}H_{20}$$
$$+ 0.0006\ 2\text{-}C_5H_{10} + 0.0012\ C_6H_{12}s + 0.0013\ C_7H_{14}s$$
$$+ 0.0005\ C_8H_{16}s + 0.0100\ C_{10}H_{20}s$$

This stoichiometry is directly obtained at 1,040 K by solving the initial continuity equations for all the radical species. A better insight into this approach is given in Appendix 1 which also refers to the specific MAMA code. The MAMA program is actually much more than a simple automatic reaction mechanism generator. It may generate the reaction path, namely the isomerization and decomposition reactions of the intermediate species, but it also solves the overall system and evaluates the primary distribution products in accordance with the hypothesis of the autonomous fate of the large intermediate radicals. The net result is the generation of equivalent or lumped reactions that drastically reduce the overall dimension of the reacting system, in terms of both species and reactions. This equivalent stoichiometry lumps all the elementary reactions of Fig. 2 and, as already mentioned, it is only very slightly temperature dependent at the usual steam cracking process temperature (900–1,100 K). This simplification was also strongly supported by the recent findings of McGivervy et al. (2004) who analysed the decomposition of n-octyl radicals.

H radical formation occurs as a result of the dehydrogenation reactions of all the intermediate radicals. These dehydrogenation reactions, not reported in Fig. 2 as well as in Table III, also explain the drop in the formation of 2-pentene, 2- and 3-hexenes and so on.

This lumped reaction matches the general lumping definition discussed at the "Workshop on Combustion Simulation Databases for Real Transportation Fuels" (Hudgens, 2003) very well. The lumped reaction is a collection of elementary reactions expressed as an "equivalent" apparent single step reaction. A lumped reaction may be a stepwise reaction, involving a consecutive set of elementary reactions, the simplest example being A→I→B, where I is a steady-state intermediate. A lumped reaction may also be a set of related but dissimilar reactions, operating as alternative or parallel pathways, or A→{In}→B where {In} represents the set of parallel pathways. Finally, a lumped reaction may also be a set of similar reactions, operating collectively on similar molecular species, or {A}→{I}→{B} where {A}, {I} and {B} are sets of similar reactants, intermediates and products, respectively.

2. Lumping of Species

The sets of similar reactants, intermediates and products in the previous lumped reaction are also an initial example of lumped species. Thus, lumped species {A} are defined as a weighted mixture of several isomers:

$$\{A\} = \sum x_j A_j$$

Lumped species react in a similar way and their internal distribution remains almost unchanged with the reactions.

Thus, all the isomers of large hydrocarbon species should be conveniently grouped into a single lumped component. A large number of alkene isomers are formed during the pyrolysis of heavy alkanes. For instance, dehydrogenation reactions of branched alkyl radicals generate all the different branched alkenes. In order to avoid an unnecessary number of species, all these isomers are conveniently grouped and considered as a single lumped compound. The clear advantage of the automatic generation is that the internal splitting of the real components is stored and the impact on the final product distribution is correctly taken into account.

Apart this "horizontal" lumping among species with the same molecular weight, there is also another very convenient simplification: the lumping of homologous species with a different number of carbon atoms. Let us refer to this as "vertical" lumping, i.e. lumping of homologous species with different molecular weights. The reactivity and the product distribution of large hydrocarbons with n_C carbon atoms can be correctly and conveniently estimated with a linear combination of the reactivity and product distribution of the homologous species with (n_C-1) and (n_C+1) carbon atoms. This fact is clearly demonstrated in Fig. 4 where the selectivities of small radicals and alkenes obtained from the H-abstraction reactions of n-alkanes are reported as a function of the carbon number. A quasilinear tendency of the different products with the carbon number can be observed.

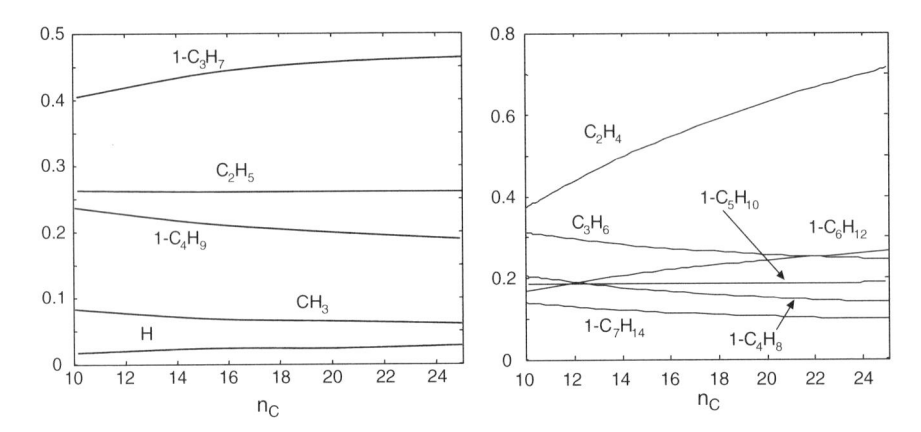

FIG. 4. Stoichiometric coefficients of small radicals and α-alkenes from the H-abstraction reactions of n-alkanes at 1,040 K, as a function of the carbon number.

Two other considerations can be drawn from the analysis of this figure:

- Ethylene selectivity increases with carbon number due to the importance of the β-decomposition of large intermediate radicals.
- 1-Pentene and, more importantly, 1-hexene selectivities also increase due to the internal (1–4) and (1–5) H-transfer reactions. 1-Propyl radical also increases for the same reason.

B. Pyrolysis of Large Hydrocarbons

It is possible to analyse and to generate the primary pyrolysis reactions of the various hydrocarbon classes using the aforementioned kinetic parameters as a basis. Normal and branched alkanes, *cyclo*-alkanes and aromatics are briefly analysed in order to discuss the analogies and similarities, as well as the differences in their pyrolysis reactions. The previously mentioned assumptions relating to the autonomous fate of large alkyl radicals and the consequent lumping of reactions are consistently applied. The automatic generation of the primary pyrolysis reactions is obtained by applying the same rules and criteria already discussed when analysing the primary reactions of *n*-decane and once again using the reference kinetic parameters from Table II (Ranzi et al., 2001).

1. Normal Alkanes

The clear advantage of this lumping approach is that the analysis of new components only requires the definition of the primary pyrolysis reactions. For large alkanes, only the initiation and H-abstraction reactions need to be defined.

KINETIC MODELLING OF PYROLYSIS PROCESSES

Thus, the lumped stoichiometry of the initiation and H-abstraction reactions of n-eicosane, again evaluated by the MAMA program at 1,040 K, is simply:

$$nC_{20}H_{42} = 0.0357 \; H\bullet + 0.1791 \; CH_3\bullet + 0.5068 \; C_2H_5\bullet + 0.8955 \; 1\text{-}C_3H_7\bullet$$
$$+ 0.3829 \; 1\text{-}C_4H_9\bullet + 1.2145 \; C_2H_4 + 0.3766 \; C_3H_6$$
$$+ 0.1944 \; 1\text{-}C_4H_8 + 0.2768 \; 1\text{-}C_5H_{10} + 0.0010 \; 2\text{-}C_5H_{10}$$
$$+ 0.3905 \; 1\text{-}C_6H_{12} + 0.0017 \; 2\text{-}C_6H_{12} + 0.0005 \; 3\text{-}C_6H_{12}$$
$$+ 0.1066 \; 1\text{-}C_7H_{14} + 0.0816 \; 1\text{-}C_8H_{16} + 0.1025 \; 1\text{-}C_9H_{18}$$
$$+ 0.1022 \; 1\text{-}C_{10}H_{20} + 0.0517 \; 1\text{-}C_{11}H_{22} + 0.0448 \; 1\text{-}C_{12}H_{24}$$
$$+ 0.0424 \; 1\text{-}C_{13}H_{26} + 0.0353 \; 1\text{-}C_{14}H_{28} + 0.0237 \; 1\text{-}C_{15}H_{30}$$
$$+ 0.0197 \; 1\text{-}C_{16}H_{32} + 0.0073 \; 1\text{-}C_{17}H_{34} + 0.0009 \; 1\text{-}C_{18}H_{36}$$
$$+ 0.0004 \; 1\text{-}C_{19}H_{38} + 0.0025 \; C_7H_{14}s + 0.0016 \; C_8H_{16}s$$
$$+ 0.0015 \; C_9H_{18}s + 0.0015 \; C_{10}H_{20}s + 0.0015 \; C_{11}H_{22}s$$
$$+ 0.0013 \; C_{12}H_{24}s + 0.0012 \; C_{13}H_{26}s + 0.0011 \; C_{14}H_{28}s$$
$$+ 0.0011 \; C_{15}H_{30}s + 0.0010 \; C_{16}H_{32}s + 0.0009 \; C_{17}H_{34}s$$
$$+ 0.0007 \; C_{18}H_{36}s + 0.0007 \; C_{19}H_{38}s$$

$$R\bullet + nC_{20}H_{42} = RH + 0.0286 \; H\bullet + 0.0615 \; CH_3\bullet + 0.2562 \; C_2H_5\bullet$$
$$+ 0.4539 \; nC_3H_7\bullet + 0.1997 \; 1\text{-}C_4H_9\bullet + 0.6331 \; C_2H_4$$
$$+ 0.2627 \; C_3H_6 + 0.1620 \; 1\text{-}C_4H_8 + 0.1922 \; 1\text{-}C_5H_{10}$$
$$+ 0.2496 \; 1\text{-}C_6H_{12} + 0.1130 \; 1\text{-}C_7H_{14} + 0.0972 \; 1\text{-}C_8H_{16}$$
$$+ 0.1045 \; 1\text{-}C_9H_{18} + 0.1047 \; 1\text{-}C_{10}H_{20} + 0.0813 \; 1\text{-}C_{11}H_{22}$$
$$+ 0.0763 \; 1\text{-}C_{12}H_{24} + 0.0749 \; 1\text{-}C_{13}H_{26} + 0.0714 \; 1\text{-}C_{14}H_{28}$$
$$+ 0.0660 \; 1\text{-}C_{15}H_{30} + 0.0625 \; 1\text{-}C_{16}H_{32} + 0.0588 \; 1\text{-}C_{17}H_{34}$$
$$+ 0.0550 \; 1\text{-}C_{18}H_{36} + 0.0244 \; 1\text{-}C_{19}H_{38} + 0.0011 \; 1\text{-}C_{20}H_{40}$$
$$+ 0.0005 \; 2\text{-}C_5H_{10} + 0.0009 \; 2\text{-}C_6H_{12} + 0.0002 \; 3\text{-}C_6H_{12}$$
$$+ 0.0011 \; C_7H_{14}s + 0.0008 \; C_8H_{16}s + 0.0007 \; C_9H_{18}s$$
$$+ 0.0008 \; C_{10}H_{20}s + 0.0008 \; C_{11}H_{22}s + 0.0006 \; C_{12}H_{24}s$$
$$+ 0.0006 \; C_{13}H_{26}s + 0.0006 \; C_{14}H_{28}s + 0.0005 \; C_{15}H_{30}s$$
$$+ 0.0005 \; C_{16}H_{32}s + 0.0005 \; C_{17}H_{34}s + 0.0002 \; C_{18}H_{36}s$$
$$+ 0.0100 \; C_{20}H_{40}s$$

Of course, α-linear-alkenes ($1C_nH_{2n}$) prevail but all the remaining linear alkenes ($C_nH_{2n}s$) are also formed as a result of the dehydrogenation reactions of all the intermediate radicals.

For the sake of simplicity and because their quantities are so tiny, these linear non α-alkenes are conveniently grouped. Table IV shows the temperature effect on the molar selectivities of the primary products from the H-abstraction

TABLE IV
TEMPERATURE EFFECT ON THE PRIMARY PRODUCT DISTRIBUTION FROM H-ABSTRACTION REACTIONS OF
n-EICOSANE (MOLAR SELECTIVITY)

T/K	940	990	1,040	1,090
H•	0.0172	0.0224	0.0286	0.0360
CH$_3$•	0.0645	0.0633	0.0615	0.0596
C$_2$H$_5$•	0.2790	0.2666	0.2562	0.2469
1-C$_3$H$_7$•	0.4358	0.4473	0.4539	0.4558
1-C$_4$H$_9$•	0.2035	0.2004	0.1997	0.2018
C$_2$H$_4$	0.4036	0.5082	0.6331	0.7759
C$_3$H$_6$	0.3043	0.2832	0.2627	0.2430
1-C$_4$H$_8$	0.1883	0.1749	0.1621	0.1500
1-C$_5$H$_{10}$	0.1957	0.1939	0.1922	0.1905
2-C$_5$H$_{10}$	0.0004	0.0005	0.0005	0.0006
1-C$_6$H$_{12}$	0.2418	0.2476	0.2496	0.2474
C$_6$H$_{12}$s	0.0008	0.0009	0.0011	0.0013
1-C$_7$H$_{14}$	0.1260	0.1195	0.1130	0.1069
C$_7$H$_{14}$s	0.0009	0.0011	0.0013	0.0014
1-C$_8$H$_{16}$	0.1049	0.1009	0.0972	0.0938
C$_8$H$_{16}$s	0.0006	0.0007	0.0008	0.0009
1-C$_9$H$_{18}$	0.1087	0.1070	0.1045	0.1014
C$_9$H$_{18}$s	0.0005	0.0006	0.0007	0.0009
1-C$_{10}$H$_{20}$	0.1113	0.1087	0.1047	0.1000
C$_{10}$H$_{20}$s	0.0005	0.0007	0.0008	0.0009
*1-C$_{11}$H$_{22}$	0.1244	0.1219	0.1194	0.1169
*C$_{11}$H$_{22}$s	0.0007	0.0009	0.0011	0.0013
*1-C$_{13}$H$_{26}$	0.1869	0.1850	0.1827	0.1798
*C$_{13}$H$_{26}$s	0.0010	0.0012	0.0015	0.0018
*1-C$_{16}$H$_{32}$	0.2098	0.2090	0.2080	0.2067
*sC$_{16}$H$_{32}$s	0.0010	0.0013	0.0015	0.0018
*1-C$_{20}$H$_{40}$	0.0597	0.0606	0.0616	0.0626
*C$_{20}$H$_{40}$s	0.0064	0.0082	0.0102	0.0125

*Lumped or equivalent compounds.

reactions of n-eicosane. Predicted ethylene selectivities move from 40.36% at 940 K up to 77.59% at 1,090 K, clearly indicating the prevailing importance of the decomposition reactions at high temperatures. The ethylene increase actually corresponds to a reduction in the heavier alkenes, particularly propene and butene. The higher ethylene selectivity at high temperature is mainly due to the increase in decomposition reactions with respect to isomerization ones. Heavy alkenes are also grouped using vertical lumping. H radical production is very low due to the high activation energy of the dehydrogenation reaction.

2. Branched Alkanes

Branched alkane decomposition becomes increasingly difficult, mainly as a result of the large number of possible isomers with the same molecular weight.

TABLE V
AVERAGE CARBON NUMBER, BOILING TEMPERATURES AND THE NUMBER OF PARAFFIN ISOMERS OF
DIFFERENT PETROLEUM FRACTIONS. ADAPTED FROM ALTGELT AND BODUSZYNSKI (1994)

Carbon no.	Boiling temperature (°C)	Paraffin isomers	Petroleum fraction
8	126	18	Naphtha and gasoline
10	174	75	Kerosene
12	216	355	Jet fuels
15	271	4,347	Diesel fuels
20	344	3.66×10^5	Light gasoil
25	402	3.67×10^7	Gasoil
30	449	4.11×10^9	Heavy gasoil
35	489	4.93×10^{11}	Atmospheric residue

TABLE VI
RELATIVE AMOUNT OF BRANCHED ISOMERS OF C_8H_{18} (WT%)

Origin	Ponca	Occidental	Texas	Internal weights
Isomers				
2-methylheptane	46.3	36.9	42.1	45.8
3-methylheptane	15.4	28.5	23.4	22.9
4-methylheptane	10.3	10.2	9.3	11.5
2,3-dimethylhexane	3.6	5.4	6.3	3.4
2,4-dimethylhexane	3.1	5.5	4.2	3.4
2,5-dimethylhexane	3.1	5.7	4.0	3.4
3,4-dimethylhexane	6.7	2.6	3.7	3.4
2,2-dimethylhexane	0.5	–	0.3	–
3,3-dimethylhexane	1.5	1.7	0.4	
2,3,4-trimethylpentane	0.3		1.1	1.2
2,2,3-trimethylpentane	0.2	–	–	–
2,3,3-trimethylpentane	0.3	–	0.6	–
3-ethylhexane	4.6	3.5	3.1	3.8
2-methyl-3-ethylpentane	3.1	–	1.5	1.2
3-methyl-3-ethylpentane	1.0	–	–	–

Several simplifications, whose importance increases with carbon numbers, are thus introduced. The difficulties involved in a detailed description of feed components are primarily caused by the very large number of compounds in the petroleum fractions, as clearly outlined in Table V.

One convenient simplification becomes clear when considering relatively simple or light components such as the branched C_8H_{18}. As shown in Table VI, only a few isomers describe the whole fraction of branched alkanes with eight carbon atoms: three monomethyl-heptanes, ethyl-hexane and four dimethyl-hexanes with a tertiary C structure. In spite of the different origins of these feeds, there is clear regularity with regard to their composition. In fact, monomethyl-heptanes prevail on dimethyl-hexanes and monoethyl-hexanes. Trimethyl-pentanes are less abundant and quaternary C-atoms are of very limited importance. On this basis, it is possible to empirically derive an internal distribution of the isomer mixture and to derive a lumped component {$isoC_8$}, as shown in the last column of Table VI. Of course, this internal distribution is a peculiarity of the virgin fractions and can change drastically after thermal or catalytic refinery processes.

Thus, for virgin feeds, a single equivalent or lumped component {$isoC_8$} is defined by grouping all the different isomers with their relative weights, where the three mono-methyl-heptanes sum up to about 80% of the whole fraction. The corresponding lumped H-abstraction reaction, obtained by simply mixing the primary distribution products generated by MAMA program, has the following stoichiometry:

$$R\bullet + \{isoC_8\} = RH + 0.0167\ H\bullet + 0.1842\ CH_3\bullet + 0.2181\ C_2H_5\bullet$$
$$+ 0.1835\ nC_3H_7\bullet + 0.1296\ i\text{-}C_3H_7\bullet + 0.0671\ i\text{-}C_4H_9\bullet$$
$$+ 0.1415\ 1\text{-}C_4H_9\bullet + 0.0592\ 2\text{-}C_4H_9\bullet + 0.1628\ C_2H_4$$
$$+ 0.2499\ C_3H_6 + 0.107\ 1\text{-}C_4H_8 + 0.0414\ 2\text{-}C_4H_8$$
$$+ 0.1417\ i\text{-}C_4H_8 + 0.0704\ 1\text{-}C_5H_{10} + 0.0326\ 2\text{-}C_5H_{10}$$
$$+ 0.0510\ 2me\text{-}1C_4H8 + 0.0599\ 3me\text{-}1C_4H_8$$
$$+ 0.0103\ 2me\text{-}2C_4H_8 + 0.009\ 1\text{-}C_6H_{12} + 0.0214\ 2\text{-}C_6H_{12}$$
$$+ 0.0047\ 3\text{-}C_6H_{12} + 0.1012\ me\text{-}C_5H_{10}s + 0.0155\ 1\text{-}C_7H_{14}$$
$$+ 0.1402\ C_7H_{14}s + 0.0137\ C_8H_{16}s$$

It is important to stress that this lumped reaction allows a significant reduction in both feed and intermediate species as well as in reactions, yet still retaining a very high level of accuracy in the prediction of the various products of pyrolysis.

It is quite difficult to find accurate and detailed analyses of single isomers for branched alkanes with more than 10 C-atoms in the literature. Nevertheless, GC analysis of heavy naphtha, kerosene and light gasoils indicates the prevailing presence of isoprenoid structures characterized by an average probability of methyl substitution of about 0.20 (Altgelt and Boduszynski, 1994).

One possible use of kinetic generators and the advantages associated with them can be demonstrated by considering the lumped component {$isoC_{15}$}. There are more than 4,000 different isomers, as reported in Table V: 12 different methyl-tetradecanes, 55 dimethyl-C_{13} (without quaternary C-atoms), 120 tri-methyl-C_{12}, 126 tetra-methyl-C_{11} and so on. These numbers of different isomers are significantly reduced by accounting for the symmetry of the molecule. By accounting for mono-, di-, tri-, tetra- and penta-methyl isomers with the proper relative weight, the lumped reaction of {$isoC_{15}$} (well-defined mixture of different branched alkanes $C_{15}H_{32}$) becomes:

$$R\bullet + \{isoC_{15}\} = RH + 0.0217\ H\bullet + 0.2086\ CH_3\bullet + 0.2033\ C_2H_5\bullet$$
$$+ 0.2138\ nC_3H_7\bullet + 0.1239\ i\text{-}C_3H_7\bullet + 0.0534\ i\text{-}C_4H_9\bullet$$
$$+ 0.1120\ 1\text{-}C_4H_9\bullet + 0.0633\ 2\text{-}C_4H_9\bullet + 0.2208\ C_2H_4$$
$$+ 0.3302\ C_3H_6 + 0.092\ 1\text{-}C_4H_8 + 0.0677\ 2\text{-}C_4H_8$$
$$+ 0.0864\ i\text{-}C_4H_8 + 0.0538\ 1\text{-}C_5H_{10} + 0.0184\ 2\text{-}C_5H_{10}$$
$$+ 0.0378\ 2me\text{-}1C_4H_8 + 0.0259\ 3me\text{-}1C_4H_8$$
$$+ 0.0187\ 2me\text{-}2C_4H_8 + 0.0443\ 1\text{-}C_6H_{12} + 0.0209\ C_6H_{12}s$$
$$+ 0.0847\ me\text{-}C_5H_{10}s + 0.0226\ 1\text{-}C_7H_{14} + 0.1195\ C_7H_{14}s$$
$$+ 0.1168\ C_8H_{16}s + 0.0115\ 1\text{-}C_8H_{16} + 0.0070\ 1\text{-}C_9H_{18}$$
$$+ 0.1041\ C_9H_{18}s + 0.0044\ 1\text{-}C_{10}H_{20} + 0.0865\ C_{10}H_{20}s$$
$$+ 0.0028\ 1\text{-}C_{11}H_{22} + 0.1313\ C_{11}H_{22}s + 0.0010\ 1\text{-}C_{13}H_{26}$$
$$+ 0.1760\ C_{13}H_{26}s + 0.0414\ C_6H_{12}s$$

Figure 5 demonstrates the sensitivity of the primary products of this lumped H-abstraction reaction by varying the probability of methyl substitution, i.e. by varying the relative amount of the different classes of isomers (mono-, di-, tri-, tetra-methyl and so on). While ethylene and 1-butene selectivities decrease with the increase in degree of methyl substitution, methyl radical, 2-butene and isobutene formation is enhanced.

The isoprenoid structure hypothesis corresponds to an average of three methyl substitutions along the carbon chain. By comparison, the equivalent H-abstraction reaction of the 2,6,10-tri-methyl-dodecane, with a regular tri-isoprene structure, becomes:

$$R\bullet + 2,6,10\text{-}tri\text{-}me\ C_{12} = RH + 0.0220\ H\bullet + 0.2254\ CH_3\bullet + 0.2181\ C_2H_5\bullet$$
$$+ 0.0673\ nC_3H_7\bullet + 0.1666\ i\text{-}C_3H_7\bullet + 0.0826\ i\text{-}C_4H_9\bullet$$
$$+ 0.0654\ 1\text{-}C_4H_9\bullet + 0.1527\ 2\text{-}C_4H_9\bullet + 0.3300\ C_2H_4$$
$$+ 0.2208\ C_3H_6 + 0.0632\ 1\text{-}C_4H_8 + 0.0699\ 2\text{-}C_4H_8$$
$$+ 0.1441\ i\text{-}C_4H_8 + 0.0192\ 1\text{-}C_5H_{10} + 0.0070\ 2\text{-}C_5H_{10}$$
$$+ 0.1125\ 2me\text{-}1C_4H_8 + 0.0590\ 3me\text{-}1C_4H_8$$
$$+ 0.2me\text{-}2C_4H_8 + 0.0002\ 1\text{-}C_6H_{12} + 0.0049\ C_6H_{12}s$$
$$+ 0.1447\ me\text{-}C_5H_{10}s + 0.0023\ 1\text{-}C_7H_{14}$$
$$+ 0.1042\ C_7H_{14}s + 0.1028\ C_8H_{16}s + 0.1458\ C_9H_{18}s$$
$$+ 0.1110\ C_{10}H_{20}s + 0.1147\ C_{11}H_{22}s$$
$$+ 0.1625\ C_{13}H_{26}s + 0.0483\ C_6H_{12}s$$

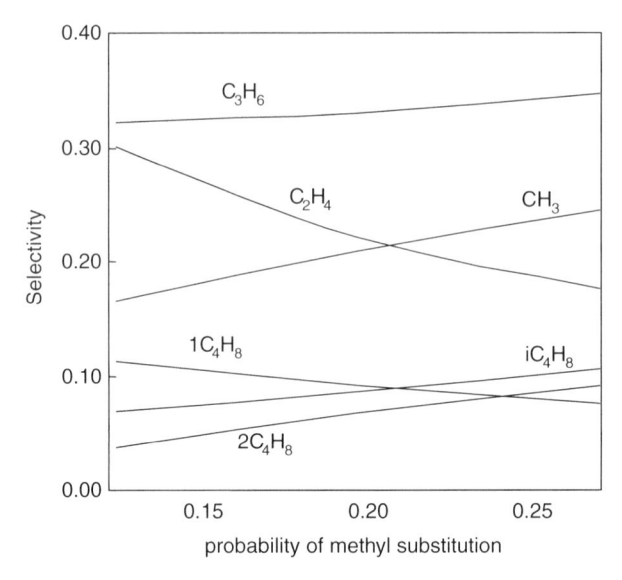

FIG. 5. Pyrolysis of lumped component of branched alkanes {$C_{15}H_{32}$}. Selectivity of methyl radical and small alkenes as a function of the probability of methyl substitution.

Due to the regular branched structure of this isomer, linear 1-alkenes heavier than 1-heptene are not present and the relative amount of propyl and butyl radicals is significantly different too. In other words, the lumped H-abstraction reaction of a single model component loses the variety of primary products obtained from the previous lumped {$isoC_{15}$}. It seems relevant to observe that to improve ethylene selectivity prediction, alkene components heavier than hexenes can be conveniently described with two different species, respectively corresponding to the true component 1-C_nH_{2n} and to a lumped mixture of the remaining normal and branched isomers.

As previously mentioned, the difference in the primary product distributions from homologous species decreases as the number of carbon atoms (n_c) becomes larger than 10–12. Therefore, branched alkanes heavier than C_{10} are also lumped species that contain not only isomers (with the same n_c) but also adjacent homologous species. For instance, all the different branched alkanes with 11 C-atoms are split equally between the two lumped components {$isoC_{10}$} and {$isoC_{12}$}. As a result of this vertical lumping, only a few reference and lumped components are selected inside each family thus further decreasing the computation effort involved while maintaining prediction accuracy.

3. Cyclo-Alkanes and Alkenes

There is a great abundance of groundwork in the scientific literature on both the fundamental and applied chemical kinetics of pyrolysis reactions relating to normal and branched alkanes. The pyrolysis of *cyclo*-alkanes, on the other hand, has received very little attention. Nevertheless, naphthas and liquid feeds contain large quantities of *cyclo*-alkanes, which consist of both alkyl *cyclo*-pentane and alkyl *cyclo*-hexane components.

Cyclo-alkane pyrolysis involves a more complex overall reacting system due to the presence of new relevant reaction classes, such as the ring opening and the *cyclo*-addition reactions (Dente et al., 2005; Green et al., 2001; Matheu et al., 2003; Pierucci et al., 2005). While the ring opening can take advantage of the reference kinetic values of the β-decomposition reactions of alkyl radicals, new reference kinetic parameters need to be singled out for the internal isomerization of *cyclo*-alkyl radicals as well as the *cyclo*-addition reactions of alkenyl radicals. The result of the successive decomposition and *cyclo*-addition reactions of these radicals gives rise to several different isomers. Starting, for instance, from *cyclo*-hexyl radical, various methyl-cyclopentyl as well as various normal and branched hexenyl radicals are obtained.

Furthermore, the pyrolysis of *cyclo*-alkanes is a chain radical process where several molecular reactions also play a significant role. Specifically, the molecular isomerizations between *cyclo*-alkanes and alkenes are typical examples of four- and six-centre concerted reactions.

The complexity rises due to the number of intermediate components in the primary decomposition paths rapidly increasing with the molecular weight. The MAMA kinetic generator becomes very useful and practically essential in evaluating the net result of these reacting systems. With the focus on a reliable evaluation of the ethylene yields in commercial reactors, the complexity of these reacting system demands major simplifications if it is to describe primary product distribution and successive decomposition paths without requiring too great a computational effort.

Cyclo-alkanes are typically *cyclo*-pentanes and *cyclo*-hexanes with a certain degree of methylation and a single more or less long alkyl side chain. The C–C cleavage in the side chain follows the same rules and applies the same reference kinetic parameters as the initiation reactions of normal and branched alkanes.

The formation of *cyclo*-alkyl radicals with the radical position on the ring is the result of a C–C bond cleavage in the alkyl substituted position. For instance, the following model reaction:

$$CH_3\text{-}cyclo\text{-}C_6H_{11} \rightarrow CH_3\bullet + cyclo\text{-}C_6H_{11}\bullet$$

is similar to the initiation reaction of *iso*-butane, and it is possible to refer to the same activation energy and to assume the following rate expression:

$$k = 5.0 \times 10^{16} \exp\left(\frac{-85,000}{RT}\right) \quad [\text{s}^{-1}]$$

This value agrees reasonably well with the rate suggested by Tsang (1978a)

$$k = 1.09 \times 10^{26} \, T^{-2.6} \exp\left(\frac{-90,400}{RT}\right) \quad [\text{s}^{-1}]$$

for *iso*-butane, although it is quite a lot faster than the one proposed by Brown and King (1989) for the demethylation reaction of methyl-*cyclo*-hexane:

$$k = 1.41 \times 10^{16} \exp\left(\frac{-88,000}{RT}\right) \quad [\text{s}^{-1}]$$

New reactions involve the cleavage of a C–C bond inside the ring to form biradical intermediates, which quickly rearrange themselves into molecular products: initiation, in this case, acts as an isomerization reaction to form alkenes. On the basis of available thermochemical evaluations (Benson, 1976), the suggested reference kinetic parameters are

$$\text{Cyclo-}C_5H_{10} \rightarrow \text{1-pentene} \quad k_{cy5} = 1.6 \times 10^{16} \exp\left(\frac{-83,500}{RT}\right) \quad [\text{s}^{-1}]$$

$$\text{Cyclo-}C_6H_{12} \rightarrow \text{1-hexene} \quad k_{cy6} = 1.6 \times 10^{16} \exp\left(\frac{-86,000}{RT}\right) \quad [\text{s}^{-1}]$$

These parameters agree with the information in the literature:

$$\text{Cyclo-}C_5H_{10} \rightarrow \text{1-pentene} \quad k = 1.4 \times 10^{16} \exp\left(\frac{-85,400}{RT}\right) \quad [\text{s}^{-1}]$$

(Tsang, 1978b)

$$\text{Cyclo-}C_6H_{12} \rightarrow \text{1-hexene} \quad k = 5.0 \times 10^{16} \exp\left(\frac{-88,800}{RT}\right) \quad [\text{s}^{-1}]$$

(Tsang, 1978a)

$$CH_3\text{-}Cyclo\text{-}C_6H_{11} \rightarrow \text{1-}C_7H_{14} \text{ and 2-}C_7H_{14}$$

$$k = 2.5 \times 10^{16.4} \exp\left(\frac{-83,000}{RT}\right) \quad [\text{s}^{-1}]$$

(Brown and King, 1989)

The differences in the activation energies are due both to the strain energy of the *cyclo*-pentane ring and to the presence of methyl and alkyl substitutions on the ring.

From the above consideration it can be deduced that the initiation of *cyclo*-alkanes then turns into the initiation reactions of alkenes. In this situation, the weakest C–C bonds in the alkene skeleton are those forming allyl type radicals.

The following reactions are typical examples in this class:

$$1\text{-}C_4H_8 \rightarrow aC_3H_5\bullet + CH_3\bullet \quad k = 10^{16} \exp\left(\frac{-72,750}{RT}\right) \quad [s^{-1}]$$
(Dean, 1985)

$$1\text{-}C_5H_{10} \rightarrow aC_3H_5\bullet + C_2H_5\bullet \quad k = 10^{16} \exp\left(\frac{-71,200}{RT}\right) \quad [s^{-1}]$$
(Tsang, 1978b)

$$1\text{-}C_6H_{12} \rightarrow aC_3H_5\bullet + nC_3H_7\bullet \quad k = 10^{16} \exp\left(\frac{-70,500}{RT}\right) \quad [s^{-1}]$$
(Tsang, 1978a)

For alkenes heavier than 1-pentene, the selected reference value for this initiation reaction is

$$k = 10^{16} \exp\left(\frac{-71,200}{RT}\right) \quad [s^{-1}]$$

The activation energy adopted here is obtained by correcting the bond energy with the contribution of the formation of a resonantly stabilized radical. This correction agrees with the data proposed for the formation of two allyl radicals as in the case of 1,5-hexadiene initiation reaction

$$CH_2=CH-CH_2-CH_2-CH=CH_2$$
$$\rightarrow CH_2=CH-CH_2\bullet + CH_2=CH-CH_2\bullet$$

Dean (1985) suggests $1.58 \times 10^{15} \exp(-57,000/RT)$ which is also confirmed as $6.3 \times 10^{14} \exp(-57,630/RT)$ by Roth et al. (1991).

Initiation through the decomposition of various C–C bonds is significantly less important and is similar to alkane species decomposition:

$$1\text{-hexene} \rightarrow CH_2=CH-CH_2-CH_2\bullet + C_2H_5\bullet$$
$$k = 5.0 \times 10^{16} \exp\left(\frac{-82,000}{RT}\right) \quad [s^{-1}]$$

H-abstraction reactions of *cyclo*-alkanes follow the same rules and apply the same reference kinetic parameters as the analogous reactions of normal and branched alkanes. For example, Fig. 6 shows the main *cyclo*-hexyl radical pyrolysis pathways. For simplicity's sake, most of the dehydrogenation reactions are not reported.

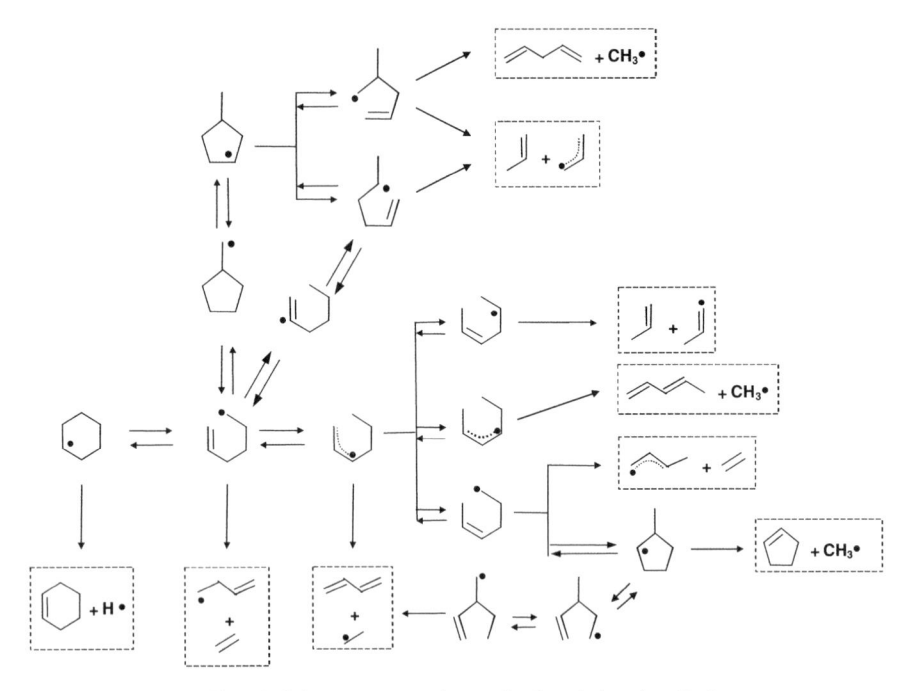

FIG. 6. Primary propagation path of *cyclo*-hexyl radical.

Once again, these reaction paths follow elementary steps with very simple and predefined rules. They may be simply classified as: isomerization and decomposition reactions.

Isomerization reactions include:

- Ring decomposition and the reverse *cyclo*-addition reactions
- 1–4, 1–5, 1–6 H-transfer reactions via five-, six- and seven-membered ring intermediates. These reactions are favoured when forming allyl or similar resonantly stabilized radicals (RSR).

Decomposition reactions include:

- β-decomposition reactions
- Dehydrogenation reactions.

A simple inspection of the scheme reported in Fig. 6 demonstrates two things. Firstly, each reaction step falls into the list reported above, and secondly, the more the initial radical carbon number increases, the larger the number of possible reactions and the greater the pool of all the intermediate and final species.

Several kinetic parameters of these propagation reactions are taken directly from the analogous reactions of alkyl radicals. For instance, the kinetic parameters of the dehydrogenation reaction of cyclohexyl radical to form cyclohexene

$$k_1 = 2.0 \times 10^{14} \exp\left(\frac{-39,500}{RT}\right) \quad [s^{-1}]$$

are the same parameters used for the dehydrogenation reaction of 3-pentyl radical to form 2-pentene:

$$CH_3-CH_2-CH\bullet-CH_2-CH_3 \rightarrow CH_3-CH=CH-CH_2-CH_3 + H\bullet$$

Similarly, the kinetic parameters of the decomposition reaction of the 5-hexen-1-yl radical to form C_2H_4 and the 3-butenyl radical

$$k_2 = 10^{14} \exp\left(\frac{-30,000}{RT}\right) \quad [s^{-1}]$$

are the same parameters used for the β-decomposition reactions of primary alkyl radicals to form ethylene and smaller primary radicals. The presence of a double bond inside the radical does not affect the reaction itself until the double bond is far enough from the reaction zone.

The isomerization reaction to form the resonantly stabilized hexenyl radical:

$$\bullet CH_2-CH_2-CH_2-CH_2-CH=CH_2$$
$$\rightarrow CH_3-CH_2-CH_2-CH\bullet-CH=CH_2$$

applies the same estimation rules common to all alkyl radicals. It is thus possible to estimate the following rate parameters for the direct and reverse reactions, respectively

$$k_{for} = 1.58 \times 10^{11} \exp\left(\frac{-15,500}{RT}\right) \quad [s^{-1}]$$

$$k_{rev} = 3.16 \times 10^{11} \exp\left(\frac{-29,000}{RT}\right) \quad [s^{-1}]$$

These kinetic parameters take into account the formation of a five-membered ring intermediate, the H-abstraction reaction of two H-atoms of allyl type and the formation of the resonantly stabilized 1-hexen-3-yl radical. These facts explain why the reverse isomerization reaction requires greater activation energy. As clearly shown in Fig. 6, there is a new class of important reactions, i.e. ring decomposition (e.g., *cyclo*-hexyl to form hexenyl radical) and the reverse *cyclo*-addition reaction. The activation energies of ring decomposition to form primary radicals are 31,500 and 28,000 kcal/kmol respectively for the six- and five-membered ring cleavage. The reverse *cyclo*-addition reactions (through the addition of a primary radical) require 8,000 and 13,500 kcal/kmol for the formation of a five- and six-membered ring respectively (Matheu *et al.*, 2003). Successive isomerization and β-decomposition reactions of these radicals increase the complexity of the overall mechanism and explain the primary formation of propene, butadiene, pentadiene and cyclopentene from *cyclo*-hexane.

The complete set of the reference kinetic parameters for *cyclo*-alkane and alkene pyrolysis is reported in Table VII.

The overall *cyclo*-hexyl radical isomerization and decomposition mechanism, already shown in Fig. 6, is once again reduced to a single equivalent reaction which lumps all the intermediate propagation and decomposition steps:

$$R\bullet + \text{cyclo-hexane} \rightarrow RH + \{cy\text{-}C_6H_{11}\} \rightarrow$$
$$\rightarrow RH + 0.34\,(C_2H_5\bullet + C_4H_6) + 0.11\,(cy\text{-}C_6H_{10} + H\bullet)$$
$$+ 0.02\,(C_6H_{10} + H\bullet) + 0.23\,(C_5H_8 + CH_3\bullet)$$
$$+ 0.20\,(C_2H_4 + pC_4H_7\bullet) + 0.01(cy\text{-}C_5H_8 + CH_3\bullet)$$
$$+ 0.01(C_2H_4 + aC_4H_7\bullet) + 0.09\,(C_3H_6 + aC_3H_5\bullet)$$

This initial product distribution, evaluated by MAMA at 1,040 K, shows the importance of butadiene formation through the decomposition of 1-hexen-3yl radical, and also confirms that this stoichiometry is only slightly temperature dependent in this case too.

The analysis of the primary decomposition products from the decomposition of methyl-*cyclo*-pentane requires the addition of only a few reactions to the scheme shown in Fig. 6. Three methyl-*cyclo*-pentyl radicals were already formed from the successive isomerization reactions of *cyclo*-hexyl-radical, and thus only the tertiary radical, with its isomerization and decomposition reactions to form methyl-allyl radical and ethylene, needs to be included

The net result of the H-abstraction, propagation and decomposition reactions of methyl-*cyclo*-pentane becomes

$$R\bullet + CH_3\text{-}cyclo\text{-pentane} \rightarrow RH + \text{Mix}\,[CH_3\text{-}cyclo\text{-pentyl}]$$
$$\rightarrow RH + 0.34(C_3H_6 + aC_3H_5\bullet) + 0.25(C_2H_4 + iC_4H_7\bullet)$$
$$+ 0.32(C_2H_4 + aC_4H_7\bullet) + 0.03(C_2H_5\bullet + C_4H_6)$$
$$+ 0.02(cy\text{-}C_5H_8 + CH_3\bullet) + 0.02(me\text{-}cy\text{-}C_5H_8 + H\bullet)$$

TABLE VII
REFERENCE KINETIC PARAMETERS FOR THE PYROLYSIS OF CYCLO-ALKANE AND ALKENES (UNITS ARE: KCAL, KMOL, L, S)

H-abstraction reactions	Primary H-atom	Secondary H-atom	Primary allyl H-atom
Primary alkyl radical	$10^8 \exp(-13,500/RT)$	$10^8 \exp(-11,200/RT)$	$10^8 \exp(-10,500/RT)$
Primary allyl radical[a]	$3.16 \times 10^8 \exp(-22,500/RT)$	$3.16 \times 10^8 \exp(-19,000/RT)$	$3.16 \times 10^8 \exp(-18,500/RT)$
Isomerization reactions of allyl radicals (Transfer of a primary H-atom)			
	1–4 H-transfer	1–5 H-transfer	1–6 H-transfer
Primary radical[b]	$10^{11} \exp(-28,000/RT)$	$1.58 \times 10^{10} \exp(-23,500/RT)$	$3.16 \times 10^9 \exp(-28,500/RT)$
	5-membered ring	6-membered ring	
Cyclo-addition reactions	$10^{11} \exp(-13,500/RT)$	$1.58 \times 10^{10} \exp(-8,000/RT)$	
Cyclo-alkyl radical decomposition Reactions (to form primary radicals)	$5 \times 10^{13} \exp(-28,000/RT)$	$10^{14} \exp(-31,500/RT)$	
Allyl radical decomposition reactions (to form conjugated dienes)		$5 \times 10^{13} \exp(-35,000/RT)$	

[a]Corrections for secondary and tertiary radicals are the same as for alkyl radicals.
[b]Rate values for secondary and tertiary radicals are obtained by adding 1,500 and 2,000 kcal/kmol to the activation energies.

The lumped radical Mix [CH_3-*cyclo*-pentyl] is the mixture of the four different alkyl radicals derived from the H-abstraction reactions on methyl-*cyclo*-pentane

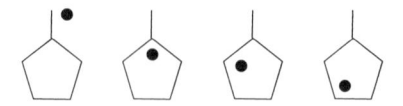

Figure 6 also describes the 1-hexene reactions. Thus, the H-abstraction reactions of 1-hexene simply become

$$R\bullet + \text{1-hexene} \rightarrow RH + \text{Mix [1-hexenyl]}$$
$$\rightarrow RH + 0.37(C_2H_5\bullet + C_4H_6) + 0.37(C_5H_8 + CH_3\bullet)$$
$$+ 0.15(C_3H_6 + aC_3H_5\bullet) + 0.07(C_6H_{10} + H\bullet)$$
$$+ 0.03(C_2H_4 + C_4H_7's\bullet)$$

The system becomes increasingly complex as the molecular weight rises. Figure 7a and 7b show the distributions of small alkenes and radicals from the decomposition, at 1,040 K, of the series of 1-methyl-4-alkyl-*cyclo*-hexanes vs. the carbon number (from 1,4-dimethyl-cyclohexane up to 1-methyl-4-heptyl-cyclohexane).

The analysis of the primary decomposition and isomerization of *cyclo*-alkanes is limited to 14 C-atoms due to the large number of possible isomers and to the resulting dimension of the overall problem. Nevertheless, it should be pointed

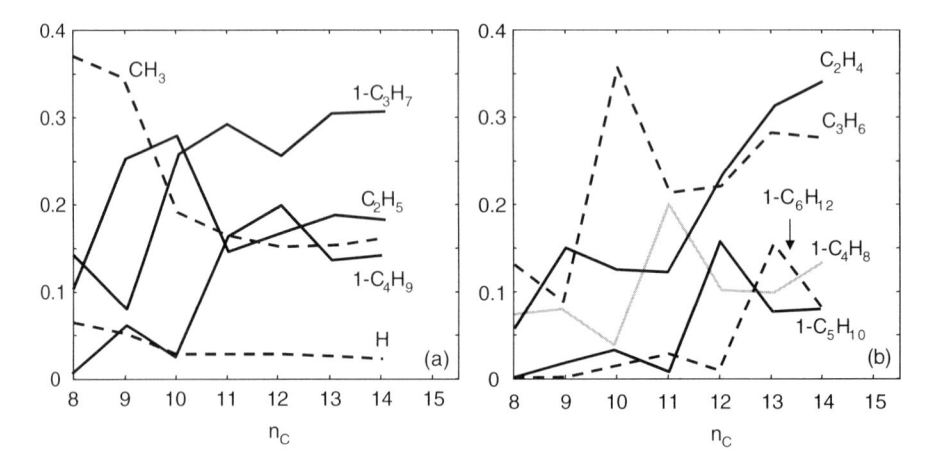

FIG. 7. Panel (a): primary selectivity of small radicals from the decomposition of 1-methyl-4-alkyl-*cyclo*-hexanes vs. carbon number; panel (b): primary selectivity of alkenes from the decomposition of 1-methyl-4-alkyl-*cyclo*-hexanes vs. carbon number.

out that the trend for small alkenes and radicals is asymptotic for C-atoms higher than 12–13. A complete and detailed description of heavier species soon becomes unmanageable. For this reason, the stoichiometry of the primary reactions of heavier homologous species is obtained using simpler extrapolation rules.

The detailed analysis of the primary product distribution from 1-methyl-4-propyl-cyclohexane (10 C-atoms) is useful for a better understanding of the relative importance of intermediate species. It can also provide useful information regarding the choice of intermediate lumped components.

Thus, MAMA gives the following result for the H-abstraction reaction:

$R\bullet$ + 1methyl-4propyl-cyclohexane = RH + 0.10 H$\bullet$ + 0.21 CH$_3\bullet$
+ 0.29 C$_2$H$_5\bullet$ + 0.23 1-C$_3$H$_7\bullet$ + 0.03 2-C$_3$H$_7\bullet$ + 0.03 1-C$_4$H$_9\bullet$
+ 0.01 2-C$_4$H$_9\bullet$ + 0.001 iC$_4$H$_9\bullet$ + 0.02 aC$_3$H$_5\bullet$ + 0.03 aC$_4$H$_7\bullet$
+ 0.02 pC$_4$H$_7\bullet$ + 0.04 iC$_4$H$_7\bullet$ + 0.119 C$_2$H$_4$ + 0.307 C$_3$H$_6$
+ 0.02 1-C$_4$H$_8$ + 0.006 2-C$_4$H$_8$ + 0.007 iC$_4$H$_8$ + 0.03 1-pentene
+ 0.002 2-pentene + 0.01 1-hexene + 0.002 2-hexene
+ 0.002 3-hexene + 0.005 2-me-1-pentene + 0.07 1,3-C$_4$H$_6$
+ 0.04 1,3-C$_5$H$_8$ + 0.003 1,4-C$_5$H$_8$ + 0.08 isoprene
+ 0.03 1,3-hexadiene + 0.004 1,4-hexadiene + 0.004 1,5-hexadiene
+ 0.004 2-me-1,3-pentadiene + 0.02 3-me-1,3-pentadiene
+ 0.01 4-me-1,3-pentadiene + 0.01 2,4-hexadiene
+ 0.01 ethyl-butadiene + 0.01 *cyclo*-hexene
+ 0.06 diaC$_7$ + 0.05 diaC$_8$ + 0.02 diaC$_9$ + 0.004 diaC$_{10}$ (C7 – C10 dialkenes)
+ 0.17 Cy$_6$C$_7$ + 0.19 Cy$_6$C$_8$ + 0.10 Cy$_6$C$_9$ + 0.023 Cy$_6$C$_{10}$ (C7 – C10 *cyclo*-hexenes)
+ 0.002 Cy$_5$C$_7$ + 0.002 Cy$_5$C$_8$ (C7 – C8 *cyclo*-pentenes)

In order to simplify this stoichiometry, several species heavier than C6 are conveniently grouped into families of homologous species.

The following considerations can be singled out:

- Unsaturated radicals constitute more than 10% of the radical pool
- There is a significant production of conjugated di-alkenes and alkyl-*cyclo*-hexenes.

Both *cyclo*-hexane and *cyclo*-pentane components must be considered. As in the case of branched alkanes, the virgin feeds for *cyclo*-alkanes are relatively regular, once again due to the same diagenesis process. *Cyclo*-hexane rings are more abundant than *cyclo*-pentane ones. The probability of methyl substitutions is about 20%, on both the side chain and the naphthenic carbon. Moreover, experimental distributions of these components indicate the prevailing presence of a single long alkyl chain and several methyl substitutions.

On this basis, the fraction of the C10 *cyclo*-alkane isomers has been assumed as a single equivalent component with a default internal composition. MAMA analyses the primary propagation reactions of the real isomers of the mixture, groups and lumps all the stoichiometries, and generates the equivalent stoichiometry of the lumped component, once again using the same reference kinetic parameters. After processing by MAMA, the following lumped reaction represents the H-abstraction reaction of the lumped component {NAFC$_{10}$}:

$R\bullet$ + {NAFC$_{10}$} = RH + 0.0916 H$\bullet$ + 0.2878 CH$_3\bullet$ + 0.1667 C$_2$H$_5\bullet$
+ 0.1073 nC$_3$H$_7\bullet$ + 0.0959 iC$_3$H$_7\bullet$ + 0.0216 1-C$_4$H$_9\bullet$ + 0.0342 2-C$_4$H$_9\bullet$
+ 0.0064 iC$_4$H$_9\bullet$ + 0.0651 aC$_3$H$_5\bullet$ + 0.0478 aC$_4$H$_7\bullet$ + 0.0135 pC$_4$H$_7\bullet$
+ 0.0619 iC$_4$H$_7\bullet$ + 0.1800 C$_2$H$_4$ + 0.2385 C$_3$H$_6$ + 0.0298 1-C$_4$H$_8$
+ 0.0213 2-C$_4$H$_8$ + 0.0053 iC$_4$H$_8$ + 0.0185 1-C$_5$H$_{10}$ + 0.0022 2-C$_5$H$_{10}$
+ 0.0142 1-C$_6$H$_{12}$ + 0.0179 C$_6$H$_{12}$s + 0.0002 1-C$_7$H$_{14}$ + 0.0340 C$_7$H$_{14}$s
+ 0.0438 me-1-C$_5$H$_8$ + 0.0044 me-2-C$_5$H$_8$ + 0.0109 2me-1-C$_4$H$_8$
+ 0.0103 3me-1-C$_4$H$_8$ + 0.0008 2me-2-C$_4$H$_8$ + 0.0545 C$_4$H$_6$
+ 0.0437 pentadiene + 0.0899 isoprene + 0.0286 1, 3-C$_6$H$_{10}$
+ 0.0094 1, 5-C$_6$H$_{10}$ + 0.0601 hexadienes + 0.0677 heptadienes
+ 0.0399 octadienes + 0.0045 cyC$_5$H$_8$ + 0.0201 cyC$_6$H$_{10}$
+ 0.0229 me-cyC$_5$H$_8$ + 0.1256 me-cyC$_6$H$_{10}$ + 0.0557 dime-cyC$_6$H$_{10}$
+ 0.1786 NAF$_9$H$_{16}$ + 0.0105 NAF$_{11}$H$_{20}$

Analysis of this product distribution quite clearly reveals the importance of diolefins and unsaturated *cyclo*-alkanes. These components must therefore be described along with their successive pyrolysis reactions which soon move toward the formation of aromatic species.

4. Aromatics

The alkyl aromatic propagation reactions are automatically generated by the MAMA program in this case also and only a few new reactions need to be discussed. The formation of RSR is the first important feature. In fact, the H-abstraction reactions to form benzyl and benzyl-like radicals are the favoured ones with respect to H-abstraction on aromatic sites.

Similarly, the formation of RSR is also favoured in the initiation reactions. The kinetic parameters $6.1 \times 10^{15} \exp(-75,000/RT)$ [s^{-1}] suggested by Baulch *et al.* (1992) for the model reaction:

$$CH_3-CH_2-C_6H_5\bullet \rightarrow Benzyl + CH_3\bullet$$

were used. The same authors suggest the following kinetic parameters:

$$k = 0.398 \times T^{3.44} \exp\left(\frac{-3,100}{RT}\right) \quad [\text{m}^3/\text{kmol s}]$$

for the H-abstraction reactions of H radical on toluene, to form benzyl radical:

$$\text{H}\bullet + \text{Toluene} = \text{H}_2 + \text{Benzyl}\bullet$$

These values are very similar to the ones of the H-abstraction from propene to form allyl radical (Tsang, 1992). Successive reactions of aromatics include the substitutive addition reactions that progressively favour the dealkylation reactions of aromatic species. The kinetic parameters for the reference reaction H• + Toluene → CH$_3$• + Benzene

$$k = 5.78 \times 10^{10} \exp\left(\frac{-8,090}{RT}\right) \quad [\text{m}^3/\text{kmol s}]$$

are taken from Baulch *et al.* (1994) whilst the kinetic parameters for the reverse reaction become (Robaugh and Tsang, 1986):

$$k = 1.2 \times 10^9 \exp\left(\frac{-16,000}{RT}\right) \quad [\text{m}^3/\text{kmol s}]$$

Other important reactions refer to vinyl and allyl radical addition reactions on alkyl aromatic species. These substitutive addition reactions also contribute to a progressive dealkylation of the aromatic species. The reference kinetic parameters of these reactions are respectively

$$k_{\text{vinyl}} = 1 \times 10^9 \exp\left(\frac{-6,000}{RT}\right) \quad [\text{m}^3/\text{kmol s}]$$

$$k_{\text{allyl}} = 2 \times 10^9 \exp\left(\frac{-25,000}{RT}\right) \quad [\text{m}^3/\text{kmol s}]$$

The differences in activation energies caused by the additions of the different radicals reflect the different levels of stability of the attacking radicals.

β-decomposition and chain cleavages of alkyl side substitution still refer to previously discussed kinetic parameters. On this basis and once again using the MAMA program, the following stoichiometry is obtained for the H-abstraction reaction of the lumped component {C$_{15}$AR}. This component is a mixture of several isomers C$_{15}$H$_{24}$ with the prevailing presence of only one long side chain with the usual methyl substitutions considered once again also.

$$R\bullet + \{C_{15}AR\} \rightarrow RH + 0.0892\ H\bullet + 0.1469\ CH_3\bullet + 0.1303\ C_2H_5\bullet$$
$$+ 0.0149\ nC_3H_7\bullet + 0.1774\ iC_3H_7\bullet + 0.0036\ 1\text{-}C_4H_9\bullet$$
$$+ 0.0799\ 2\text{-}C_4H_9\bullet + 0.1645\ iC_4H_9\bullet + 0.1618\ rC_8H_9\bullet$$
$$+ 0.0314\ rC_9H_{11}\bullet + 0.2199\ C_2H_4 + 0.2539\ C_3H_6$$
$$+ 0.0481\ 1\text{-}C_4H_8 + 0.0306\ 2\text{-}C_4H_8 + 0.0694\ iC_4H_8$$
$$+ 0.0035\ 1\text{-}C_5H_{10} + 0.0039\ 2\text{-}C_5H_{10} + 0.0427\ 2me\text{-}1\text{-}C_4H_8$$
$$+ 0.0275\ 3\text{-}me\text{-}1\text{-}C_4H_8 + 0.0323\ me\text{-}C_5H_{10}s + 0.0844\ C_7H_{14}s$$
$$+ 0.2296\ me\text{-}Styrene + 0.3652\ di\text{-}me\text{-}Styrene + 0.1395\ ArC_{12}H_{16}$$
$$+ 0.0725\ ArC_{15}H_{22}$$

The end primary products of this reaction include a significant quantity of RSR, referred to here as $rC_8H_9\bullet$ and $rC_9H_{11}\bullet$, as well as methyl- and di-methyl-styrene together with larger unsaturated aromatic components, once again lumped into $ArC_{12}H_{16}$ and $ArC_{15}H_{22}$. As clearly set out by Tsang (2004) with regard to the aromatic compounds, a resonance stabilized radical is formed once the alkyl side chains have been reduced to one carbon, and there is general agreement that the condensation of such long-lived radicals play an important role in the formation of PAH. As it will be discussed in the next paragraph, these species are important precursors of carbon deposition in the pyrolysis coils and are also important intermediates in the ultimate formation of soot particles in combustion processes. This is why there has been considerable work on the benzyl radical (Braun-Unkhoff *et al.*, 1990; Ellis *et al.*, 2003) (Frisch *et al.*, 1995) and it is clear that isomerization processes, quite rapidly releasing H-atoms, are competitive with condensation reactions.

C. Detailed Characterization of Liquid Feeds

As previously mentioned, pyrolysis feedstocks are usually made up of complex hydrocarbon mixtures derived from the refinery. An example of this complexity has been already given in Table V in which boiling temperatures and the number of paraffin isomers are compared with the carbon numbers of various petroleum fractions (Altgelt and Boduszynski, 1994).

1. Naphtha, Kerosene and Light Gasoil Fractions

Commercial naphthas are complex mixtures of a large number of different isomers and are generally characterized by specific gravity and boiling curves (TBP curves or ASTM D86). The relevant properties of different straight run naphthas are reported in Table VIII. The sulphur content is usually lower than 2% and nitrogen is in the order of a few ppm. Light and heavy naphthas are

TABLE VIII
Major Properties of Straight-Run Naphthas Adapted from Aalund (1976a, b)

Origin	Arabian	Iranian	U.S.S.R.	Indonesia	Algeria	Nigeria	Lybia	Ecuador
Property								
Boiling range (°C)	100–150	93–149	77–155	110–193	95–175	75–175	65–110	77–154
Gravity	0.7366	0.7389	0.7389	0.7861	0.7523	0.7792	0.7624	0.7459
P	70.3	53	60	43.8	63.8	27.5	77	49.4
N	21.4	34	32.4	29.1	25	58.5	18	43.9
A	8.3	13	7.6	27.1	11.2	14	5	6.7
Sulphur (wt%)	0.0028	0.13	0.03	0.01	–	0.01	0.0008	0.015
Yield (vol %)	6.8	9.6	12.58	17.9	18.11	8.7	11.67	11.91

TABLE IX
Regularity of Naphtha Composition Adapted from Smith (1968)

Ratio (vol %) of hydrocarbons to *n*-hexane	Minas	Beaver Lodge
n-pentane	0.69	0.82
n-hexane	1.00	1.00
n-heptane	1.23	1.10
iso-pentanes	0.53	0.52
iso-hexanes	0.81	0.68
iso-heptane	0.81	0.61
cyclo-pentanes	0.41	0.31
cyclo-hexanes	0.99	0.86
Benzene plus toluene	0.06	0.38

mainly aliphatic but they also contain a significant quantity of *cyclo*-alkanes. The aromatics may account for as much as 10–20% and most of these are alkyl-benzenes. When naphtha with a high aromatic content is cracked in a pyrolysis plant, alkene production decreases as it is difficult to convert aromatic hydrocarbons.

Naphtha can constitute from 10 to 35 volume percent of crude oil. Despite this difference, considerable similarities were observed in different crude oils if the comparison uses the ratios of each hydrocarbon to the volume percent of *n*-hexane (Smith, 1968). Table IX gives a very simple example of this by demonstrating the similarity between two different naphthas.

The fact that the relative presence of alkane and *cyclo*-alkane hydrocarbons in naphtha fractions is so similar suggests that the most of the oil has undergone the same sequence of reactions or diagenesis. The difference in the naphtha volume fraction inside the oil as well as the larger difference in the aromatic components, however, may indicate the extent to which the diagenesis has proceeded. This is why it is important to define the PONA index, i.e. the relative amount of paraffins, olefins, naphthenes and aromatics, so that the relative sizes of the different fractions can be characterized. It is also important to specify both normal and branched alkanes. Alkenes are rarely contained in straight-run petroleum fractions; however, substantial amounts are present in certain refinery streams.

Naphtha feed is often characterized using PINA analysis that simply is the weight % of *n*-paraffin, *iso*-paraffin, naphthene and aromatic compounds. If the typical commercial indexes (specific gravity, PINA analysis and TBP curves or ASTM D86) are used properly, it is possible to empirically derive detailed naphtha composition by referring to the four different hydrocarbon classes and only to a limited number of reference components within each class. In fact, the PINA information indicates the relative abundance of the four different classes directly. The specific gravity and boiling curve allow the specification of the initial and final cuts of the hydrocarbon mixture as well as the relative presence and distribution of the reference pseudo components inside each fraction.

n-Alkanes from *n*-butane up to *n*-decane can be assumed as real components. However, *iso*-alkanes, *cyclo*-alkanes (heavier than C_6–C_7) and aromatics (from xylenes) are described with lumped components for each carbon number for the sake of convenience. As already mentioned, the analysis and comparison of different virgin feeds shows that significant regularities in the internal distribution of the different isomers can be singled out fairly independently of the crude origin. This fact has already been observed in Table VI for the isomers of the branched octanes. The same approach and similar internal distributions can be defined for all the lumped components as well as for *cyclo*-alkanes and aromatics. This means that the feed is described on the basis of a limited number of reference components defined as fixed mixtures of real isomers. This simplified but still reliable composition of the naphtha feed reduces the total number of species and reactions and makes it possible to deal with the complexity of the pyrolysis system.

Kerosene is the generic name for the lighter end of a group of petroleum substances known as middle distillates, the heavier end being gasoils. Kerosene can be divided into two basic types of kerosenes (straight-run and cracked) but subsequent treatments often blur this simple distinction. Kerosenes obtained from crude oil by atmospheric distillation are known as straight-run kerosenes. The typical distillation range of 145–300°C for kerosenes is such that benzene and *n*-hexane concentrations are always below 0.01% by mass. The main components of kerosenes are normal, branched and *cyclo*-alkanes. They normally account for at least 70% by volume of a process stream. Aromatic hydrocarbons, mainly alkylbenzenes and alkylnaphthalenes, will not normally exceed 25% by volume of kerosene streams (CONCAWE, 1995).

In the case of kerosene and light gasoil also, density, TBP or ASTM D86 distillation curves and PINA analysis (if available) are used, and empirical correlations are derived to yield a detailed picture of hydrocarbon mixture composition. The aromatic fraction can vary greatly for the different feeds, and information, such as the H/C of the feed, is important when PINA is not available.

Vertical lumping is also applied for components with more than 10 carbon atoms. The MAMA program generates the primary propagation reactions, i.e. the lumped initiation and H-abstraction reactions, for all these reference components. The overall resulting stoichiometry of these lumped species is simply obtained by averaging and weighting the lumped stoichiometries of the individual isomers, once again with regard to their internal composition.

2. Heavy Gasoils

The first difficulty in extending the kinetic model to very heavy feeds, and then vacuum oils and crude distillates containing feed components up to 35 C-atoms, involves a proper description of the cracking mixture. Altgelt and Boduszynski's book (Altgelt and Boduszynski, 1994) made an important contribution to this and remains a highly relevant reference work. Specific gravity, boiling points, viscosities, aniline points, UOP factor and hydrogen contents of the fractions are the usual characterization indexes. As Watson et al. (1935) discovered in their pioneering study, if two of these properties are known for any particular feedstock, the others can be more or less accurately approximated. This fact supports the idea that it is possible to derive some information on the internal composition of the fraction from these data. Much higher concentrations of certain petroleum compounds are found than others of similar structures. These survived the diagenesis of petroleum with fewer changes than the remainder. They are known as "biomarkers" or "molecular fossils" and are a further index of the regularities inside the oil fractions. They also support to some extent the idea that the internal detailed composition of the oil fractions can be derived by using only a few industrial indexes such as gravity, boiling curve and H/C data.

The n-d-M method (refractive index, density and molecular weight) is another way of calculating the distribution of carbon atoms in paraffinic, aromatic and naphthenic groups. Simple empirical correlations established by Tadema allow this group type characterization (Van Nes and Van Westen, 1951) to be completed. NMR, particularly the combination of 1H- and 13C-NMR and elemental analysis, allows the determination of numerous average structural groups in a petroleum fraction, such as all the aromatic and aliphatic hydrocarbons. Oka and co-workers (Chang et al., 1982; Oka et al., 1977) assessed relevant steps towards so-called computer-assisted molecular structure construction (CAMSC), i.e. an accurate selection of reference components capable of describing the average properties of the mixture.

Parallel to this, it is also important to define the description level required by the model for it to be reliable and effective. Liguras and Allen (1989, 1992) quite correctly observe that one of the key issues in developing ideal characterization and kinetic models for hydrocarbon fractions is determining the point at which increasing the number of model compounds no longer enhances the model predictions. According to Krambeck (1992), the most important application of detailed compositional analysis is in developing the reaction networks and the kinetic models of hydrocarbon processes. This means that a lumped composition of the feed must be derived. Adequacy and accuracy are the two key features of this lumped composition. Adequacy means that the lumped composition must have sufficient detail to determine all product properties of interest. Different feedstocks with the same lumped composition must yield reaction products with the same lumped composition. This is the first guarantee of accuracy.

On this basis, the detailed characterization of heavy oil fractions first requires the selection of model components. The following relevant classes of model components were selected:

n- and *iso*-alkanes		C_nH_{2n+2}
cyclo-alkanes:	mono-naphthenes	C_nH_{2n}
	di- and poly-	C_nH_{2n-2}–C_nH_{2n-4}–C_nH_{2n-6}
aromatics:	alkyl-benzenes	C_nH_{2n-6}
	alkyl-tetralines and -indanes	C_nH_{2n-8}
	alkyl-naphthalenes, -phenanthrenes	C_nH_{2n-12}–C_nH_{2n-18}
	alkyl-crysenes	C_nH_{2n-24}

Once again using vertical lumping, only one reference component every five C-atoms can be selected in the range between 20 and 35 carbon atoms, for each family. These components are schematically reported in Fig. 8. Primary distribution products for all these species are generated by the MAMA program, using the same rules and procedures previously discussed.

Figure 9 drastically simplifies the major reaction paths of alkyl-naphthalene components. Via H-abstraction and successive decomposition reactions, they can easily form, either naphthalenes with unsaturated side chains (vinyl, allyl or alkenyl side chains) or RSR and smaller decomposition products. The preferential radical attack on the alkyl side chain is in the benzyl position due to the weak hydrogen bond. This makes it easy to justify either the formation of RSR or the successive β-decomposition reaction to form vinylnaphthalene. The net result of the successive recombination and condensation reactions of these aromatic species is the formation of PAH of increasing molecular weight with a progressively lower hydrogen to carbon ratio.

Similar reaction mechanisms are also considered for the other classes of aromatic species. The reacting system gradually moves towards more stable species. These are light gases (H_2 and CH_4) and increasingly heavier polycondensed aromatic species with a low degree of side substitutions and progressively lower hydrogen to carbon ratios. Thus, heavy gasoils at high severities, i.e. after long residence times at high temperatures, produce a fuel

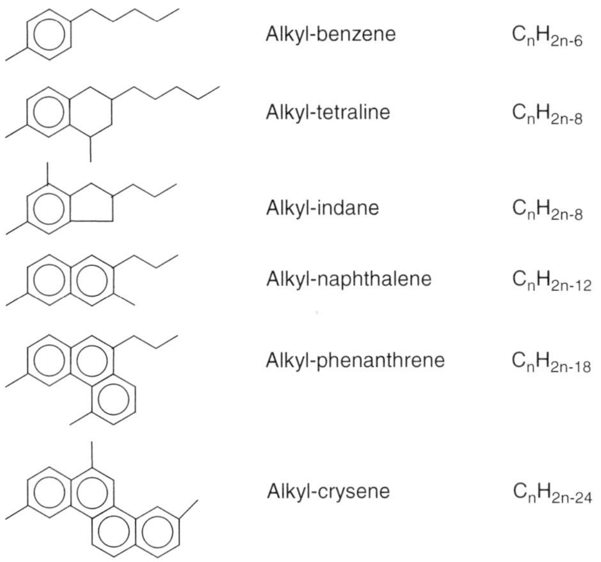

Fig. 8. Classes of model components in aromatic fractions of gasoil feeds.

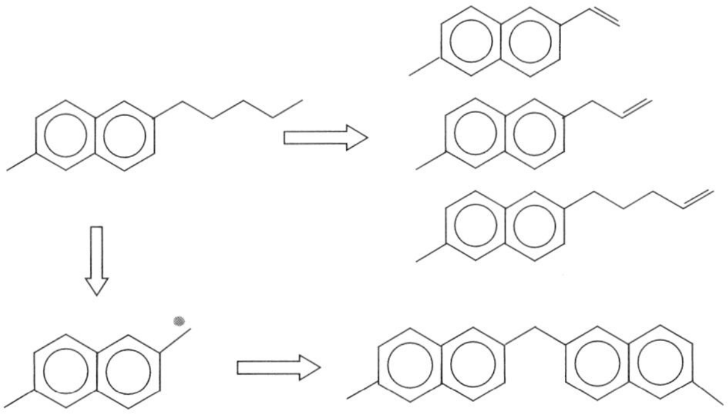

	Alkyl-benzene	C_nH_{2n-6}
	Alkyl-tetraline	C_nH_{2n-8}
	Alkyl-indane	C_nH_{2n-8}
	Alkyl-naphthalene	C_nH_{2n-12}
	Alkyl-phenanthrene	C_nH_{2n-18}
	Alkyl-crysene	C_nH_{2n-24}

Fig. 9. Major reaction paths of alkyl-naphthalenes.

oil fraction ($C12^+$) largely consisting of naphthalene, phenanthrene, crysenes and heavier aromatic condensed species too.

The description of the fuel oil fraction is also a first step towards the understanding of the fouling phenomena in cracking coils and in transferline exchangers. Fouling phenomena and carbon deposition on the pyrolysis coils mostly follow a heterogeneous process. They require the description of the liquid-phase transformation of the carbon macro-polymer structure and its progressive degradation and dehydrogenation to coke, as will be discussed in the coming paragraphs.

3. Crude Distillation Residues

There are analogies between the characterization of crude distillation residues and the previous feeds. The main difference is that for atmospheric and vacuum distillation residues of crude PINA and H/C can be estimated only through the kind of analysis (such as NMR) not normally available in refineries. Moreover, their final boiling point is not defined and the internal distribution of the different hydrocarbon classes of alkanes, *cyclo*-alkanes and aromatics has to be deduced in a different way. Inside each macro-class, the relative amount of the components can be derived from the following statistical distribution:

$$\frac{df}{dn_c} = \frac{1}{n_{c,av} - n_{c,min}} \exp\left[-\frac{n_c - n_{c,min}}{n_{c,av} - n_{c,min}}\right]$$

where f is the fraction of the components with n_c C-atoms, $n_{c,av}$ and $n_{c,min}$ are the average and the minimum number of C-atoms, respectively. Lumped components group sections of 5 to 10 carbon atoms, while a few pseudo components are assumed to represent the heavier fractions up to 200 C-atoms. This statistical distribution, which is also useful for heavy gasoils, was obtained after careful analysis and processing of the data available in the literature (Ali *et al.*, 1985).

Only a small number of experimental data are used to define the feed characteristics: atmospheric equivalent initial boiling point (AEBPI), kinematic viscosity, specific gravity, sulphur content and CCR (Conradson Carbon Residue). AEBPI defines $n_{c,min}$ and the viscosity of the residue $n_{c,av}$. Specific gravity and CCR are essential to defining the relative fraction of the three hydrocarbon classes. Sulphur content is used for correcting the effective specific gravity of the hydrocarbon fraction. The specific gravity of different hydrocarbon classes is defined by means of group contribution methods. In the case of heavy components, experimental data are not available; when the number of C-atoms and of rings increases over a certain amount, specific group contributions have to be derived.

Figures 10 and 11 show some examples of model predictions of the feedstock characterization of crude distillation residues. Figure 10 compares model predictions with the experimental distillation curves of three Arabian atmospheric residues. Figure 11 shows the model ability in predicting the aromatic carbon content and the H/C of different feeds in comparison with some NMR data. A more detailed description and discussion of this residue characterization is reported elsewhere (Bozzano *et al.*, 1995, 1998).

D. EXTENSION OF THE MODELLING TO LIQUID-PHASE PYROLYSIS

The interest here is focused mainly on heavy hydrocarbons feedstocks, as in the case of certain refinery processes, and on polymer thermal degradation. A radical chain mechanism is also involved in the liquid- or condensed-phase pyrolysis. This is once again characterized by initiation, radical recombination

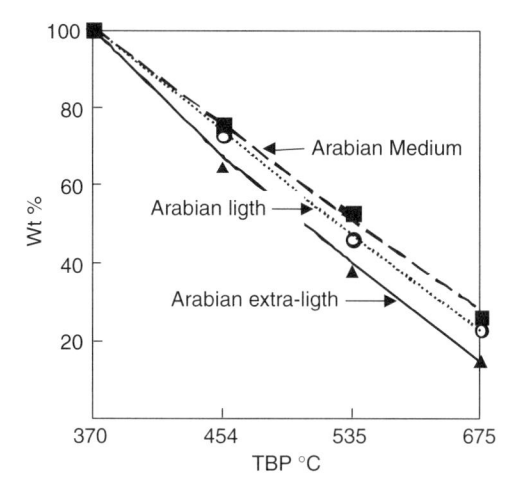

FIG. 10. Predicted and experimental distillation curves of Arabian residues (Ali et al., 1985).

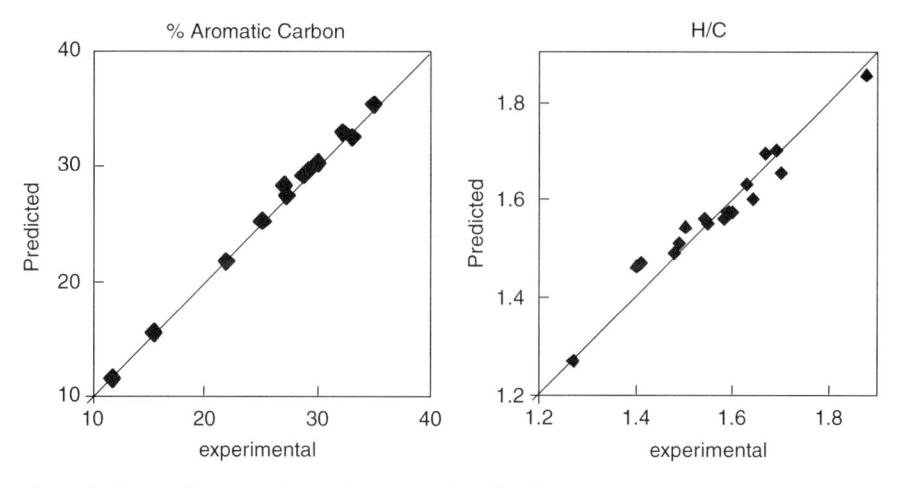

FIG. 11. Scatter diagrams of experimental and predicted values of aromatic carbon atoms and H/C in several residues.

and propagation reactions (i.e. β-scission, H-abstraction, radical isomerization, and substitutive addition of radicals onto unsaturated molecules). Thus, liquid-phase pyrolysis presents close analogies with the gas-phase pyrolysis. Because of the sterical hindrance in the liquid phase, the molecular rotation of large C–C segments is greatly inhibited and internal isomerization reactions of alkyl radicals are less important, in terms of the evaluation of the overall degradation process. For this reason, concerted path molecular reactions can be neglected too.

When the same reaction is "moved" from the gas to the liquid phase, frequency factor and activation energy need to be modified in principle because of the condensed state. In fact, the transition state entropy and enthalpy change in passing from gas to the liquid phase. The transposition to the liquid phase of rate constants can be evaluated (Benson, 1960) from

$$\frac{k_{\text{liq}}}{k_{\text{gas}}} = \exp\left(\frac{\Delta S^{\#}_{\text{liq}} - \Delta S^{\#}_{\text{gas}}}{R}\right) \exp\left(-\frac{\Delta H^{\#}_{\text{liq}} - \Delta H^{\#}_{\text{gas}}}{RT}\right)$$

where $\Delta S^{\#}_{\text{liq}} - \Delta S^{\#}_{\text{gas}}$ = entropy variation of the transition state in passing from liquid to gas phase (this variation is typically negative). $\Delta H^{\#}_{\text{liq}} - \Delta H^{\#}_{\text{gas}}$ = enthalpy variation of the transition state in passing from liquid to gas phase (this is typically negative and about equivalent to the hypothetical heat of evaporation of the transition state).

The corrections are significant if the absolute value of reaction energy is very large; thus, they mainly affect initiation reaction and radical recombinations. The first consideration regards initiation reactions. Unlike the case of gas phase, the entropy change is related to the fact that when the two radicals are formed, they remain "caged" and cannot fully develop their translational and external rotational degrees of freedom (internal rotations and vibrational frequencies remain more or less the same in the reactant and in the transition state).

Just to give few examples, some initiation reactions related to C–C bond cleavage in heavy hydrocarbon and/or polymer decomposition in the liquid phase are reported (consequent to the application of the rule of simulating a fictitious condensation of the transition state).

1. Secondary–secondary carbon bond cleavage:

$$k = 7.94 \times 10^{14} \exp\left[-\frac{81,000 + F(n_c)}{RT}\right] \quad [\text{s}^{-1}]$$

2. Allyl (or Allyl-like and Benzyl-like)–primary carbon bond cleavage:

$$k = 3.16 \times 10^{13} \exp\left[-\frac{72,000 + F(n_c)}{RT}\right] \quad [\text{s}^{-1}]$$

Activation energies are corrected by means of the following approximated function:

$$F(n_c) = 1,140\sqrt{\frac{200 \cdot n_c}{(200 + n_c)}} \quad [\text{kcal/kmol}^{-1}]$$

where n_c is the number of carbon atoms of the heavy component or simply the number of carbon atoms characterizing the so-called flow unit (Van Krevelen, 1976) for the diffusion in the molten polymer.

Rate constants of propagation reactions can be evaluated as in the case of gas phase because the "caging effect" previously mentioned is negligible.

For termination reactions, it is more convenient to make use of the collision theory with an orientation factor. The molecule mobility in liquid phase intuitively is reduced. When radicals approach with the proper orientation, their recombination is very fast but their mobility is low; therefore, the radical recombination is diffusionally limited. The number of collisions between radicals per unit time (Benson, 1960) is provided by $4\pi r_A D_{AS} C_{AS}$ (where r_A is the distance among the atoms containing the radical positions, typically C-atoms, D_{AS} is the diffusion coefficient of the radical in the surrounding area, C_{AS} is the radical concentration in the surrounding area).

The previous expression represents the number of collisions for the single radical A. By multiplying for the number of molecules per unit volume, the number of collisions is obtained

$$v_{coll} = \frac{1}{2} 4\pi r_A D_{AS} \left(\frac{N_{AV}}{1,000}\right) C_{AS}^2 = k_{coll} C_{AS}^2 \quad [kmol/m^3 \, s]$$

where N_{AV} is the Avogadro number and the 1/2 factor is included to avoid counting the collisions twice. k_{coll} is the collision constant (m³/kmole/s). The fraction of collision with orientation favourable to the coupling is the same as in the gas phase. The self-recombination constant is therefore provided by

$$k_{AA} = k_{coll} \cdot \Phi_{AA}^2 \quad [m^3/kmol \, s]$$

where Φ_{AA}^2 is the orientation factor of the collision for effective recombination. The traditional symmetry rule is valid for the recombination A–B

$$k_{AB} = 2\sqrt{k_{AA} \cdot k_{BB}} \quad [m^3/kmol \, s]$$

This rule has the obvious advantage of drastically reducing the number of useful recombination constants simply related to the self-recombinations. The k_{coll} evaluation deserves some further consideration. First of all, the Stokes–Einstein theory suggests that when molecules are not too "long" (the exclusion regards the case of polymers or that one of solvents with extended linear backbones), the diffusion coefficient is

$$D_{AS} \cong \frac{kT}{3\pi\mu_S r_A} \quad [m^2/s]$$

where k is the Boltzmann constant. According to the Eyring theory, the liquid viscosity μ_S is estimated as

$$\mu_S \cong \frac{N_{AV} h}{\tilde{V}_S} \exp\left(\frac{E_V}{RT}\right) \quad [kg/m \, s]$$

where h = Planck constant, $\tilde{V}_S$ = liquid molar volume, E_v = energy required for the viscous motions. The latter can be approximately estimated as $E_v \approx 0.36 \Delta H_{ev}$ where ΔH_{ev} is the heat of vaporization of the liquid at its normal boiling point, T_b. When ΔH_{ev} is not available, the simple Trouton rule $\Delta H_{ev} \approx 21\, T_b$ can be applied.

In the case of molten polymers, oligomers or solvents characterized by highly extended linear backbones, the proposed expression is the same but E_v corresponds to the so-called unit of flow energy, i.e. the energy needed to move the proper segment of the chain in the viscous flow (Van Krevelen, 1976). The same concept is applied also to $\tilde{V}_S$, which becomes the molar volume of the "segment".

Therefore, by combining the given expressions, it follows that

$$k_{coll} \approx \frac{2}{3}\frac{kT}{h}\tilde{V}_S \exp\left(-\frac{E_v}{RT}\right) \quad [\text{m}^3/\text{kmol s}]$$

where $\tilde{V}_S$ = the molar volume of the liquid or of the polymer segment = W_S/ρ_S (with W_S = the average molecular weight of the liquid (kg/kmole), ρ_S = liquid density (kg/m^3)). Therefore

$$k_{coll} \approx 10^{10.2} T \cdot \tilde{V}_S \exp\left(-\frac{E_v}{RT}\right) \quad [\text{m}^3/\text{kmol s}]$$

Note that the self-recombination kinetic constant follows an Arrhenius-type formula where E_V assumes the role of activation energy.

By applying the given formulas at 450 K, the recombination constant of macro-radicals in molten polymethylene becomes

$$k_{AA} \approx 10^{10.3} \exp\left(-\frac{5,800}{RT}\right) \quad [\text{m}^3/\text{kmol s}]$$

where the estimated orientation factor is $\Phi_{AA}^2 = 10^{-2.2}$.

III. Fouling Processes: Formation of Carbonaceous Deposits and Soot Particles

The formation of carbonaceous deposits and the formation of soot particles are two different aspects of the same pyrolysis reactions involving a complex sequence of condensation and dehydrogenation reactions.

Fouling processes on surfaces are common to all equipment involving hydrocarbon feeds at temperatures over ~300°C. They are therefore widespread inside pyrolysis coils because they both operate at high temperatures and are fed by hydrocarbon fractions, which are responsible for the deposit formation. Fouling in pyrolysis coils has always had a negative impact on the behaviour of the process. Deposit formation reduces heat transfer and increases tube skin temperatures so that the maximum allowable value is approached, limiting the total run-time. Moreover, the furnace thermal efficiency as well as the reaction volume inside the coils is progressively reduced. The pressure drop increases with a detrimental effect on process selectivity. The thermal efficiency of the downstream transfer line heat exchangers (TLE) is reduced too due to the progressive fouling of the tubes. The experimental information from industrial furnaces usually refers to the evolution of pressure drops and external tube skin temperature vs. on-stream time.

However, fouling is present not only in gas- and liquid-phase pyrolysis processes, but also in several other systems. In fact, the mechanisms involved in the formation and growth of the deposit are quite similar in coils, heat exchangers surfaces and drums of delayed coking, in the formation of PAH and soot in combustion processes as well as in the formation of nano and micro-fibres. In all these cases, despite different temperature levels ranging from 300°C up to more than 1,500°C, the continuous degradation towards char and coke formation plays a common role (Bozzano *et al.*, 2002). It is clear from this scenario that an understanding of the phenomena and their modelling is fundamental to improving not only the design of pyrolysis coils and cracking furnaces but also the operation of the heat exchangers and other process units.

The deposit grows as a consequence of the continuous addition of new species, molecules or radicals from the process fluid; these form a polymeric layer at the process fluid side. This layer has a relatively soft consistency (rubbery or glassy, depending on the temperature and time). It is then progressively dehydrogenated, causing it to release gases and its specific volume to shrink in many cases. This gives rise to morphological and structural modifications of the deposit, which becomes increasingly porous because of the formation of micro- and nano-fractures. As a consequence of these phenomena, the bulk of the deposit progressively assumes the characteristics of disordered graphite with time and temperature, and only then it can be defined as coke. While the deposit is progressively transformed into coke, the fouling process continuously increases its thickness. Thus, the deposit changes in nature, composition and properties not only with the time but also along the thickness.

On this basis, the fouling mechanisms must be discussed first and only then can the successive transformation of the deposit be analysed, with particular attention to the evolution of its structure and properties.

A. FOULING AND COKING MECHANISMS IN PYROLYSIS COILS AND TLE'S

Feed composition, operating conditions, equipment geometry, materials and their treatments are the main variables influencing the fouling rate. Firstly, qualitative information on deposit formation is obtained by a direct analysis of the morphology of the material. Coke samples from both industrial and

FIG. 12. Coke deposit in tubes and tubesheet of TLE of an ethane cracker.

lab-scale devices often show two different regions inside the deposit. There is a very thin polymeric layer close to the metal surface (50–80 μm) characterized by a fibrous structure with large void fractions and low thermal conductivity. Figure 12 gives a sample of the coke deposit from the tubes and the tubesheet of a TLE of an ethane cracker. Micro-fibres are present on the metal side of these deposits. The typical diameters of these fibres are in the order of a few μm with a length of some tens of μm. The second layer, close to the process side, is more compact and does not have this fibrous structure. Albright et al. (Albright and Marek, 1988, 1992; Albright et al., 1979) discussed the morphology of several coke samples not only from steam cracking coils and TLE's but also from laboratory devices feeding ethylene, acetylene and butadiene at moderate temperatures. Similar studies and results were also obtained on coke samples coming from the pyrolysis coils and TLE's of crackers fed with naphtha and ethane (Dente and Ranzi, 1983).

After the pioneering work done by Dente and Ranzi (1983), the different mechanisms and features of fouling processes were more fully understood and clearly singled out (see for instance Albright, 2002; Albright and Marek, 1988; Blaikley and Jorgensen, 1990; Cai et al., 2002; Chan et al., 1998; Dente et al., 1983; Froment, 1990; Kopinke et al., 1988; Wauters and Marin, 2002; Zou et al., 1987).

The fouling process consists of at least two mechanisms that contribute to the formation of the deposit: an initial catalytic mechanism, followed by a radicalic

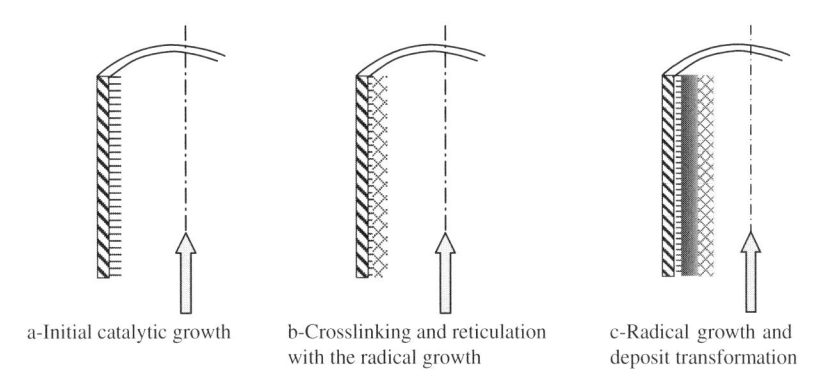

FIG. 13. Relevant steps in the fouling process and deposit formation.

one, which becomes more relevant when crosslink and reticulation reactions inside the deposit, increase the diffusion resistances of the monomers. This hinders their access to the catalytic sites on the metal surface. A simple outline of the evolution of the carbonaceous deposit over time is shown in Fig. 13. During continuous radical growth, the existing deposit progressively develops into more dehydrogenated structures (panel c).

1. Catalytic Mechanism and Initial Growth

Several researchers have studied the fouling of metal surfaces at moderate temperatures by analysing the effect of different feed components. The pyrolysis coil metals (and their oxides) act as a heterogeneous poly-addition catalyst, forming an initial fouling deposit very similar in its morphology to the polymers initially formed with traditional Ziegler–Natta catalysts. The transition metals of the coil surfaces are very good polymerization catalysts and promote this initial deposit formation. All the molecules containing vinyl groups can polymerize on the active sites. When a relatively uniform layer is formed on the metal surface, the catalytic mechanism turns out to be controlled by the diffusion of the monomer into the polymeric layer and a different radical mechanism rules the further fouling process. This now generally accepted conclusion means that the deposit growth can be described by means of at least two different contributions: first a catalytic mechanism and then a radicalic or pyrolytic one (Bach *et al.*, 1995; Barendregt *et al.*, 1981; Goossens *et al.*, 1978; Ranzi *et al.*, 1985; Zimmerman *et al.*, 1990).

This model is also supported by experimental data indicating a larger growth rate in aged coils (already decoked) with respect to new ones. The greater roughness and the higher metal oxide content increase both the metal surface and the number of active sites so that the catalytic mechanism is enhanced. Catalytic growth consists of the continuous addition of vinyl and vinyl-like compounds on the metal sites; these form polymer fibres with a "metallic head" at their end, which eventually detaches from the wall surface. This is probably due to the initial formation of carbides, which can cause nano-fractures in the surface and generate catalyst particles that encourage carbonaceous fibre formation (Lobo and Trimm, 1973). This mechanism is supported by the analysis of coke layers near the metal surfaces where the metal atoms remain quite dense up to about 50 μm from the surface (Albright and Marek, 1988). Metal inclusions can be also responsible for fibre branching. Very similar findings and structures are also observed in the formation of carbon nano-tubes and microfibres in counter-flow diffusion flames (Helveg and Hansen, 2006; Helveg et al., 2004; Merchan-Merchan et al., 2003).

At the beginning of this catalytic growth, the metal surface is not limited by diffusion of monomer species; typical monomers are unsaturated molecules with a terminal vinyl group and are not only aromatics. With the proper sterical orientation, the monomers can add to the growing polymer chains. Polymer fibres then start to cover the metal surface. They grow and assemble so that diffusion resistances limit the further growth. The catalytic mechanisms progressively slow down, and dealkylation and dehydrogenation reactions increase in the formed polymer.

This situation can be clearly seen when observing the time evolution of the tube metal temperature of the pyrolysis coils: there is a fast initial increase and then a reduced asymptotic slope. Note that although the initial slope is initially related to the catalytic rate, it is also due to the relatively low thermal conductivity of the initial fibrous material as a result of the large void fraction. The thickness of this layer is in the order of 20–40 μm. The evolution of the fluid temperature over time either at the TLE outlet or in visbreaking processes and in delayed coking furnaces shows a very similar behaviour.

As recently summarized by Cai et al. (2002), this mechanism requires the presence of catalytic sites on wall surfaces at proper temperatures. On both new and aged metallic surfaces, there are significant populations of catalytic sites where ethylene and olefins can be preferentially adsorbed. Fe and Ni on metal surfaces help coke formation/deposition. The effect of tube materials on coke deposition under TLE conditions was also clearly demonstrated by Bach et al. (1995) who measured and compared the coking rates of 12 different kinds of steels. The steels can be further divided into three categories: low-alloyed and unalloyed steels consisting almost completely of iron; alloyed but nickel-free steels and commonly used high-temperature CrNi steels. The coking rates were the highest for unalloyed and low-alloyed steels, and lowest for CrNi steels. The high iron content and absence of Cr in the low or unalloyed steels are the primary reasons for rapid coke formation. Zychlinski et al. (2002) confirm this catalytic effect with their analysis of fouling processes on conventional and coated HP 40 materials. The results show that the Cr-based coating decreases coke formation in the radiant zone of a steam cracker by up to 90% and 80%, using ethane and naphtha, respectively, as feedstock.

2. Radical and Concerted Path Growth

After the catalytic formation of a basis substrate, a radical mechanism (also referred to as the pyrolytic mechanism) takes place with a reduced growth rate. This second mechanism is active in the bulk of the formed polymer, close to the fluid interface, and becomes more important at high temperatures. The Diels–Alder or concerted path reactions also give a major contribution.

Three main classes of addition reactions contribute to this mechanism:

1. Additions of radicals contained in the process fluid on unsaturated sites of the deposit surface (see Fig. 14)
2. Addition of acetylene, dienes and unsaturated molecules on the deposit surface: Diels–Alder reactions (see Fig. 15)
3. Addition of aromatic and polycyclic aromatic molecules with alkyl side chains on radical positions of the deposit surface (see Fig. 16).

These three classes of reactions consist of two consecutive steps: an initial addition of species from the process side on the coke surface and a successive release of small species (H_2, alkyl side chain, CH_3 or H) with the continuous growth of the original aromatic structure. Successive dealkylation and dehydrogenation reactions further contribute to this coke formation mechanism.

The relative importance of the three reaction classes depends on the operating conditions. The first reaction class is more important at high temperatures as the

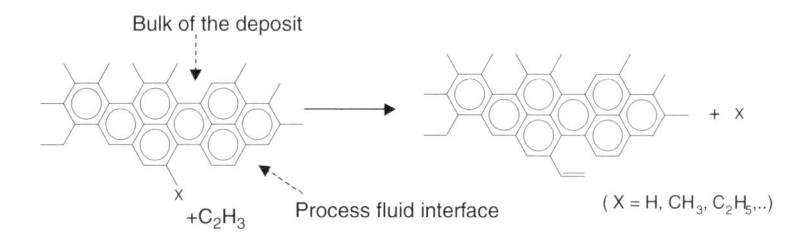

FIG. 14. Addition of unsaturated radicals on the surface of the polymer layer.

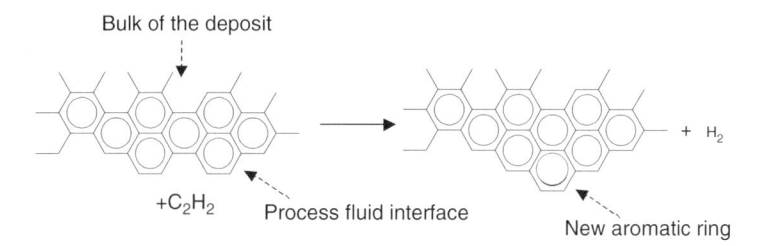

FIG. 15. Addition of unsaturated molecules on the surface of polymer layer (Diels–Alder reaction).

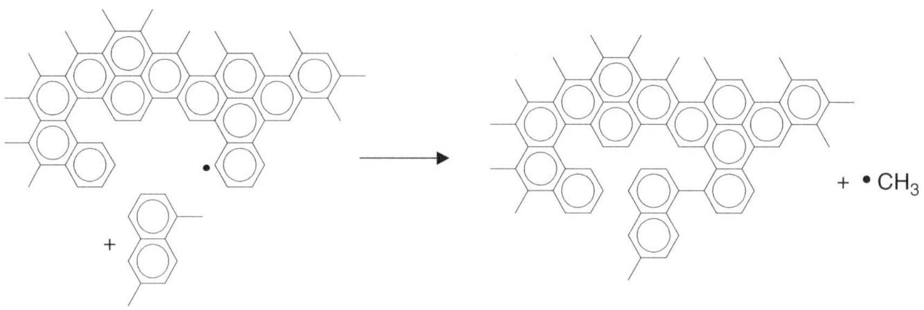

FIG. 16. Addition of an aromatic molecule on radical positions of the deposit surface.

radical concentration increases with temperature. Vinyl and vinyl-like radicals, aromatic and poly-aromatic phenyl-like radicals are the major precursors and are responsible for the growth.

Unsaturated hydrocarbons can add to the deposit-forming cyclic structures with five or six C-atoms through Diels–Alder reactions. The successive dehydrogenations produce polycyclic aromatic structures. In the third reaction class, aromatic species add to macro-radicals of the surface. Note that these radicals are greatly stabilized by the aromatic resonance. Thus, the third reaction class is more important at lower temperatures, typically, in TLE, visbreaking and delayed coking units.

The growth rate depends on the concentration values of the various reacting species at the interface. For gas phase pyrolysis, these concentrations are controlled by their solubility inside the soft layer of the deposit. In fact, only the external layer can be considered as a polymer in a molten or rubbery state. It has not yet been subject to sufficient dealkylation or dehydrogenation reactions, which create a progressive stiffness. For this reason it is important to properly evaluate the solubility of gaseous reactants in the polymer. For both vapour and liquid processes, the reactant concentrations inside the deposit rapidly decrease due to the increasing diffusion limitations. In fact, moving from the external to the inner layers, the polymer becomes increasingly cross-linked, approaching a glassy to solid state. The deposit structure becomes similar to disordered graphite. The only reactants able to reach the inner layers of the deposit are H and methyl radicals.

Finally, it is relevant to observe that this dissolution presents strong analogies with a condensation process discussed and stressed by several authors (Cai et al., 2002) as being responsible for coke formation/deposition in the TLE tube outlet section at operating temperatures of 350–450°C. Indeed this mechanism can be explained on the basis of the solubility of heavy species of the process fluid phase in the soft polymer. There has also been research into the computer generation of a network of elementary steps for coke formation during steam cracking process (Wauters and Marin, 2002).

B. EVOLUTION OF STRUCTURE AND PROPERTIES OF THE DEPOSIT

Once the initial and successive formation of the deposit is analysed, a further important aspect in the overall fouling process is the transformation of the deposit. Due to the thermal degradation of the polymer, the deposit goes from rubbery to glassy and solid state with structural variations over time. The part of the deposit being considered here is the layer between the initial catalytic polymer (near the metal wall) and the one just generated by the radical mechanism close to the process interface (see Fig. 13 panel c). The morphology and properties of the deposit in this intermediate layer are similar to those of a glassy or solid polymer; the material is amorphous and rather rigid. Relatively disordered graphite with cata-condensed aromatic rings is the asymptotic structure, characterized by condensed aromatic rings, with methyl substitution on the free boundary. This is due to the fact that aromatic poly-alkenes are highly reactive even at low temperatures; polystyrene (PS), for instance, will decompose in about 1 h at 360°C. This decomposition process is further enhanced if the aromatic poly-alkenes also contain alkyl or methyl substitutions. So when they are formed or dissolved in the rubber part of the deposit, they are promptly subjected to dealkylation reactions. Only methyl-aromatic structures are present in the inner layers of the deposit. Delayed coking and visbreaking processes' modelling has shown that the average probability of methyl substitution on the free boundary of the aromatic sheets is about 50%. A simple example of a typical poly-aromatic structure with methyl substitutions and methylene bridges is given in Fig. 17.

These aromatic structures are roughly the same for all the fouling processes with the exception of ethane pyrolysis. It is convenient to distinguish and define the different C-atoms and groups inside the deposit. As clearly shown in the previous figure, methyl (P_{CH_3}), methylene (P_{CH_2}), aromatic (P_A) and fused aromatic (P_{FUS}) positions can be identified. Inside the rubbery polymer, initiation, H-abstraction and substitutive addition reactions form the corresponding

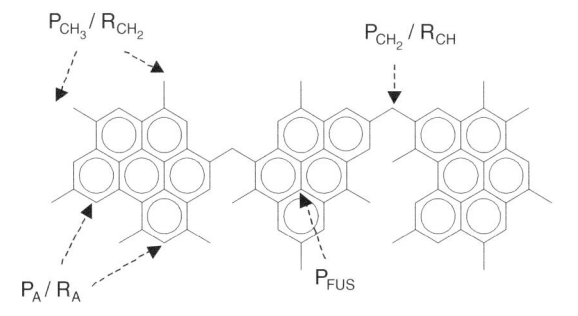

FIG. 17. Typical poly-aromatic structure of the deposit.

radical positions. Three different types of macro-radicals are formed: benzyl-like (R_{CH_2}), phenyl-like (R_A) and radicals on the methylene bridge (R_{CH}). Macro-radicals R_{CH_2} and R_{CH} are present in large amounts due to aromatic resonance but cannot move freely in the polymer. H and CH_3 radicals, on the other hand, can diffuse and react in the deposit porosity via the usual radical mechanism. H_2, CH_4, C_2H_6 are released in the gas phase as a result of H-abstraction, substitutive addition and termination reactions which form radical positions also in the glassy layers.

The H/C ratio of the deposit gradually decreases and, at high temperatures, the hard aromatic C–C and C–H bonds can be broken also. Initially, P_{CH_3} and P_{CH_2} are the preferred positions for H-abstraction reactions with the formation of R_{CH_2} and R_{CH} macro-radicals. Less stable radicals are generated next. All the radicals are active either in H-abstraction or in substitutive additions; otherwise, they recombine with each other.

The deposit from ethane cracking is more compact than the ones derived from other feeds. Practically, all of the growth is due solely to the very ordered additions of ethylene and vinyl radicals. Successive cyclization and dehydrogenation reactions complete the formation of well-organized aromatic structures. This leads to the formation of a more ordered and compact deposit: unlike the case of different feedstocks, large molecules or radical additions are of less importance. Thus, the material produced during ethane pyrolysis is equivalent to disordered graphite, and P_{CH3} positions are practically absent even though the methylene bridges and a similar aromatic structure are maintained.

1. Kinetic Modelling

Thermal conductivity, H/C ratio, specific volume and specific heat vary during the chemical evolution of the deposit. Unfortunately, there is very small quantity of data in the literature on thermal conductivity. In fact, what little there is refers to coke or bitumen and provides limited or sometimes contradictory information because of the high dependency on the structure and composition of the solid. More reliable data refer to disordered graphite, similar to an aged deposit, without hydrogen and with a low porosity. The available experimental data on the time evolution of pressure drop and tube metal temperature in pyrolysis coils of ethylene crackers only permit rough estimates of the overall and average thickness and thermal conductivity of the deposit.

For this reason, the evolution of the thermal conductivity can be obtained on the basis of a group contribution method, mainly based on the concentration of P_{CH_3}, P_{CH_2}, P_A and P_{FUS} positions.

It is quite clear that the thermal conductivity is heavily dependent on the H/C ratio and structural properties, such as the void fraction. Porosity (i.e. free volume) is an important characteristic. Dehydrogenation and dealkylation reactions progressively reduce the molar volume of the deposit, with the formation of fused aromatic carbons. This reduction brings the density of the deposit to about 1,600 kg/m^3. Further volume contractions generate fractures and increase the void fraction but the apparent density remains about constant. The temporal and spatial evolution of the deposit depends mainly on temperature, but temperature evolution and distribution are difficult to describe too. The different equipment and operating conditions greatly influence the thickness and physico-chemical properties of the deposit. Even a rough initial description of this evolution would involve a large number of species and reactions. Consequently, the appropriate lumping techniques are applied once again in this case. Important reactions are grouped into different classes, to be considered as equivalent reactions between positions and radicals, with their kinetic constants. The evolution of the deposit is the result of a radical degradation mechanism based on the following steps:

- Initiation reactions: $P_i \rightarrow R_1 \bullet + R_2 \bullet$
- Propagation reactions (H-abstraction or substitutive addition): $R_1 \bullet + P_1 \rightarrow R_2 \bullet + P_2$
- Termination reactions: $R_1 \bullet + R_2 \bullet \rightarrow P_i$

P_i and R_j are molecular species (or positions) and radicals, respectively.

In order to give an example of the main reactions involved in this modelling, let us consider the conditions when the local evolution in the bulk of the deposit is such that the H/C ratio is less than about 0.8. The deposit mostly consists of polycyclic aromatic structures connected by methylenic bridges and only a few methylated positions are present on the aromatic edge. As already mentioned, these structures are quite stiff (glassy or solid), and the movements of the large macromolecules are more or less limited to vibrations and small-width oscillations. Under these conditions, only small species radicals and molecules react effectively, i.e. $H\bullet$, $CH_3\bullet$, H_2 and CH_4. The average free volume in the layer (about 30% of the total deposit) is made up of "nanochannels" in which the gas radicals and molecules diffuse. Consequently, the mobility of the $H\bullet$ and $CH_3\bullet$ radicals is very similar to that typical of the gas phase. The same thing happens for their H-abstraction or substitutive addition reactions on P_{CH_3}, P_{CH_2} and P_A macromolecular positions.

Radical chain terminations are dominated by $H\bullet$ and $CH_3\bullet$ recombinations into the void volume: these are typical three-body reactions where the third body is generally made up of the close rigid walls of the polymeric material.

Due to the lack of reliable experimental data on fouling rates, kinetic parameters of these reactions were first obtained by means of *a priori* prediction methods. The H/C data published by Lohr and Dittman (1978) relating to several TLE deposit samples provided the first kinetic validation of the model. A further confirmation of these kinetic data relied on experimental data on bituminous coals (Van Krevelen, 1961).

A sample of the kinetic constants involved in this process includes:

Initiation

$$P_{CH_2} \to H\bullet + R_{CH} \quad 1.0 \times 10^{15} \exp\left(\frac{-78,000}{RT}\right) \quad [s^{-1}]$$

$$P_{CH_2} \to R_A + R_{CH_2} \quad 2.0 \times 10^{17} \exp\left(\frac{-94,000}{RT}\right) \quad [s^{-1}]$$

$$P_{CH_3} \to H\bullet + R_{CH_2} \quad 2.0 \times 10^{15} \exp\left(\frac{-96,000}{RT}\right) \quad [s^{-1}]$$

$$P_{CH_3} \to CH_3\bullet + R_A \quad 1.0 \times 10^{17} \exp\left(\frac{-100,000}{RT}\right) \quad [s^{-1}]$$

$$P_A \to H\bullet + R_A \quad 1.0 \times 10^{15} \exp\left(\frac{-110,000}{RT}\right) \quad [s^{-1}]$$

Propagation

$$H\bullet + P_{CH_2} \to R_{CH_2} + P_A \quad 2.0 \times 10^{10} \exp\left(\frac{-6,000}{RT}\right) \quad [m^3/kmol\ s]$$

$$CH_3\bullet + P_{CH_3} \to CH_4 + R_{CH_2} \quad 1.0 \times 10^9 \exp\left(\frac{-10,000}{RT}\right) \quad [m^3/kmol\ s]$$

High-pressure limits are assumed for the termination reactions of small radicals.

The local H and C content of the deposit is directly obtained from the concentration of the different positions (P_A, P_{CH_2} and P_{CH_3}). After the first evolution steps, P_{CH_2} and P_{CH_3} concentrations decrease. As P_A positions become dominant, the successive reactions transform P_A into P_{FUS}. The concentration of aliphatic C-atoms drastically decreases during degradation. Figure 18 shows a typical evolution of the H/C ratio as predicted by the model at about 900 K.

Ethane and methane are released in the first part of this evolution. With a rapid decrease in P_{CH_3} positions, the H/C ratio reaches an intermediate value of about 0.5. The successive dehydrogenation process evolves at a lower kinetic rate. Of course, methane decreases and the gaseous phase is made up mainly of hydrogen.

The initial H/C ratio in the more ordered deposit obtained from ethane pyrolysis is lower than that in a non-compact structure. Due to its more ordered structure, the kinetic rate of successive dehydrogenation reactions of the ethane deposit is lower too.

The evolution of the H/C ratio of the bulk of deposit is extremely important because it affects the void fraction, specific gravity and thermal conductivity of the solid. For ethane cracking, an asymptotic value for H/C, slightly larger than the one yielded by heavy feeds, is obtained at low and intermediate temperatures (700–900 K). At low temperatures (700 K), the deposit mostly loses methyl

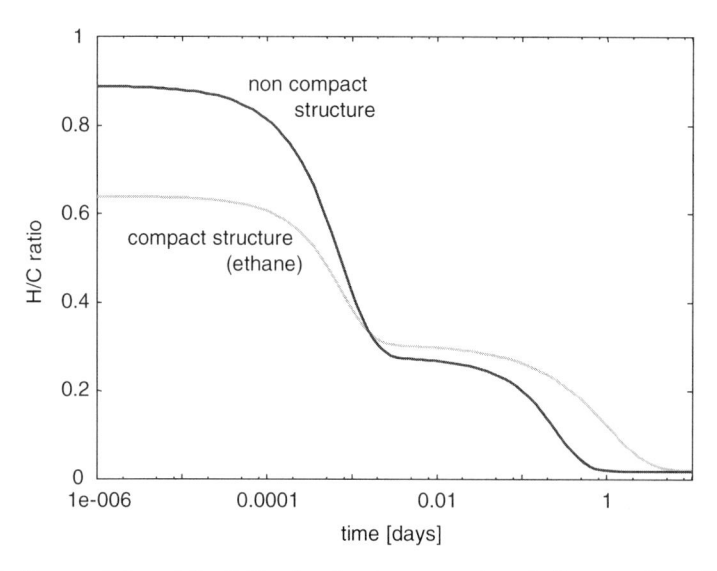

FIG. 18. Time evolution of the H/C ratio of carbonaceous deposits from pyrolysis of naphthas and ethane feed.

positions after about 100 days. The dehydrogenation of P_A positions in favour of P_{FUS} formation requires an higher activation energy. A more or less extended flat region can then be observed. The value of H/C ratio in the flat region and its time dependence vary with local temperature. The total carbon concentration is larger in the ordered deposit because of the different formation mechanism and asymptotic specific gravity. In addition to this, the carbon making up the deposit is mainly aromatic so that the carbon loss is limited.

The evolutions of thermal conductivity vs. H/C ratio for the compact and non-compact deposit morphologies are reported in Fig. 19. The asymptotic value for compact deposit, after deep dehydrogenation, is about 4 kcal/m/h/K (similar to the value of disordered graphite) while for the non-compact deposit, the maximum value is about 2.2 kcal/m/h/K, typical of coke.

It is interesting to observe that the gas released during the aging of the deposit generates micro-bubble in the softer and rubber-like external layers. Some of these bubbles are then trapped in the material and leave micro-spherical cavities and voids close to the process fluid interface. This microscopic aspect of the deposit is strictly connected to the "*bubbling*" phenomenon.

As a consequence of the more ordered and compact structure, the volume contractions due to the dehydrogenation and dealkylation reactions of carbon deposits from ethane crackers are less important. The deposit maintains a limited void fraction, the density is higher and decoking process is more difficult.

Of course, during the decoking process, the void fraction increases with the time because of the decrease in the solid fraction.

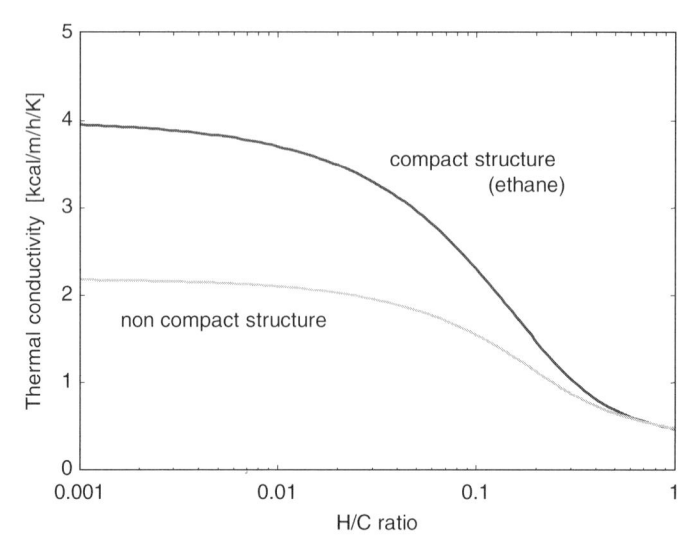

FIG. 19. Thermal conductivity as a function of the H/C ratio for compact and non-compact deposit.

2. Comparisons with Experimental Data

Lohr and Dittman (1978) studied the fouling of TLE in heavy gasoil pyrolysis and their experimental data provided the H/C ratios for three different deposit samples. Sample A at TLE inlet near the process fluid interface (H/C about 0.024); Sample B at TLE outlet near the metal wall (H/C about 0.365) and Sample C at TLE outlet near the process fluid interface (H/C about 0.224). The on-stream time of the TLE was 60 days. The deposit close to the process fluid was exposed throughout the running time at a temperature of about 730–750°C, similar to the furnace outlet temperature; Sample A was exposed to these temperatures for only 3–4 h, after which time the temperature dropped due to the shutdown of the furnace. Sample B was exposed to a more or less constant temperature for 60 days; in this case the temperature was a few degrees higher than the cooling process fluid (boiling water near critical temperature). Sample C, on the other hand, was exposed to rising temperatures due to the progressive thickening of the deposit. Nevertheless, the temperature of Sample C was also fairly constant and equal to that of the process fluid interface (about 600–630°C when the process fluid is made up of heavy gasoil and in the range of 400–430°C in the case of naphtha cracking).

The model predictions for Sample A after 3 h at 1,000 K are very close to the experimental value. Assuming a deposit growth of 5 mm/month, 20 µm are obtained after 3 h; this value is compatible with the scraped amounts taken for the analysis.

There are some uncertainties in the case of sample B, mostly relating to the size of the sample drawn for analysis. Near the metal surface, the deposit is grown via the catalytic mechanism with micro-fibres and highly porous structures. The drawn material is therefore limited and the sample had to be quite thick to ensure that an adequate amount had been obtained. Furthermore, the low thermal conducibility of this fibrous layer induces a sharp increase in temperature near the tube wall. In the case of a typical Borsig heat exchanger, a temperature increase of about 25 K is estimated in the 50 µm near the wall; the temperature to be assumed is therefore in the region of 650 K. With this temperature constant for the 60 days of run length, the model prediction (H/C ≈ 0.38) agrees quite well with the experimental measurements. Lastly, the model predicts the proper H/C value for sample C too when 3 h of exposure at about 600°C is assumed.

A second set of comparisons refers to the H/C ratio of the coke obtained in a delayed coking process. This material remains inside the coking drums for about 24 h and its final H/C ratio becomes about 0.5 (de Freitas Sugaya, 1999). The model is able to predict this ratio both by maintaining about 700 K for 24 h and also by decreasing the temperature profile (from 715 K to 690 K) according to the temperature evolution of the coking drum.

Further experimental data and further model comparisons relate to the rapid pyrolysis of different coals. In the absence of air, this experimental device heats and converts small coal particles (10–200 µm) in gas and distillates. Figure 20 shows a very satisfactory agreement between experimental data relating to a bituminous coal and model results at 1,260 K. It is noteworthy that despite the strong differences between carbon deposit and bituminous coal, the characteristic times for the dehydrogenation processes are practically the same. Further data on this subject, as well as a detailed model for the analysis of the pyrolysis and devolatilization process of coal particles, are available in a recent paper (Migliavacca *et al.*, 2005).

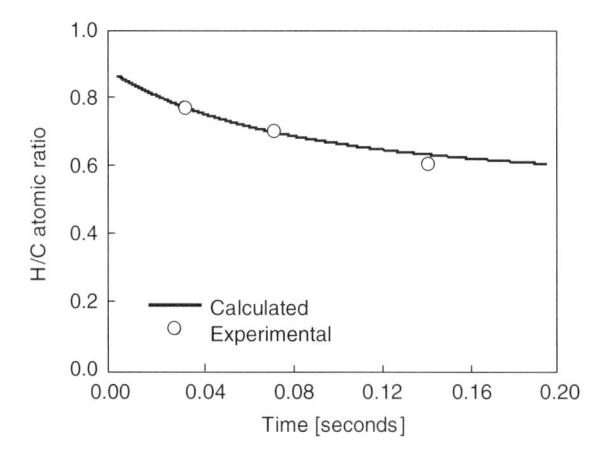

FIG. 20. Comparison with "Rapid Pyrolysis" data.

C. SOOT FORMATION

High-temperature pyrolysis reactions of hydrocarbons are responsible for the production of PAH and solid carbon black particles, soot. This phenomenon is common in diffusion flames where, at high temperatures and without oxygen, hydrocarbon fuel aggregates follow pyrolysis and condensation paths with the formation of heavy aromatic structures. Many PAH's identified in aerosols have been found to be mutagenic and are certainly important soot precursors. This formation of carbonaceous particles has recently become one of the main topics in chemical reaction engineering, especially in the field of pyrolysis and combustion of hydrocarbon fuels. This interest rises from environmental concerns about PAH and soot particle emissions because of their dangerous impact on the human health (Oberdorster et al., 2004).

As already reported in Glasman's book (Glasman, 1996), soot characteristics were clearly described by Palmer and Cullis (1965). On an atomic basis, the carbon formed in flames generally contains a considerable proportion of hydrogen and corresponds to an empirical formula C_8H. When examined under the electron microscope, the deposited carbon consists of a number of roughly spherical particles strung together rather like pearls on a necklace. Figure 21 provides a representative image of the soot formed in rich flames.

Despite minor differences (Alfè et al., 2005), young soot particles formed from ethylene and from benzene flames clearly show very similar morphologies. The diameters of the small spherical particles range from 10 to 20 nm. Each particle is made up of a large number of crystallites. Each crystallite consists of several aromatic sheets with interlayer spacing of about 0.34 nm, similar to the spacing of ideal graphite (0.335 nm).

Several kinetic mechanisms have been proposed in the literature for modelling the formation and growth of soot particles (Appel et al., 2000; Balthasar and Frenklach, 2005; D'Anna et al., 2001a; Frenklach and Wang, 1994; Richter et al., 2005). All these mechanisms agree that the process can be described in terms of the following major steps:

– formation of the first aromatic rings
– growth of PAH structures
– homogeneous nucleation of particles
– particle coagulation
– particle surface growth
– particle agglomeration

Figure 22 (Bockhorn, 1994) clearly summarizes the evolution from light molecules to heavy aromatic structures and finally to carbon particles and aggregates. The transition of the species in the gas phase to aerosols and solid particles is possibly the least understood process.

YOUNG SOOT

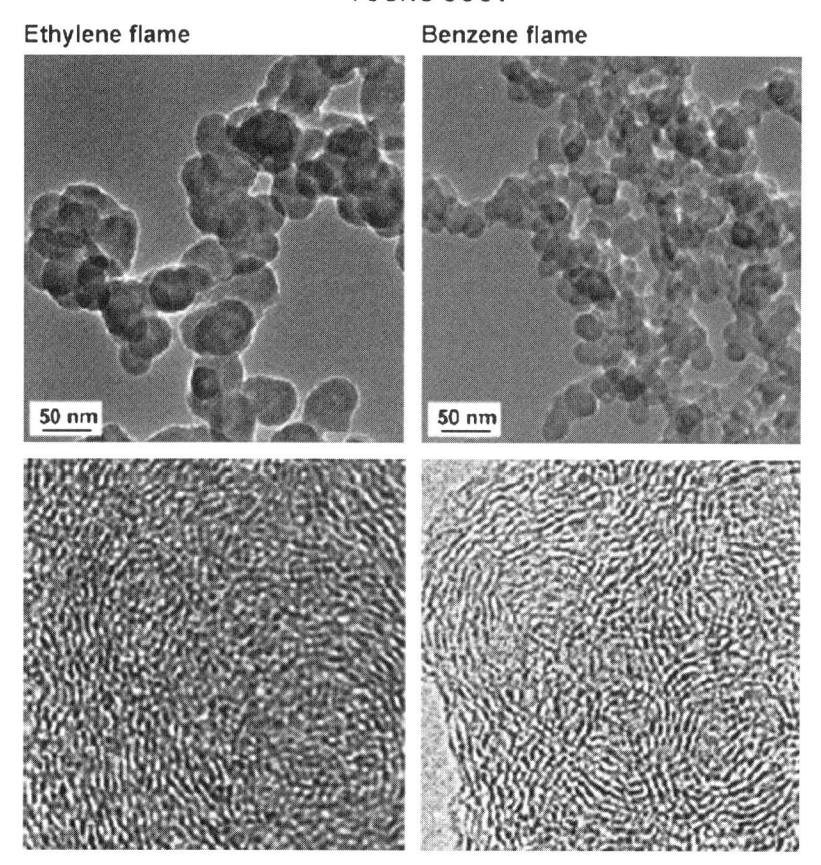

Ethylene flame　　　　Benzene flame

Fig. 21. TEM and HR-TEM of soot particles from ethylene and benzene flames (Alfè et al., 2005).

The very first hypotheses regarding soot inception attributed a key role to ionic (Calcote, 1981) or polyacetylene species (Homann and Wagner, 1967). Recent studies all agree that PAH are the precursors of soot formation (Haynes and Wagner, 1981). Several alternative mechanisms were also discussed as part of the hypothesis. These considered different paths of PAH growth, involving condensed rings (Frenklach and Wang, 1994), aromatic sheets with methylenic or aliphatic bridges (D'Anna et al., 2001b) or, lastly, polyyne pathways (Krestinin, 2000).

There are several analogies between the soot formation mechanism and the fouling mechanism, even though the typical temperatures are higher, and, as a result, more dehydrogenated species and different reactions dominate the process.

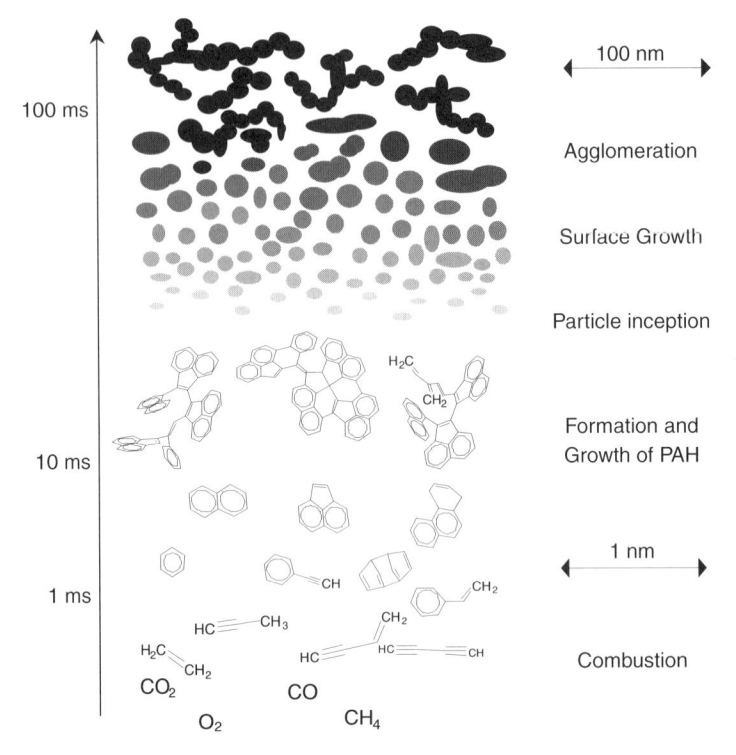

FIG. 22. Reaction steps toward soot formation (Adapted from Bockhorn, 1994).

1. Formation of the First Aromatic Rings and Aromatic Growth

The first step in the soot formation mechanism is the formation of the first aromatic ring when aromatic species are not directly present in the feed. Thus, the formation mechanisms of benzene and the first aromatic species in hydrocarbon pyrolysis, particularly at high temperatures, have been investigated in great depth. Nevertheless, several uncertainties about these formation paths remain. Due to the high temperatures, acetylene addition reactions on unsaturated radicals play a key role (Frenklach and Wang, 1994):

$$n\text{-}C_4H_3\bullet + C_2H_2 \rightarrow C_6H_5\bullet$$

$$n\text{-}C_4H_5\bullet + C_2H_2 \rightarrow C_6H_6 + H\bullet$$

Similarly, at lower temperatures, the following reaction also contributes to benzene formation when there is a significant amount of ethylene in the reacting system:

$$n\text{-}C_4H_5\bullet + C_2H_4 \rightarrow C_6H_6 + H\bullet + H_2$$

In addition to these reactions of even carbon atoms, odd pathways involving the resonantly stabilized propargyl radical (C_3H_3) have also been identified as relevant contributions to benzene formation (Miller and Melius, 1992):

$$C_3H_3\bullet + C_3H_3\bullet \to C_6H_6$$

$$C_3H_3\bullet + C_3H_3\bullet \to C_6H_5\bullet + H\bullet$$

Although propargyl recombination is not an elementary reaction and most probably passes through the linear C_6H_6 compound (Miller, 2001), it can be conveniently lumped into a single equivalent reaction (Appel et al., 2000). Similarly, C_3H_3 can also react with methyl-allenyl (n-C_4H_5) to form toluene or H and benzyl radical:

$$n\text{-}C_4H_5\bullet + C_3H_3\bullet \to C_7H_8$$

$$n\text{-}C_4H_5\bullet + C_3H_3\bullet \to C_7H_7\bullet + H\bullet$$

At intermediate temperatures, vinyl, allyl, methyl-allyl (aC_4H_7), cyclopentadienyl radicals as well as cyclopentadiene are involved in a relevant subset of addition reactions (Faravelli et al., 1998). Vinyl addition to ethylene, through the intermediate pC_4H_7 and aC_4H_7 radicals, easily forms butadiene.

$$C_2H_3\bullet + C_2H_4 \to \text{p-}C_4H_7\bullet$$

$$\text{p-}C_4H_7\bullet \to \text{a-}C_4H_7\bullet$$

$$\text{a-}C_4H_7\bullet \to C_4H_6 + H\bullet$$

$$\text{p-}C_4H_7\bullet \to C_4H_6 + H\bullet$$

The allyl type and more stable a$C_4H_7\bullet$ can also add to C_2H_4 giving rise to cyclopentene (cyC_5H_8) and methyl-cyclopentene. Successive dehydrogenation and dealkylation reactions explain the formation of cyclopentadiene. cyC_5H_6 is also formed by the addition of vinyl radical to butadiene

$$C_2H_3\bullet + C_4H_6 \to CH_3\bullet + cyC_5H_6$$

This reaction, which is very important to butadiene disappearance, competes with benzene formation.

$$C_2H_3\bullet + C_4H_6 \to H_2 + H\bullet + C_6H_6$$

At moderate temperatures, further important sources of benzene are the *cyclo*-addition of ethylene to cyC_5H_6 (Dente et al., 1979) and the radical recombination reaction between cyclopentadienyl and methyl radicals (Ikeda et al., 2000; Moskaleva et al., 1996)

$$C_2H_4 + cyC_5H_6 \rightarrow C_6H_6 + H\bullet + CH_3\bullet$$

$$C_5H_5\bullet + CH_3\bullet \rightarrow H\bullet + H\bullet + C_6H_6$$

Cyclopentadienyl ($C_5H_5\bullet$) is another important resonantly stabilized radical originating from cyclopentadiene or formed by the *cyclo*-addition of propargyl on C_2H_2

$$C_3H_3\bullet + C_2H_2 \rightarrow C_5H_5\bullet$$

This initial cyclization is similar to the one from propargyl recombination. However, due to the formation of a five-membered ring, it is less favoured from the kinetic point of view: the rate constant is about 30 times lower (Frenklach, 2002). Depending on the operating conditions, this lower reactivity may often be compensated for by the larger amount of C_2H_2 with respect to $C_3H_3\bullet$.

For several years now (Dente et al., 1979), the critical role of C_5H_5 radicals in the formation of heavy species, and typically of naphthalene has been defined as

$$C_5H_5\bullet + C_5H_5\bullet \rightarrow C_{10}H_8 + H\bullet + H\bullet$$

The kinetic parameters of the reactions involved in the formation of the first aromatic rings are summarized in Table X.

A reliable and well-defined model for describing the successive growth of aromatic species comes from Frenklach and Wang and is the so-called HACA mechanism (Frenklach and Wang, 1990). This acronym stands for the sequence of two main repetitive reactions, i.e. H-abstraction and C_2H_2 addition. This mechanism explains the successive growth of aromatic structures. The H-abstraction from the generic aromatic with *i* condensed rings (A_i) forms the corresponding aromatic radical ($A_i\bullet$). The successive addition reaction of $A_i\bullet$ on acetylene is the first growing step.

$$A_i + H\bullet \rightarrow A_i\bullet + H_2$$

$$A_i\bullet + C_2H_2 \rightarrow \text{products}$$

The successive repetition of these reactions with parallel cyclization and dehydrogenation reactions easily yields condensed aromatic structures with high molecular weights

TABLE X
RATE CONSTANTS OF REACTIONS INVOLVED IN FIRST AROMATIC RING FORMATION (UNITS ARE KMOL, m^3, K, KCAL)

n-$C_4H_3\bullet + C_2H_2 \to C_6H_5\bullet$	7.01×10^{11}	-0.86	6,379	Westmoreland et al. (1989)
n-$C_4H_5\bullet + C_2H_2 \to C_6H_6 + H\bullet$	1.60×10^{13}	-1.33	5,400	Appel et al. (2000)
n-$C_4H_5\bullet + C_2H_4 \to C_6H_6 + H\bullet + H_2$	3.00×10^8	0.00	7,000	Faravelli et al. (1998)
$C_3H_3\bullet + C_3H_3\bullet \to C_6H_5\bullet$	3.00×10^8	0.00	0.00	Miller and Melius (1992)
$C_3H_3\bullet + C_3H_3\bullet \to C_6H_5\bullet + H\bullet$	3.00×10^8	0.00	0.00	Miller and Melius (1992)
n-$C_4H_5\bullet + C_3H_3\bullet \to C_7H_8$	1.00×10^9	0.00	0.00	Faravelli et al. (1998)
n-$C_4H_5\bullet + C_3H_3\bullet \to C_7H_7 + H\bullet$	1.00×10^9	0.00	0.00	Faravelli et al. (1998)
$C_2H_3\bullet + C_2H_4 \to p$-$C_4H_7\bullet$	4.00×10^8	0.00	6,500	Dente et al. (1979)
p-$C_4H_7\bullet \to a$-$C_4H_7\bullet$	2.50×10^{12}	0.00	36,000	Dente et al. (1979)
a-$C_4H_7\bullet \to C_4H_6 + H\bullet$	1.00×10^{14}	0.00	51,000	Dente et al. (1979)
p-$C_4H_7\bullet \to C_4H_6 + H\bullet$	2.50×10^{13}	0.00	38,000	Dente et al. (1979)
$C_2H_3\bullet + C_4H_6 \to CH_3\bullet + cyC_5H_6$	3.40×10^9	0.00	3,000	Dente et al. (1979)
$C_2H_3\bullet + C_4H_6 \to H_2 + H\bullet + C_6H_6$	6.00×10^8	0.00	3,000	Dente et al. (1979)
$C_2H_4 + cyC_5H_6 \to C_6H_6 + H\bullet + CH_3\bullet$	3.00×10^8	0.00	30,000	Dente et al. (1979)
$C_5H_5\bullet + CH_3\bullet \to H\bullet + H\bullet + C_6H_6$	2.00×10^9	0.00	0.00	Ikeda et al. (2000)
$C_3H_3\bullet + C_2H_2 \to C_5H_5\bullet$	1.00×10^8	0.00	0.00	Frenklach (2002)
$C_5H_5\bullet + C_5H_5\bullet \to C_{10}H_8 + H_2$	5.00×10^8	0.00	0.00	Dente and Ranzi (1983)

This recursive sequence of reactions progressively increases the number of rings in the poly-aromatic structure whilst always referring to a fixed number of similar species, and makes the adoption of a method of moments for the modelling of this process convenient (Frenklach and Wang, 1994).

Besides acetylene, which is important due to its great abundance in pyrolytic environments at high temperatures, there are also several other components which contribute to aromatic growth. In fact, the RSR (like C_3H_3, C_4H_5 or C_4H_3, C_5H_5 etc. ...) can either recombine with aryl radicals or they can add on the five- or six-membered rings (mainly five) and contribute to the growth of these aromatic structures. Analogous to C_5H_5 recombination, different cyclopentadienyl-like structures also contribute to the successive growth of

PAH (D'Anna et al., 2001b)

Moreover, aryl radicals can add on unsaturated molecules forming radical intermediates whose successive fate is dehydrogenation and cyclization.

A further important reaction class refers to the substitutive addition reactions which were already discussed in the pyrolytic mechanism of deposit formation. During PAH formation and growth, there are alkyl substitutions (mainly methyl) on the edge surface too. Active radicals add on the aromatic structures with a simultaneous dealkylation. This results in an increase in the molecular weight and dehydrogenation of the aromatics.

Successive addition and condensation reactions of large aromatic molecules and radicals or recombination of large aromatic radicals further contribute to the growth of aromatic structures. As clearly shown in Fig. 23, five-membered ring compounds also play a significant role in this process (D'Anna et al., 2001b).

This mechanism involves the formation of ring–ring aromatics, interconnected by aliphatic chains, which after successive dehydrogenation reactions progressively organize ordered graphite structures. Vlasov and Warnatz (2002) propose a competitive role for the polyyne molecules as important soot precursors. The polyyne pathway is derived from the works of Krestinin (2000) and assumes that these species grow rapidly with successive additions of C_2H radicals

$$C_{2n}H_2 + C_2H\bullet \rightarrow C_{2n+2}H_2 + H\bullet$$

Polyyne molecules were not supposed to grow above certain mass ranges in line with the experimental observations of Homan and Wagner (1967) who analysed the composition of large PAH and soot. Their data on average soot composition indicated H/C ratios of about 0.1–0.12 with molecular weights in

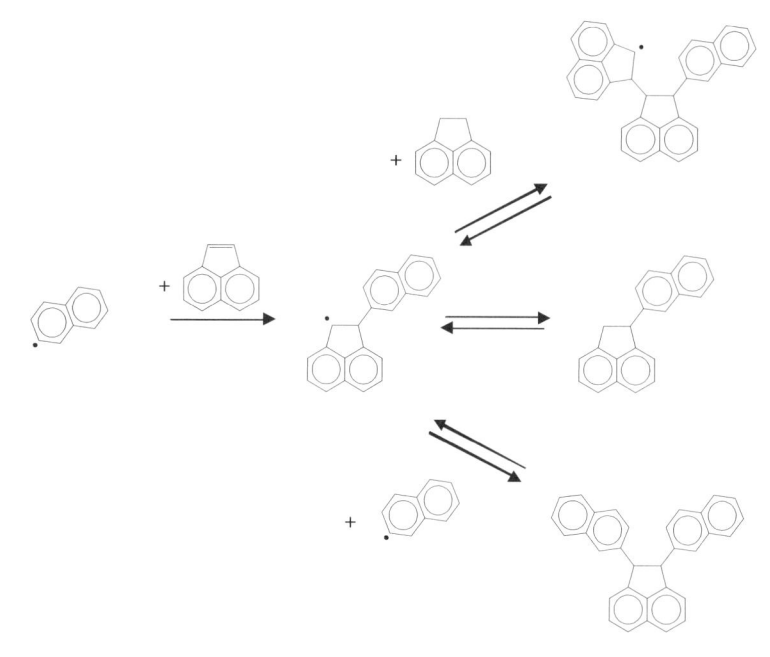

Fig. 23. Growth of PAH structures (D'Anna et al., 2001b).

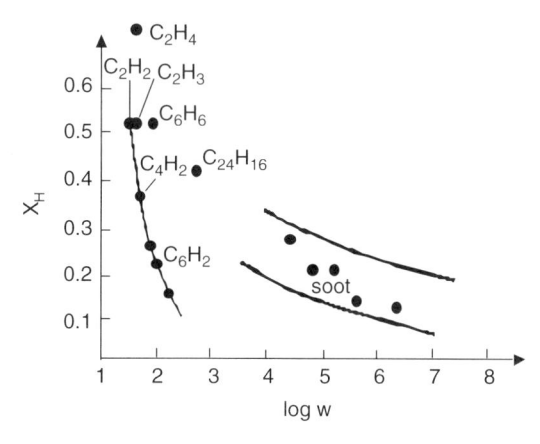

Fig. 24. Hydrogen content in hydrocarbons and soot as function of molecular weight adapted from Homan and Wagner (1967).

the order of 10^5–10^6 while polyyne species quickly move to lower hydrogen content, as shown in Fig. 24. In the Krestenin mechanism, polyyne molecules can generate heavier PAH structures through radical addition on unsaturated bonds and the formation of cross-linked structures. The polyynes are also thought to be successively responsible for major soot growth via surface reactions (Krestinin, 2000).

2. Soot Inception and Growth

Nucleation is a crucial step in the whole process of carbonaceous particle formation. According to Frenklach and Wang (1990, 1994), nucleation is controlled mainly by the sticking of PAH sheets during their collisions. Physically bound clusters of PAH are then formed and successively evolve toward aerosol, solid particles and crystallites. As shown in Fig. 25, different polycyclic aromatic layers can form more or less regularly ordered graphite structures, all of which have interlayer distances of about 0.35 nm. These two to four-layer structures are assumed as the threshold of the formation of the solid phase; particle inception typically takes place at molecular masses of 1,000–2,000 amu.

A partially different approach assumes that the transition from the gas to the condensed phase simply occurs as a sequence of successive addition reactions, with the formation of large poly-aromatic structures. Once again, above a certain molecular weight PAH becomes aerosol or soot particles (D'Anna et al., 2001b; Richter et al., 2005; Sarofim and Longwell, 1994).

The kinetic modelling of soot formation processes refer to gross simplifications and the use of lumped pseudospecies aimed at describing the pysicochemical evolution of heavy PAH towards soot particles with molecular weight of more than 10^6–10^7. The discrete sectional method (DSM) is commonly adopted to solve this problem. Soot formation is described in the usual form of gas-phase kinetics (Granata et al., 2005; Richter et al., 2005) and a limited number of lumped pseudospecies is selected. Large PAH species and particles are grouped into sections (equivalent components, here referred to as BINs), each of these has an average mass, mass range and defined H/C ratio. The first BIN corresponds to coronene, $C_{24}H_{12}$, but either lighter (pyrene) or heavier aromatics can be adopted. The next BIN has twice the average mass and mass range, i.e. BIN_2 has an average mass of 600 amu and a mass range of

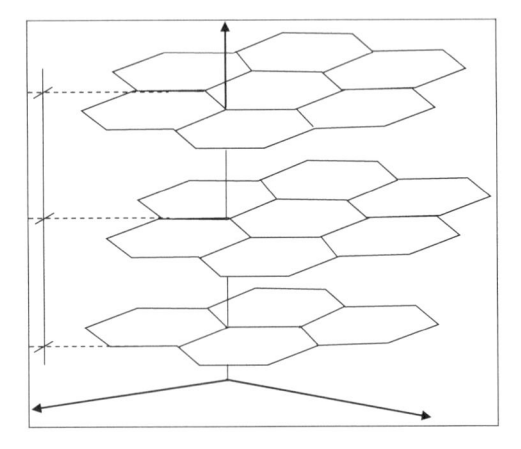

FIG. 25. Three-layer PAH cluster with ordered structure and interlayer distances of about 0.35 nm.

401–800 amu. This exponential sectional approach saves on computation time whilst still offering a good description of the low-molecular weight components. The largest BIN reaches nominal species with more than 10^6 C-atoms and H/C ratios lower than ~0.125 in line with the experimental measurements.

Large BINs, i.e. heavier than 2,000–2,500 amu, are considered to be soot or solid particles and therefore interact like aerosol particles. Kinetic rates are scaled to account for the variation in the collision frequency with reactant sizes. Particle diameter and collision interactions are usually based on the assumption of spherical structures.

Nascent soot particles are formed via reactions between PAH radicals or between stable PAHs and their radicals. The kinetic rates are evaluated on the basis of the analogy with the first phenyl radical and benzene reactions. Frequency factors are scaled on the basis of the change in collision frequencies relative to those of benzene addition on phenyl or phenyl recombination. These reactions are treated as irreversible. BIN radicals are formed via H-abstraction, mainly by H radicals, with both forward and reverse reactions assumed. Unimolecular hydrogen loss from parent BINs and the corresponding reverse reaction also occur.

Surface growth mainly takes place via reactions with acetylene, PAH and PAH radicals to soot nuclei activated by H-abstraction as well as PAH radicals to parent BINs. Soot particle coagulation is the result of addition reactions between BINs and BIN radicals, recombination reactions of BIN radicals and, lastly, interactions between heavy BIN species.

This modelling approach allows the description not only of the overall soot volume fraction but also of the distribution of the equivalent diameters of the carbon particles. Figure 26 shows a sample of comparisons between model

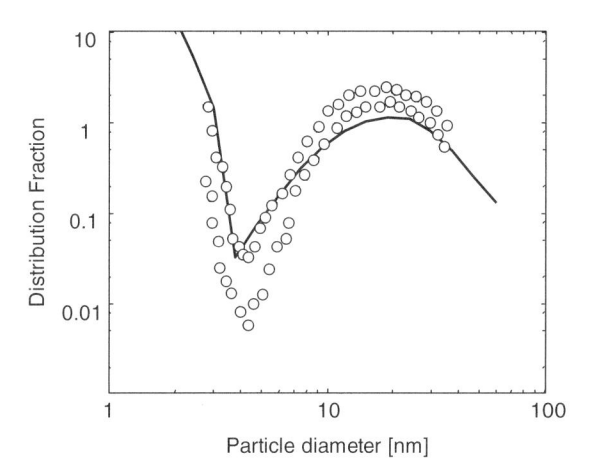

FIG. 26. Distribution of particle diameters in an ethylene sooting flame. Comparison between model predictions and the experimental data (Zhao et al., 2003).

predictions and the experimental data of Zhao *et al.* (2003) relating to an example of this distribution of particle diameters in an ethylene sooting flame.

IV. Applications

This chapter presents the industrial applications and validations of certain detailed models which refer to the kinetics analysed earlier. The steam cracking process will be analysed first followed by visbreaking and delayed coking processes. Last of all, the method will be applied to the thermal degradation of plastic waste.

A. STEAM CRACKING PROCESS AND PYROLYSIS COILS

Thermal cracking is a well-known and widely accepted method of olefin production. A steam cracker can be conveniently divided into three sections: pyrolysis, quenching and product separation. Of course, the pyrolysis section is the heart of the steam cracker process. Hydrocarbons first enter the convection section of the pyrolysis furnace and are preheated to about 600–650°C. Then, the vaporized steam is mixed with superheated steam and passed into long pyrolysis coils with typical diameters in the range of 30–125 mm. The pyrolysis reactions mostly take place in the radiant section of the furnace where tubes, usually made of chromium nickel alloys, are heated externally to about 1,100°C by floor and/or wall burners. Depending on the cracking severity, free radical reactions lead to the formation of ethylene and light olefins with different selectivities. The hot gases are then quenched in the TLE to 450–650°C to reduce the effect of successive reactions and generate high-pressure steam at the same time. As already described in the previous paragraphs, pyrolysis coils and heat exchangers are prone to fouling and therefore have to be periodically shut down (Ren *et al.*, 2006).

In the primary fractionation section, gasoline and fuel oil streams are quenched by oil and fractionated. The liquid fraction is extracted and the gaseous products are cooled in the quench tower by a circulating water stream. The gaseous fraction is then compressed and cooled down to remove acid gases, carbon dioxide and water. The final product separation essentially involves distillation, refrigeration and extraction. The equipment used includes chilling trains, where methane and hydrogen are separated at cryogenic temperatures, and a sequence of several fractionation towers. Ethane and propane are usually recycled as feedstock. The product yield from steam cracking process varies greatly according to the feed and operating conditions.

Regular decoking is required to deal with coke deposits in both pyrolysis coils and the TLE. The furnace is shut down and high-pressure steam and air are fed in to burn off the coke deposit on the inner surfaces of the walls. The coke deposit in the TLE is washed away with high-pressure water or is removed mechanically. Depending on the feedstock, coil configuration and severity, this decoking process must be carried out every 10–70 days and can take 20–40 h.

In order to maximize ethylene yield and selectivity, the pyrolysis process should take place at high temperatures, under low pressure and with low contact times. In fact, high temperatures favour decomposition reactions, low pressures inhibit condensation reactions, and low residence times avoid successive reactions of ethylene and derived products. Mostly for this last reason, the geometry of pyrolysis coils has been changed from the long, large-diameter coils used in the past to the current small-diameter and shorter coils (10–20 m). Today the focus is also on coil configurations and geometries capable of not only improving ethylene yields but also increasing the on-stream time of the furnaces.

As an example of recent efforts to improve overall coil performance, Fig. 27 shows the new coils recently patented by Kubota (Magnan *et al.*, 2004). MERT (Mixing Element Radiant Tube) is a centrifugal cast radiant tube with a helical mixing element on the inside diameter for use in cracking furnaces. It works by inducing high turbulence and disrupting the inside boundary layer of hydrocarbons, thus increasing the effective internal heat transfer coefficient. The high

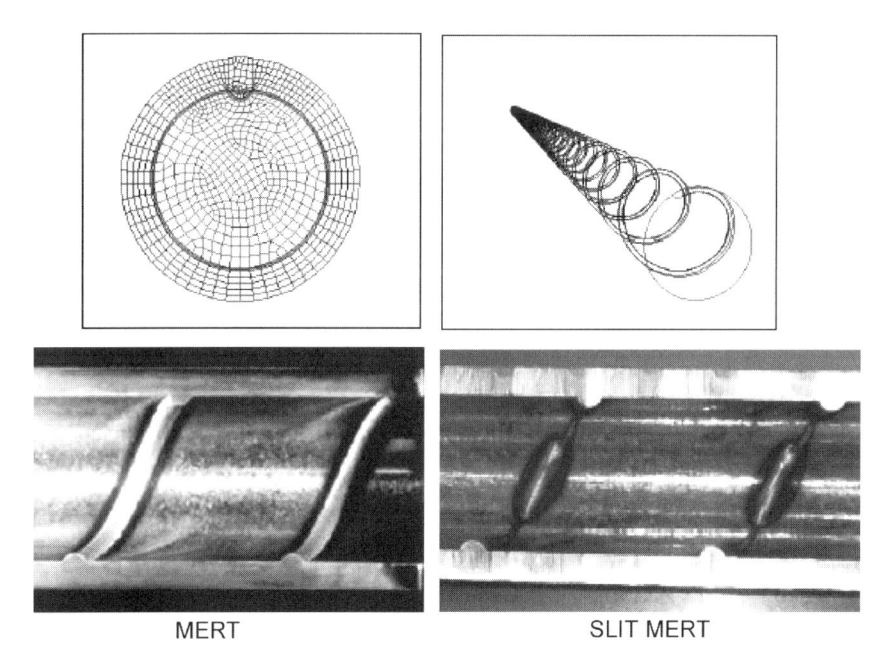

FIG. 27. (a) CFD representation of MERT. (b) Longitudinal cross-section of MERT and SLIT MERT.

turbulence and mixing improves cracking performance and provides a more uniform hydrocarbon temperature throughout the tube cross-section. However, the increased friction factor does also have certain detrimental effects. In fact, the geometry was recently modified to achieve a lower friction factor yet maintain the same mixing effect and internal heat transfer coefficient. The new coils (SLIT MERT) are similar to the previous helical finned ones except for the fact that the element is applied intermittently. The volume fraction of the element is reduced, but since the alignment of the segments is maintained, the swirl mixing effect continues throughout the length of the tube.

Engineering companies always are interested in work aimed at improving coils and furnace arrangements, regardless of whether the focus is on design or revamping. Pyrolysis coil selectivity and yields are being continuously improved by changing from older large bore tubes to smaller tubes with very short residence times. An example of these design improvements is provided by Evans et al. (2002) who observes that the key factor in minimizing investment and operation costs without sacrificing performance is larger capacity per single furnace, as would be the case with the re-introduction of the staggered or two-lane coil. The compactness of the new design decreases firebox size while maintaining high cracking furnace availability, thus lowering the investment and operating cost of the cracking furnaces. Although a staggered coil layout is not typical in modern coil design, a feasible design was obtained by minimizing the inherent drawback of the staggered design, i.e. the higher peak-to-average heat flux ratios. A new furnace design (GK6) was compared to a previous one (GK5). The staggered GK6 design has an 8–8 cracking coil configuration with the inlets and outlets of the cracking coil entering and leaving the radiant section through the roof. The tubes are arranged in two lanes: each inlet tube in one lane is connected to an outlet tube in the other lane. The bottom jump overlies the same plane as the in and outlet tubes. This means that there are no horizontal forces in the GK6 design, which avoid bowing of the radiant tubes. Feedback from operating units confirmed that the cracking performance of the GK6 is equal to or slightly better than that of a leading cracking coil arrangement, the GK5 coil design. Improvements in run length and the expected tendency to bowing have been confirmed too.

1. Coil Model and Model Validation

A mathematical model to simulate the cracking process in the pyrolysis coils contains a set of differential equations describing material, energy and momentum balances. Additional features are also included, such as the fouling process, the associated dynamic behaviour of the coke deposit in the pyrolysis coils and the rising pressure drops in the coil and in the TLE. A monodimensional model is usually adopted, although several two-dimensional description projects have been attempted (Heynderickx et al., 1992, 2001; Van Geem et al., 2004). There is less benefit to such an approach, however, than might be gained with a proper description of the pyrolysis mechanisms and the corresponding overall kinetic scheme.

The previous description of the kinetic approach to pyrolysis, as well as the fouling phenomena, have been adopted for the development of a mechanistic model and simulation program (Dente *et al.*, 1970, 1992, 2005) which has been extensively used and tested in industry since the 80s.

About 250 real and lumped components, molecules and radicals are considered in the overall kinetic model which aims to describe the pyrolysis chemistry of gaseous and liquid feeds up to heavy gasoils and condensate fractions. Figure 28 shows the distribution of these components (real and lumped species) as a function of the number of carbon atoms and of the dehydrogenation degree Z, in accordance with the formula C_nH_{2n-Z}. The families of alkanes as well as several alkyl-aromatics are singled out and the vertical lumping of the heavy homologous compounds at 25, 30 and 35 carbon atoms is also quite clear. It is worthwhile noting too that the heaviest dehydrogenated species with about 40 C-atoms maintain an H/C ratio of about 0.7. This dehydrogenation level allows the gas-phase formation of the cracking fuel oil and heavy pyrolysis fractions to be described.

Careful, in-depth validation of the overall kinetic model across a wide range of operating conditions was carried out over long periods on the basis of industrial data. The resulting pyrolysis model thus gives accurate and reliable yields predictions and is a significant reference for the ethylene producers.

Figure 29 shows a scatter diagram relating to the deviations of ethylene yields predicted by two rival models. The experimental data refer to about 300 different cases. High and low cracking severities are considered, and the data refer to

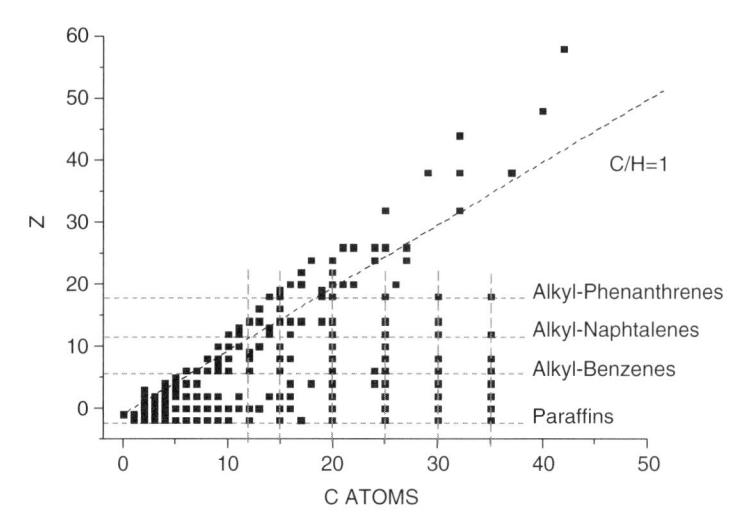

FIG. 28. Distribution of reference compounds involved in the pyrolysis model.

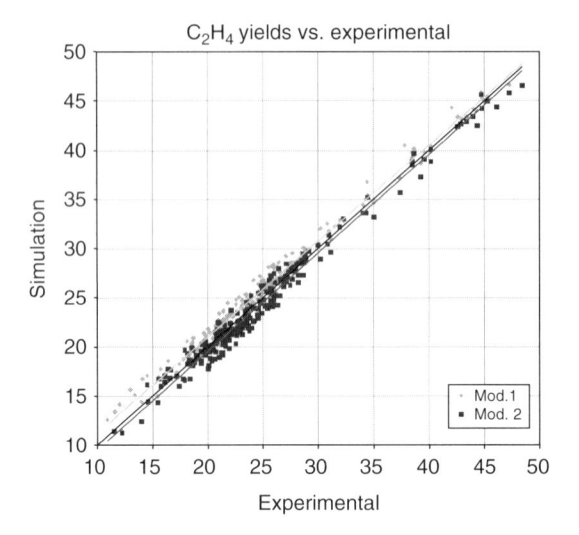

FIG. 29. Validation of the model. Comparison of two rival models. Scatter diagram of ethylene yields.

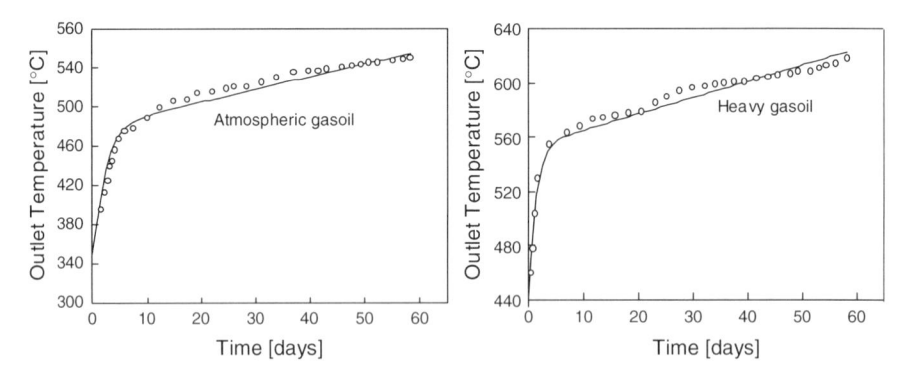

FIG. 30. Non-conventional TLE. Outlet temperature vs. time. Comparisons of predicted and experimental measurements (Lohr and Dittman, 1978).

commercial plant or pilot and laboratory-scale devices. Feedstocks range from light naphtha to virgin and hydrotreated gasoil with final boiling points of over 500°C.

Fouling process models for both pyrolysis coils and the TLE were also verified against experimental data. Figure 30 reports a comparison between model predictions and the experimental data reported by Lohr and Dittman (1978). The time evolution of the outlet temperature from the TLE indicates that the efficiency of the heat exchangers is penalized. In the case of both an atmospheric gasoil and a vacuum one, the temperature of the process gases from the TLE increases by more than 200 K in 50 days. The very sharp, net initial temperature increase is due both to the catalytic growth and to the low thermal conductivity of the initial deposit.

B. VISBREAKING AND DELAYED COKING PROCESSES

Liquid-phase pyrolysis of oil streams is widely used in refinery: visbreaking and delayed coking processes are typical examples. The feeds are made up of residues of atmospheric or vacuum crude distillation. The typical temperatures (440–490°C) and the high pressures (>10 bar), together with the nature of the feedstock, indicate that these processes essentially take place in the liquid phase. Visbreaking and delayed coking are extensively used to upgrade refinery residues or heavy streams to increase distillate production: gas, gasoline, kerosene, light, heavy and vacuum gasoils. Visbreaking is widespread in Europe and the Far East while delayed coking is more popular in the US and the other parts of the world.

As shown in Fig. 31a, a visbreaking reactor consists of very long coils inserted into a radiating furnace. The effluent often passes into the extra adiabatic

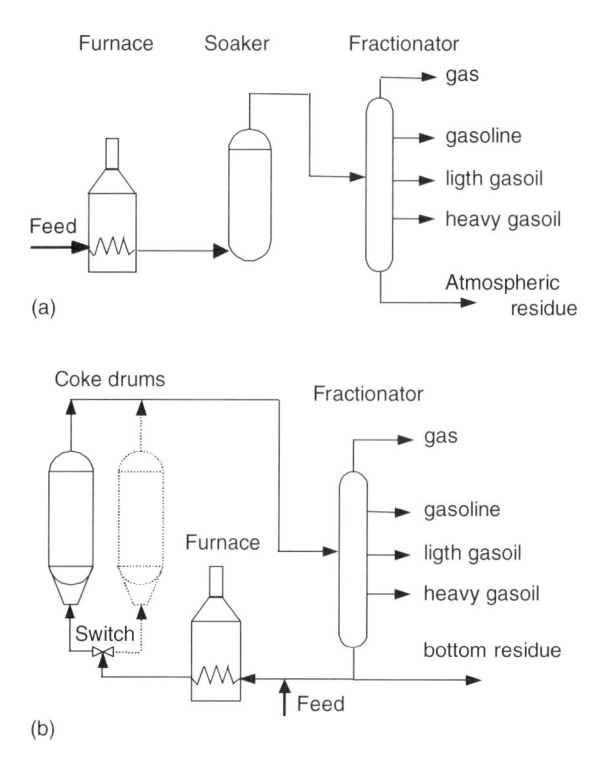

FIG. 31. (a) Visbreaking process. (b) Delayed coking process.

reaction volume of the soaker. This makes it possible to decrease the furnace outlet temperature and reduce the fouling inside the coils. One or more fractionators are aimed to the different streams recovering.

A simplified delayed coking process scheme is shown in Fig. 31b. The feed, mixed with part of the bottom of the fractionator, is preheated and partially vaporized in the coils of the furnace. The latter is connected to a system of two or more large coke drums that operate alternatively semi-batchwise, like adiabatic reactors. The liquid and the mesophase (i.e. the colloidal system precursor of coke) gradually fill the drum and the released vapours exit through the top. When the proper level of condensed phase is reached (60–70% of drum volume), the feed is progressively switched to the other drum. Steam is injected to complete the stripping and the recovery of the lighter fractions still contained into the drum. Last of all, coke removal begins.

Simplified macro-kinetic models (Del Bianco et al., 1993; Filho and Sugaya, 2001) are usually proposed in the literature: these are limited because are not able to cover the complete range of operating conditions and feedstocks. A detailed kinetic model together with reliable feed characterization, on other hand, gives a flexible representation of the process.

1. Kinetic Mechanism

The whole kinetic mechanism is essentially characterized by initiation, β-scission, H-abstraction, internal isomerization of radicals, substitutive addition of radicals onto unsaturated molecules and radical recombination reactions.

H-abstraction and β-scission reactions are the main effective propagation steps of the chain.

H-abstractions on pseudocomponents give rise to radicals which can either (re)abstract H-atoms from other molecules or undergo β-scission reactions. Due to the relative activation energies and to the high concentration of neighbouring H-atoms in the liquid phase, the H-abstractions are much, much faster. Therefore, β-scission reactions constitute the rate-determining step of the overall decomposition process. In fact, the radicals produced through β-scissions rapidly generate their products before undergoing a new β-scission. The final consequence of these considerations is that the decomposition fate of every pseudocomponent can be simply described using a lumped equivalent reaction whose stoichiometry takes into account the relative probability of the scission of different bonds. The stoichiometric coefficients, which vary slight with temperature, are conveniently evaluated at a reference temperature of 700 K.

As already mentioned, frequency factors and activation energies of all these reactions are derived from the equivalent ones in the gas phase, by applying the additive corrections of activation energies and entropies for their transposition in the condensed state.

The substitutive addition reactions of poly-aromatic radicals on methyl and alkyl positions of aromatic species produce poly-aromatic structures, as

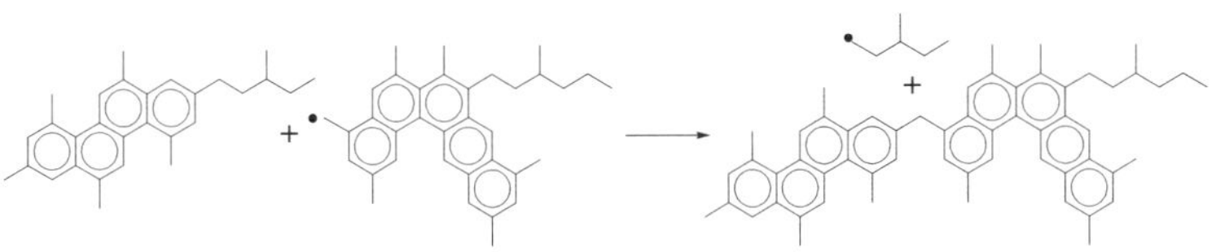

FIG. 32. Substitutive addition reaction.

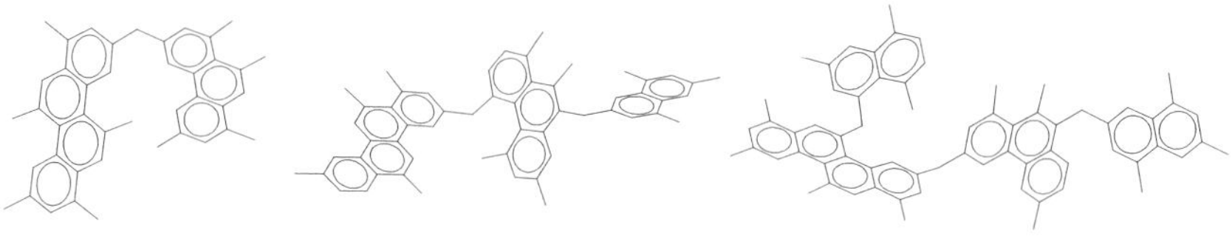

FIG. 33. Poly-aromatic species.

reported in Fig. 32. The net effect of these addition reactions with successive dealkylation is a polymerization of aromatic molecules characterized by multiple aromatic sheets connected by methylene bridges (see Fig. 33).

The index of multiplicity (i.e. the number of aromatic sheets) ranges from two to four. The maximum of four in the multiplicity index is justified mainly by two considerations. The first is that the predicted amount of the four-sheet molecules is already negligible compared with those of lower index. This simply means that further interactions, which would produce larger multi-sheet structures, become even more negligible. The second consideration refers to geometrical structure and to the increasing sterical hindrance which drastically reduces the possibility of further radical additions to the methyl sites of the molecule. During these substitutive addition reactions, the release of large alkyl radicals is more favoured than the release of methyl or even H radicals. These small radicals abstract H-atoms from the substrate forming alkanes, methane or H_2.

These reactions are very important, principally in the delayed coking process because of the high residence time in the drum of up to 24 h. There is, therefore, a clearer need for appropriate lumping in the case of delayed coking.

As a function of the different sites, reference kinetic constants for these substitutive additions in the aromatic ring, are

– alkyl site $k_{ADD} = 5 \times 10^8 \exp(-24,000/RT)$ [m^3/kmol s]
– methyl site $k_{ADD} = 5 \times 10^8 \exp(-26,000/RT)$ [m^3/kmol s]
– aromatic hydrogen position (C$_{ar}$-H)
    $k_{ADD} = 5 \times 10^8 \exp(-28,000/RT)$ [m^3/kmol s]

Further dehydrogenation and demethylation reactions contribute to the formation of coke precursors, as shown in Fig. 34.

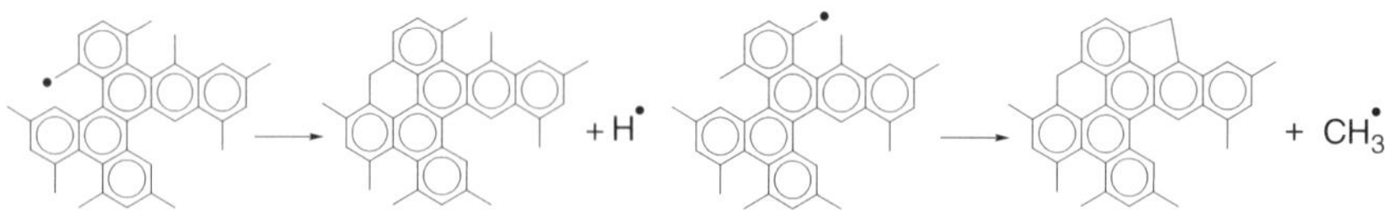

FIG. 34. Examples of internal dehydrogenation and demethylation of poly-aromatic radicals.

Significant amounts of methane and hydrogen are thus released in the gaseous products while the H/C ratio of the heavy residue gradually decreases to 0.6–0.4 and the hydrocarbon mesophase progressively becomes coke. In fact, the liquid containing the dispersed mesophase becomes more and more viscous and termination reactions become very difficult. As a consequence, poly-aromatic radical concentrations rise continuously with time and gas production is observed. The production of light fractions increases at the expense of the other distillates, such as gasoline and gasoils. This mechanism is clearly supported by experimental data and measurements relating to the kinetics of residue degradation (Del Bianco et al., 1993). After about half an hour in the drum, the average size of the alkyl side chains in poly-aromatic species is so drastically reduced that addition, dehydrogenation and demethylation become the predominant reactions. Details of the kinetic models including the description of the several hundreds of real and lumped components and equivalent reactions contained therein are available in Bozzano et al. (1995, 2005).

2. Stability of the Visbroken Residue and Pyrolysis Severity

A further important aspect in visbreaking process modelling is the stability of the residue that rules the operating conditions or process severity. The asphaltenes, which are present in the feed and formed during the cracking reactions, can flocculate forming a sort of mud, especially in systems with a vacuum distillation unit. This mud affects residue quality and therefore constitutes a limit to visbreaking severity. Asphaltenes are defined as the hydrocarbons that precipitate after the addition of *n*-heptane or *n*-pentane. An early discussion of the chemical species representative of asphaltenes is found in Algelt and Boduszynski's very comprehensive book (Altgelt and Boduszynski, 1994). Asphaltenes are best defined as high-molecular weight poly-aromatic species, containing cata-condensed aromatic sheets with methyl or short side chains and heteroatoms (mainly S, but also small amounts of N and O). Thus, the whole aromatic fraction can be considered to consist of asphaltenes and the remaining part of poly-aromatics with long alkyl side chains which also contain polar and resin compounds. Residue stability and asphaltenes flocculation are discussed in detail elsewhere (Bozzano and Dente, 2005; Bozzano et al., 1995; Dente et al., 1997). Only the essential aspects are reported here. Poly-aromatics with side chains decompose during the pyrolysis process and contribute to increase the asphaltene content in the visbreaking residue.

The reacting mixture can thus be represented by two phases. The first, the maltenic phase, is made up of aromatics, alkanes, alkenes, *cyclo*-alkenes and naphthenes. Aromatics and asphaltenes are contained in the second phase. Reciprocal compatibility is related to the solubility parameter that is in the range of 7–8 (cal/cm^3)$^{1/2}$ for alkanes and alkenes, 8–9 (cal/cm^3)$^{1/2}$ for naphthenes and *cyclo*-alkenes, 8.8–10 (cal/cm^3)$^{1/2}$ for aromatics and 11–12 (cal/cm^3)$^{1/2}$ for asphaltenes. It is worth noting too that the closer the solubility parameter of the two phases, the more resistant the system is to flocculation. All the asphaltenic molecules are supposed to be surrounded by aromatics, deposited as "sheets" over the asphaltenic "sheet", constituting a molecular aggregate equivalent to an adsorbed phase. This coverage improves the solubility parameter and compatibility with the maltenic phase of the aggregate with respect to pure asphaltenes. The non-asphaltenic aromatics are supposed to be distributed between the maltenic and adsorbed phase. Their distribution ratio is estimated on the basis of the activity coefficients of the aromatic class inside the two phases and depends on the temperature and composition of the two phases. Activity coefficients are estimated using the solubility parameters. Precipitation occurs when, as a result of dilution (e.g., with cetane), the total concentration of aromatics in the maltenic phase is so low that it results in the desorption of the aromatics deposited on asphaltenes. *Cyclo*-alkane and *cyclo*-alkene components increase the stability of the visbreaking residue at the same severity level. In fact, their presence increases the global solubility parameter of the maltenic phase and a smaller amount of aromatics is needed to cover the asphaltenes. This phenomenological model allows the estimation of the peptisation value of the mixture commonly adopted in refineries to control the severity of the process.

3. Model Validation and Comparisons with Experimental Data

Some interesting examples of industrial visbreaker performance with comparisons between experimental data and model predictions are reported in Table XI. Three different feedstocks and process conditions are considered. About 60% of the total residence time is inside the soaker. These data also show the increase in stability due to the presence of naphthenes. Feed 1 is richer in *cyclo*-alkanes while the third feed is more aromatic.

Likewise, Table XII reports a comparison of delayed coking models with industrial plant data for five different feedstocks covering a large range of different gasoil recycle ratios. The coil outlet temperature is in the range of 490°C and coil outlet pressure is about 5 ata. The results obtained confirm the reliability of this model.

Further interesting kinetic data, obtained from a batch isothermal lab reactor across a wide range of temperatures and residence times, are also discussed (Del Bianco *et al.*, 1993). The feedstock is a Belaym vacuum residue with the following properties: specific gravity (15/4) 1.028, kinematic viscosity at 100°C

TABLE XI
COMPARISON OF EXPERIMENTAL AND CALCULATED YIELDS AND PROPERTIES FOR A VISBREAKER

		1		2		3	
Feed properties							
AEBPI (°C)		424		485		495	
Sp. Gr.15/4		1.018		1.010		1.042	
K.V. cSt (100°C)		2,086		1,200		13,300	
CCR		13.7		15.5		23	
S wt%		2.97		2.75		3.2	
Operating conditions							
Coil outlet temperature (°C)		448		448		450	
Coil outlet pressure (ata)		11.5		11.5		11.5	
Residence time (min)		15.4		16.2		17.0	
Effluents		Exp.	Model	Exp.	Model	Exp.	Model
Gas	H$_2$S wt%	0.15	0.12	0.16	0.12	0.18	0.15
	Yield wt%	0.82	0.95	1.10	1.04	1.16	1.21
	Yield wt%	5.30	4.9	5.36	5.04	6.8	6.4
	ASTM D86 (°C)	22–165	23–168	36–167	27–168	32–191	34–195
Naphtha	Sp. Gr. (15/4)	0.717	0.719	0.718	0.719	0.738	0.731
	Sulphur wt%	0.76	0.85	0.82	0.85	1.00	0.94
	Br. Number	84	86	101	87	80	88
	Yield wt%	7.86	7.96	8.27	8.69	5.0	5.7
	ASTM D86 (°C)	195/325	200/324	195/331	200/324	200/300	200/321
Gasoil	Sp. Gr. (15/4)	0.843	0.840	0.843	0.841	0.852	0.847
	K.V. at 50°C cSt	1.8	1.6	1.96	1.8	1.9	1.7
	Sulphur wt%	1.91	1.68	1.39	1.47	2.1	1.84
	Br. Number	44	44	40	44	45	47
	Yield wt%	86.02	86.19	85.27	85.23	87.04	86.69
	AEBP (°C)	>340	>340	>340	>340	>320	>320
	Sp. Gr. (15/4)	1.028	1.027	1.021	1.020	1.049	1.051
Visbroken residue	K.V. at 100°C cSt	450	644	310	427	1414	2300
	P.V.	1.2	1.20	1.10	1.15	1.20	1.10
	Sulphur wt%	2.97	3.0	2.35	2.78	3.20	3.30
	CCR wt%	24	22.0	21.0	19.0	27.0	25.0

5230 cSt, CCR 20.8, n-heptane asphaltenes 18.6 (wt%). The comparisons between the experimental and predicted distillate yields are shown in Fig. 35. The maximum observed in the yields is related to the partial pressure of the distillates increasing with the time in the batch unit, so that partial condensation of heavy fractions is possible. At high temperatures and residence times, thermal degradation of the heavy fractions takes place in the gas phase; these reactions increase the gasoline and gas at the expense of heavy gasoil.

As previously discussed, fouling is an important aspect in the visbreaking process too. The adiabatic reactor volume of the soaker allows the run length of

TABLE XII
Comparisons of Experimental and Calculated Yields and Properties in a Delayed Coking Unit

		1	2		3		4		5	
Feed properties										
AEBPI (°C)		360	360		480		370		470	
Sp. Gr. 15/4		0.998	1.044		1.065		0.98		1.026	
S wt%		4.83	8.3		8.8		3.3		4.0	
CCR		9.1	17		20		14		20	
K.V. cSt (50°C)		30	400,000		25,000,000		1,100		400,000	
Gasoil recycle (% feed)			0		6		15		6	15
Effluents		Exp. Mod	Exp.	Mod	Exp.	Mod	Exp.	Mod	Exp.	Mod
Gas ($\leq C_4$)	Yield wt%	6.72 6.9	9.6	9.2	11.3	10.7	6.6	7.0	9.0	8.5
	H_2S	1.4 1.56	3.4	3.1	3.9	3.7	1.3	1.1	1.6	1.5
	$CH_4 + H_2 + C_2^{\pm}$	3.0 3.31	3.7	2.4	4.3	2.8	3.1	2.2	4.3	2.8
	$C_3^{\pm}$	1.4 1.0	1.1	1.2	1.4	1.3	1.0	1.1	1.4	1.3
	$C_4^{\pm a}$	1.0 1.5	1.4	2.6	1.7	2.9	1.2	2.4	1.7	2.9
L. Gasoline (C_5/80°C)	Yield wt%	2.91 2.94	4.6	4.4	4.2	4.6	4.8	4.3	4.2	4.4
H. Gasoline (80/180°C)	Yield wt%	10.6 10.5	7.5	8.0	6.8	7.0	7.9	8.1	6.8	7.0
	Sp. Gr. 15/4	0.777 0.775	0.76	0.77	0.76	0.76	0.75	0.76	0.75	0.75
	Sulphur wt%	2.1 1.5	3.0	3.2	3.3	3.1	1.2	1.3	1.5	1.3
L. Gasoil (180/350°C)	Yield wt%	31.8 31.9	27.7	29.0	24.0	26.4	32.5	32.4	26.2	26.1
	Sp. Gr. 15/4	0.90 0.91	0.88	0.87	0.88	0.86	0.86	0.85	0.86	0.85
	Sulphur wt%	3.0 2.6	4.3	4.5	4.8	4.1	2.0	1.9	2.5	1.9
H. Gasoil (>350°C)	Yield wt%	30.1 30.5	24.4	23.8	22.8	22.4	26.1	26.7	22.8	22.7
	Sp. Gr. 15/4	0.96 0.96	0.96	0.97	0.96	0.97	0.94	0.94	0.94	0.95
	Sulphur wt%	3.3 3.0	4.5	5.0	5.0	4.8	2.6	2.2	3.0	2.2
Coke	Yield wt%	17.8 17.7	25.5	25.4	30.2	28.9	21.4	21.5	30.3	31.3
	Sulphur wt%	6.1 5.95	8.0	9.0	10.0	9.5	3.3	3.7	3.3	3.8

[a] In the industrial cases part of the C_4 are entering into the light gasoline fraction.

the furnace to be increased up to about one year. Figure 36 shows the comparison between experimental and predicted external tube skin temperatures at the coil outlet in the case of an industrial visbreaking furnace. Moreover, the fouling also takes place in different areas of the visbreaking plant, such as the vacuum distillation tower or the heat exchangers, via the same mechanism.

Finally, it is worth pointing out that the modelling was strongly supported by a large amount of laboratory and industrial experimental data which allowed further tuning and verification of model assumptions and kinetic parameters. Detailed results of this, together with a discussion of the kinetic parameters, are reported in several papers (Bozzano and Dente, 2005; Dente et al., 1997).

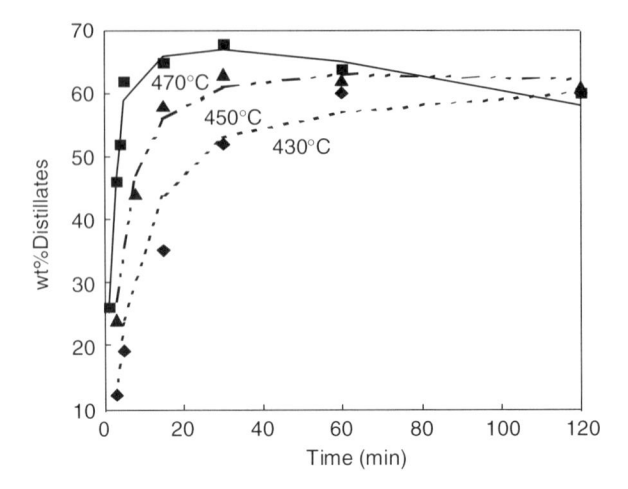

FIG. 35. Distillate yields in a batch lab reactor (Del Bianco et al., 1993). Prediction (line) vs. experimental points.

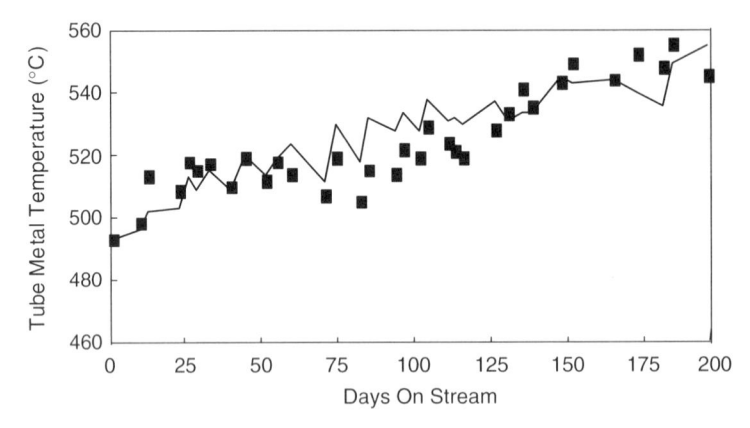

FIG. 36. External tube skin temperature at coil outlet in a visbreaking furnace. Prediction (line) vs. experimental points.

C. Thermal Degradation of Polymers

Polyvinyl polymers, such as polyethylene (PE), polypropylene (PP), polystyrene (PS) and polyvinyl chloride (PVC), are widely used in packaging and make up the main components of plastic waste in domestic refuse. Gaining a full understanding of their decomposition is therefore of great relevance to the environment. Waste recycling processes are always based on complex chain radical mechanisms taking place in the molten phase. These follow the same kinetic rules already discussed and analysed for visbreaking and delayed coking.

The scientific and technical literature on pyrolysis and gasification of plastics includes several papers, for instance, on PE (Anderson and Freeman, 1961; Miranda et al., 2001; Westerhout et al., 1997) on PP (Gambiroza-Jukic and Cunko, 1992), on PS (Lattimer and Kroenke, 1982; Madorsky, 1952) and on PVC (Bockhorn et al., 1999a; Knumann and Bockhorn, 1994; Miranda et al., 2001). The primary goal of most of these works is the evaluation of the overall rate of weight loss in thermal decomposition (Borchardt and Daniels, 1957; Coats and Redfern, 1964; Flynn and Wall, 1966; Friedman, 1963). The conventional power law model is commonly used for describing the conversion α

$$\frac{d\alpha}{dt} = k_0 \times \exp\left(-\frac{E_{att}}{RT}\right) \times (1-\alpha)^n$$

where n is the reaction order and E_{att} and k_0 are the activation energy and the pre exponential factor. The kinetic parameters relating to the same polymer and obtained from different experimental works differ significantly. The frequency factor spans several orders of magnitude whilst the activation energy can span ± 50 kJ/mol; this happens even when considering selected experiments in which transport phenomena limitations are negligible (Westerhout et al., 1997). These differences could be partially linked to either the presence of additives (which can display catalytic activity) or weak bonds in the original polymers. Consequently, a clearer understanding of the degradation process might be gained from a detailed kinetic model.

1. Mechanism Considerations

PE, PP and PS in their molten state decompose according to a liquid-phase radical chain mechanism which mainly involves the cleavage of the chain backbone while the side groups are not significantly removed from the polymer structure. Once again, the free chain radical reactions are simply described by the usual sequence of initiation, propagation and termination reactions. The same reaction classes as ever allow the pyrolysis process of the three different polymers to be explained. For the sake of simplicity, we can represent the polymers through a simplified notation in which the "side chain" (**G**) can be either H, CH$_3$ or phenyl in the case of PE, PP or PS, respectively. An exception must be considered for the specific case of PE in which the side group (**G**) undergoes the same reaction classes shown here for only one hydrogen atom.

The overall mechanism can be reduced to the usual reaction classes:

1. Initiation reactions, to produce radicals from molecules:
   (1A) Random scission of polymer backbone

(1B) Terminal scission

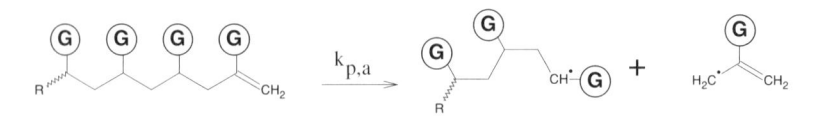

2. Propagation reactions of intermediate radicals:
   (2A) β-scission of radicals to form unsaturated molecules and smaller radicals

   (2B) β-scission in specific position, including unzipping reactions

   (2C) Alkyl radical isomerization via (1,4), (1,5) and (1,6) H-transfer

(2D) H-abstraction (metathesis) reaction on the polymer chain due to terminal radicals

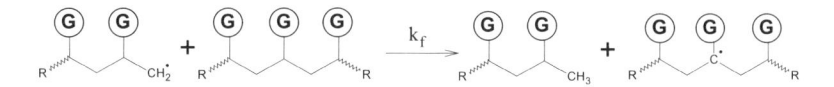

(2E) H-abstraction (metathesis) reaction on the polymer chain due to mid-chain radicals

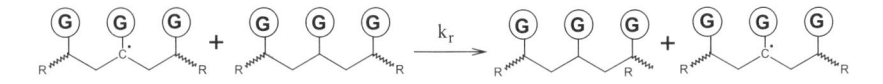

(2F) H-abstraction (metathesis) reaction on a specific position

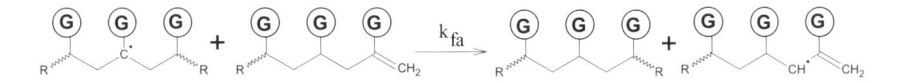

3. Termination by radical recombination reactions

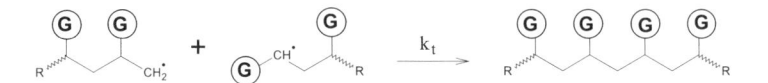

The thermal degradation of PE, PP and PS proceeds with the progressive formation of unsaturations at the end chain of the molecules. Each species in the melt, characterized by molecules or radicals, differs mostly for the type of end and tail unsaturation (Faravelli et al., 1999, 2001, 2003; Ranzi et al., 1997a). Smaller radicals and unsaturated species are formed through β-scission reactions of the radicals produced by H-abstraction reactions.

Thus, alkene backbones can be formed from alkyl radicals or also from unsaturated radicals. Similarly, α–ω dialkene backbone species are produced from parent radicals of heavier alkenes and dialkenes. By using P, O and D respectively to indicate the whole amount of alkane, alkene and α–ω dialkene backbone molecules, the resulting stoichiometry of the process is the following:

$$P_n \rightarrow O_j + P_{n-j}$$
$$O_n \rightarrow c_1(P_j + D_{n-j}) + c_2(O_j + O_{n-j})$$
$$D_n \rightarrow D_j + O_{n-j}$$

where j varies along the carbon chain of total length n. The two constants c_1 and c_2 depend on the presence of resonant position and intramolecular backbiting reactions (Faravelli et al., 1999). The average molecular weight of the melting

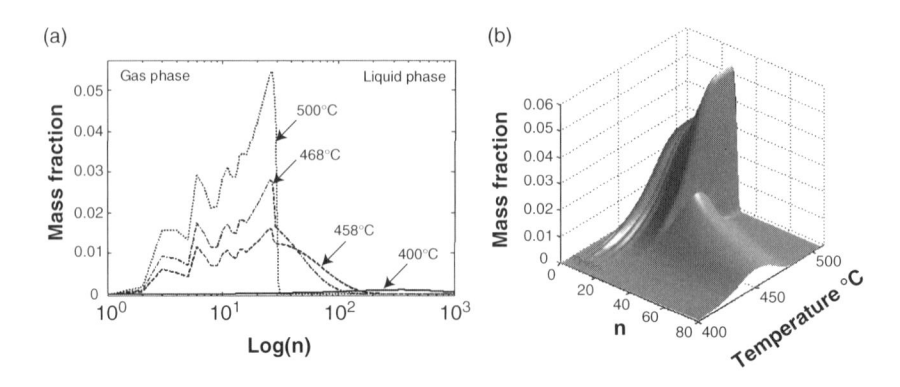

FIG. 37. Simulation of the instantaneous molecular weight distribution of PE during pyrolysis, expressed in term of repeated units n: heating rate 10°C/min, pressure 1 atm. Panel (a) Total mass fraction of chains nearby the border range of evaporable compounds at different temperature levels; panel (b) temperature profiles of components in the gas and the nearest liquid phase.

phase is progressively reduced due to the reiteration of these reactions. Ultimately, light decomposition products evaporate and leave the condensed phase without further reactions (Poutsma, 2003). The increase in the gas phase when the temperature rises and the simultaneous consumption of the melt is clear in the simulations reported in panel a of Fig. 37. Consequently, the gas released is fully described by the surface reported in Fig. 37b. During the thermal degradation process the initial molecular weight distribution (MWD) is affected by the shortening of the polymeric chains and the concentration of molecules close to the evaporation limit drastically increases. It is therefore necessary to consider a gas–liquid equilibrium at the interface and to account for mass balances in the gas and liquid phase. The simple approach considers all the components with boiling points lower than the system temperature as instantaneously entering the gas phase.

Two additional phenomena contribute to the thermal degradation mechanism and can determine the formation of the relative maximum and minimum in the characteristic surface of the gaseous MWD as shown in Fig. 1. In the backbiting process, the end of polymer chain can also react to form gaseous products with a non-random distribution through a sequence of isomerization and β-scission reactions. Finally, monomer units can also be released by an unzipping process through β-scission of the terminal radicals.

Even though PVC thermal degradation is a free chain radical mechanism in the liquid phase, it partially differs from what observed for PS, PP and PE. In fact, PVC pyrolysis involves the cleavage of the C–Cl bond of the side chain, instead of just decomposing the polymer chain. The result of this process is the formation of a large quantity of double bonds due to successive β-scissions and the formation of poly-aromatic structures as a result of the consequent cross-linking reactions.

KINETIC MODELLING OF PYROLYSIS PROCESSES 141

The polymer structures involved in PVC pyrolysis can be represented using the same previously mentioned, simplified notation in which the side chain (**G**) can be either H, Cl or PAH clusters. The free chain radical reactions are simply described by the following reaction classes:

1. Initiation reactions, to form chlorine radical:
   (1A) Random scission of C–Cl bonds

   (1B) Specific scission in allyl position

2. Propagation reactions of intermediate radicals:
   (2A) H-abstraction (metathesis) reaction on the polymer chain, mainly due to chlorine

   (2B) β-scission of radicals to form double bonds and chlorine radical

   (2C) β-scission of radicals forming tar compounds

   (2D) Radical addition on double bonds

3. Molecular reactions:
   (3A) Intramolecular Diels–Alder

   (3B) Intermolecular Diels–Alder

4. Termination or radical recombination reactions.

A sequence of H-abstractions due to the very reactive chlorine radical and β-decomposition gives rise to the formation of polyene molecules and HCl which is released from the melting phase. Cross-linking reactions between polyenes lead to alkyl aromatic intermediates which can further decompose, releasing tar compounds and/or can polymerize with char formation. These transformations are experimentally observed as a second weight loss in PVC thermo gravimetric analysis.

This mechanism is more complex with respect to other vinyl polymer ones and demands specific simplifications. One proposed approach carefully analyses all the reactions of the single monomer unit in the polymer chain (Marongiu et al., 2003). The initial PVC polymer is simply represented by the chlorinated reference unit P–(**CH₂CHCl**)–P. The successive steps of degradation form polyene molecules and these species, which have different molecular weights, are represented by the alkene reference unit: P–(**CH=CH**)–P. The reference species (reported in bold characters in the brackets) are the reacting units and are placed inside the polymer chain, represented here by the P at the beginning and end.

The resulting semi-detailed kinetic mechanism thus consists of real species and a limited number of pseudocomponents (molecules and radicals) which are representative of the whole system. They characterize the typical structures

TABLE XIII
Reference Components in the Kinetic Model of Pyrolysis of PVC

Real species	Linear pseudo species	Aromatic and char pseudo species	
C_8H_8	P-(CH_2CHCl)-P	P-($C_{10}H_{10}$)-P	P-($C_{10}H_9Cl$)-P
$C_{12}H_{10}$	P-($CH=CH$)-P	P-($C_{18}H_{16}$)-P	P-($C_{18}H_{15}Cl$)-P
$C_{14}H_{10}$	P-($CH_2 \bullet CH$)-P	P-($C_{47}H_{36}$)-P	P-($C_{47}H_{35}Cl$)-P
C_6H_5Cl	P-($\bullet CHCHCl$)-P	CHAR	CHARC

involved in the different degradation phases in terms of polymer reference pseudospecies, aromatic and char components. Table XIII provides a sample of the different components involved in PVC degradation model.

Intermediate species during the dehydrochlorination phase are other reference units which present either Cl or double bonds in the structure. H or Cl abstractions from these pseudocomponents form the corresponding radicals. Cross-linking reactions between polyene molecules lead to alkyl aromatic intermediates which can further condensate and grow into char. Reference lumped components are then introduced according to the different number of aromatic rings and the possible presence of Cl. The thermal degradation of this and similar polymers is always described on the basis of the same reaction classes, and always with a small set of reference kinetic parameters (Marongiu et al., 2003; Mehl et al., 2004).

Independently of the adopted approach for the mechanism formulation, the estimation of the rate constants for all the reaction classes involved in the pyrolysis process of PE, PP, PS and PVC is of critical importance in model development. As already mentioned, the kinetic parameters of the condensed-phase reactions are directly derived from the rate parameters of the analogous gas-phase reactions, properly corrected to take into account transposition in the liquid phase (e.g., Section II.D).

2. Balance Equations

Mass balance equations are expressed in a very general form:

$$\rho \frac{dy_n}{dt} = \sum_{j=1}^{NRC} v_{n,j} \cdot R_j \cdot W_n$$

where y is the mass fraction and NRC the number of chemical reactions. R_j is the reaction rate of disappearance or formation, $v_{n,j}$ is the corresponding stoichiometric coefficient and W_n the molecular weight of the component. The density ρ is assumed constant during the degradation. The first difficulty in developing the kinetic model for thermal degradation of polymers is to properly simplify the system to reduce and control the overall dimensions.

Three different approaches are briefly discussed here. They refer to the steady state assumption of propagating radicals, the discrete section method and the method of moments.

a. Continuously varying steady state (CVSS). The continuously varying steady state (CVSS) assumption can be conveniently applied to the pyrolysis of PE, PP and PS. The concentration of all the propagating radicals is proportional to that of a lumped single radical R•

$$[R\bullet] = \sqrt{\frac{\sum_i k_{pp,i}[BP]_i}{k_t}}$$

where k_t is the rate constant of self-recombination reaction R + R; $k_{pp,i}$ are the rate constants of the different initiation acts; $[BP]_i$ are the total concentrations of backbone reacting positions of alkanes, alkenes and dialkenes.

Because of the lower activation energy of their initiation reactions, the rising concentration of double bonds inside the polymer enhances the radical concentration and favours the pyrolysis process. This radical concentration is then used to evaluate the rate of formation and disappearance of all the species in the system. Thus, the production of the alkene O_j is simply expressed as

$$\frac{d[O_j]}{dt} = k_\beta [R_n^\bullet] = k_\beta \frac{k_r[R\bullet][S_n]}{k_f\{H\} + k_\beta} = k_{app}[R\bullet][S_n]$$

where k_{app} is the rate constant of the apparent propagation reaction. This is the proper combination of the decomposition (k_β) and abstraction rate constants (k_f and k_r). The total concentration of H-atoms is

$$\{H\} = n_H \frac{\rho}{W_0}$$

where ρ is the density of the polymer, W_0 the molecular weight of the monomer and n_H the number of available H for each monomer unit. Of course, CVSS and the assumption of the single lumped radical results in a significant reduction of the total number of balance equations (Dente *et al.*, 1979; Faravelli *et al.*, 1999, 2001, 2003; Ranzi *et al.*, 1997a).

b. Discrete section method. A further possible reduction of the overall dimension of the problem can be addressed with appropriate lumping procedures (Ebert *et al.*, 1982). This approach is equivalent to the discrete sectional model, which has been widely adopted in particle distribution research for many years and already discussed in the soot formation paragraph (Gelbard *et al.*, 1980; Wulkow, 1996). The method approximates the MWD of polymer molecules by a finite number of sections or lumped components, typically called BINs, which represent groups of species. By dividing the entire MWD into sections and dealing only with a representative pseudospecies inside each section, the number of conservation equations is reduced to the number of sections.

Using a finite element type formulation, the concentration of the discretized element B_i is

$$B_i = \sum_{n=n_{i,min}}^{n_{i,max}} S_n$$

where S_n is the molar concentration of each molecule in the section between $n_{i,min}$ and $n_{i,max}$. Below $n_{1,min}$ no clustering is introduced and each real species is taken into account. The lumped pseudospecies can be either equispaced or spaced linearly in $\ln(n)$, separated by a constant factor Δ: the exponential sectional approach saves on computation time whilst still maintaining a good description of the low molecular weight components.

During polymer decomposition, whenever a species is formed with a molecular weight not equal to that of a pseudocomponent, its amount is linearly distributed between the adjacent bins (B_i, B_{i+1}). This lumping procedure or discrete section method was successfully applied to the thermal degradation of PS (Faravelli *et al.*, 2001).

c. Method of moments. The method of moments is a very effective mathematical tool when it comes to dealing with the complexity of regular systems. In fact, the MWD of different families of species can be conveniently described by a continuous expression (Frenklach et al., 1984). Moment equations describe the time evolution of the statistical parameters characteristic of the MWD of molecules and radicals in the liquid and gas phases. The zero-order moment represents the molar concentration, the first and second moments stand for the average MW and its variance (Abramowitz and Stegun, 1968). This method involves a small number of equations and drastically reduces the computing times for the solution of the problem. McCoy and co-workers have extensively applied the method of moments and developed various models with different mechanistic details (Madras and McCoy, 2002; Madras et al., 1996; McCoy and Wang, 1994). Kruse and Broadbelt apply the method of moments to the thermal degradation of PS, PP and their binary mixture (Kruse et al., 2002, 2003, 2005).

The polymer species in the liquid phase span from L to N_{max}, where L represents the minimum number of units for species in the liquid phase and N_{max} is the corresponding maximum, thereby limiting the mass MWD to 99.9%. Monomers and oligomers up to $L-1$ units are in the gas phase. Frenklach (1985) and Bockhorn (1999b) assume the following definition of statistical moments of the size distribution of the polymer species:

$$^k\mu_{P,l} = \sum_{n=L}^{N_{max}} n^k \cdot P_n$$

The left superscript k denotes the order of moment, the right subscript represents the family of the different species (alkanes, alkenes, ...); l indicates the liquid phase and n is the number of repeated units. The distribution of volatile species is defined as follows:

$$^k\mu_{P,v} = \sum_{n=1}^{L-1} n^k \cdot P_n$$

where subscript v denotes the gas phase. Moment equations properly describe all the typical reactions involved in the thermal degradation process of the different polymers. The kinetic expression ruling the moment variation can be calculated for each reaction class, starting from the molar balances of the different species. This formulation of the system permits a mechanistic identification and description of the thermal decomposition of polymers in both isothermal and non-isothermal pyrolysis, where L changes with the time (Marongiu, 2006).

To deduce the moment equations, the balance equations of the different species must be summed up to group the contributions of the different reaction classes. The overall contribution ^kR_j for each reaction class to the different moment equations, k, is simply analytically calculated with the proper sum of all the corresponding terms for the liquid and the vapour phase:

$$^kR_{j,l} = \sum_{n=L}^{N_{max}} n^k \cdot v_{n,j} \cdot R_j \qquad ^kR_{j,v} = \sum_{n=1}^{L-1} n^k \cdot v_{n,j} \cdot R_j$$

The following system of differential equations is obtained by considering all the reaction classes and families of species in the kinetic scheme:

$$\frac{d^k \mu_{i,l}}{dt} = \sum_{j=1}^{NRC} v_{i,j} \cdot {}^k R_{j,l} \quad i = 1, \ldots, NSP$$

$$\frac{d^k \mu_{i,v}}{dt} = \sum_{j=1}^{NRC} v_{i,j} \cdot {}^k R_{j,v} \quad i = 1, \ldots, NSP$$

where NRC is the total number of reactions, $v_{i,j}$ the stoichiometric coefficient and NSP the number of families of species.

3. Model Validation and Comparisons with Experimental Data

A few comparisons between experimental measurements and model predictions are reported to show the reliability of the different models, to investigate the relative reactivity of the different polymers and, finally, to allow a brief discussion of the different numerical approaches.

Figure 38 shows the comparisons of model predictions and experimental measurements of thermal decomposition of PS (panel a), PP (panel b) and PE (panel c). These experiments refer to three dynamic TGA's with different heating rates (Anderson and Freeman, 1961; Ranzi et al., 1997a). The agreement between model predictions and experimental data is quite good. Figure 38a shows the dynamic thermal degradation of PS under vacuum 1 mmHg and heating rate 5°C/min. Thermal decomposition proceeds with a single step and the mass loss is due to the evaporation of small fragments of the polymer (monomers, dimers and trimers). Of course, for all the polymers, the shortening of the polymer chain requires the formation of double bonds before the initial weight loss. The behaviour of the three different degradation curves in Fig. 38 confirms the relative reactivity of the three polymers: PS > PP > PE, as expected from thermo kinetic analysis of the polymers structures.

As already mentioned, different modelling approaches were applied; (a) the complete system of balance equations of all the species with QSSA for the single propagating radical; (b) discrete section method; (c) the method of moment. These comparisons demonstrate the equivalence between these techniques, not only with respect to total mass loss, but also in terms of intermediate liquid components (alkanes, alkenes and $\alpha-\omega$ dialkenes).

Figure 39 presents the characteristic three phases of PVC decomposition very clearly. The initial dehydrochlorination reactions form HCl and polyene structures. During this phase, benzene and some naphthalene and phenanthrene are also formed through Diels–Alder reactions and the successive dealkylation of polyene molecules. Then, when Cl has been quantitatively released

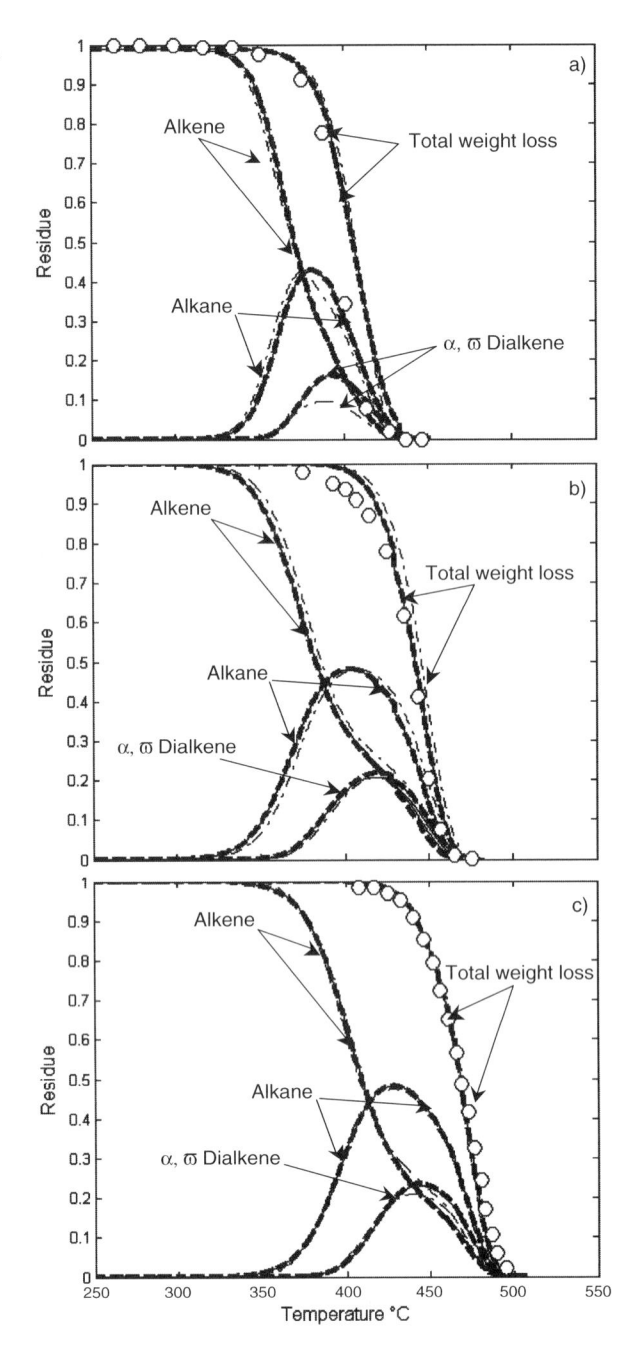

FIG. 38. Total weight loss and main species in the liquid phase during dynamic thermal decomposition. (a) Pyrolysis of polystyrene (Anderson and Freeman, 1961) (5°C/min, 1 mmHg). (b) Thermal decomposition of polypropylene (Ranzi et al., 1997a) (10°C/min, 1 atm). (c) Pyrolysis of polyethylene (Ranzi et al., 1997a) (10°C/min, 1 atm). Discrete model (Dashed) and Moment model (–).

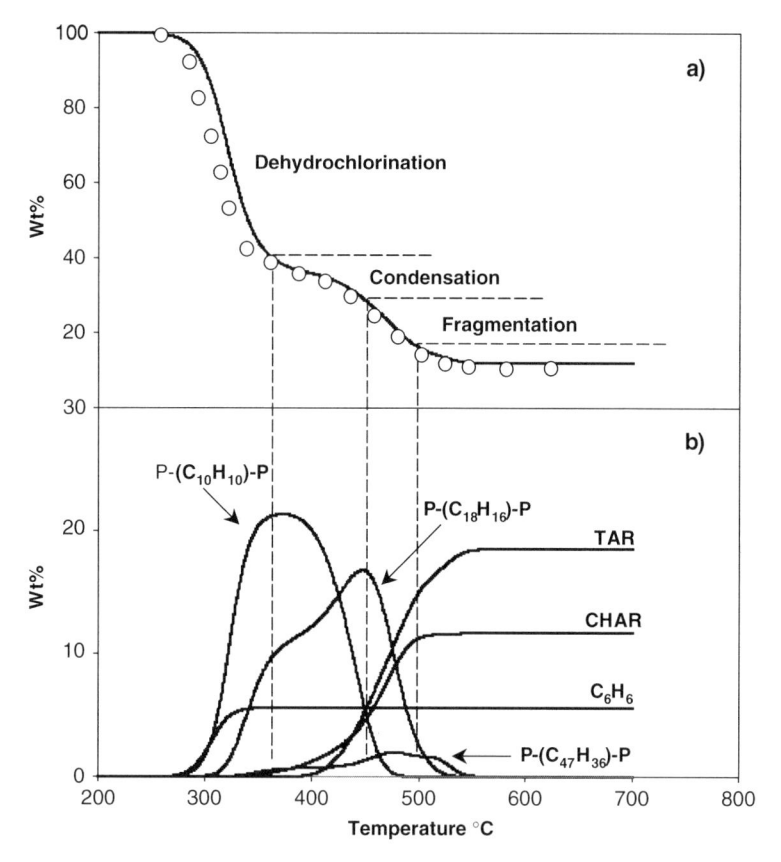

FIG. 39. Predicted dynamic TGA of PVC with an heating rate of 10°C/min, experimental data (Marks) (Montaudo and Puglisi, 1991). Panel (a) Residue (%wt) behaviour and identification of the main thermal decomposition phases. Panel (b) Benzene, PAH and char formation profiles. The TAR fraction represents the total amount of volatile aromatics.

from the melt, the polyene molecules rearrange and, through cyclization and cross-linking reactions, form alkyl aromatic hydrocarbons and char residues. The model agrees quite well with the experiments in all the different phases. Similar agreement can be found for different heating rates (Marongiu et al., 2003).

As a final consideration, it is relevant to discuss the behaviour of mixtures of different plastics. In fact, one possible process for recovering valuable chemical and petrochemical products from plastic waste is the stepwise thermal degradation of polymer mixtures. This potentially allows the step-by-step simultaneous separation of the different fractions generated by the polymers of the blend. The effect of the mixing scale of PE and PS and their interactions in the melt on the basis of several hypotheses was recently investigated (Faravelli et al., 2003). The first and simplest approach was a completely segregated model which directly combines the results from pure plastics. The opposite solution is a completely mixed model assuming a single melt phase with a homogeneous and perfectly mixed composition. Neither of these asymptotic models is correct because it is necessary to refer to the real solubility, miscibility and mass-transfer among the polymers, even though a completely segregated model is a reasonable approximation which allows easy extension to more complex mixtures.

V. Conclusions

The kinetic aspect common to all the topics discussed in this chapter is the pyrolysis reactions. The same kinetic approach and similar lumping techniques are conveniently applied moving from the simpler system of ethane dehydrogenation to produce ethylene, up to the coke formation in delayed coking processes or to soot formation in combustion environments. The principles of reliable kinetic models are then presented to simulate pyrolysis of hydrocarbon mixtures in gas and condensed phase. The thermal degradation of plastics is a further example of these kinetic schemes. Furthermore, mechanistic models are also available for the formation and progressive evolution of both carbon deposits in pyrolysis units and soot particles in diffusion flames.

Detailed kinetic schemes also consist of several hundreds of species involved in thousands of reactions. Once efficient tools for handling the correspondingly large numerical systems are available, the extension of existing kinetic models to handle heavier and new species becomes quite a viable task. The definition of the core mechanism always remains the most difficult and fundamental step. Thus, the interactions of small unsaturated species with stable radicals are critical for the proper characterization of conversion and selectivity in pyrolysis processes. Parallel to this, the classification of the different primary reactions involved in the scheme, the definition of their intrinsic kinetic parameters, the automatic generation of the detailed primary reactions and the proper simplification rules are the important steps in the successive extension of the core mechanism. These assumptions are more relevant when the interest lies in the pyrolysis of hydrocarbon mixtures, such as naphtha, gasoil and heavy residue, where a huge number of isomers are involved as reactant, intermediate and final products. Proper rules for feedstock characterizations are then required for a detailed kinetic analysis.

Due to the complexity of these kinetic schemes, an accurate model validation must always refer to different experimental devices and to the widest possible range of operating conditions and feed varieties.

In spite of the different models present in the technical literature and discussed here to some extent, not all the problems have been solved yet and much theoretical and experimental work still remains to be done. New experimental data will allow further improvements in the kinetic assumptions or suggest new hypotheses for a deeper understanding of the detailed chemistry of pyrolysis and fouling processes.

LIST OF SYMBOLS

A	matrix of the stoichiometric coefficients
B_i	concentration of the discretized element i
$[BP]_i$	total concentrations of backbone reacting positions of alkanes, alkenes and dialkenes
C_{AS}	concentration of radical A in liquid phase S
D_{AS}	diffusivity of radical A in liquid phase S
D_j	set of decomposition reactions
E_{ATT}	activation energy
f	total molar fraction of the liquid phase containing from n_{CMIN} to n_C carbon atoms
$F(n_C)$	correction of activation energy
h	Planck constant
$\Delta H^{\#}$	molar enthalpy variation in the transition state
ΔH_{ev}	molar heat of vaporization of liquid or polymer unit of flow
k	kinetic constant
k_0	frequency factor
k_{coll}	collision constant
K	Boltzmann constant
Ij	set of isomerization reactions involving radical j
n_C	number of carbon atoms
$n_{CMIN}, n_{C,AV}$	minimum and average number of carbon atoms
O_j	olefin concentration
P_j	concentration of reacting position into the deposit layer
$\dot{P}_j$	total production rate of the component j
$\vec{P}_0$	vector of the initial production rate
r_A	distance among atoms in A+A recombination
R	gas constant
R_j	set of decomposition reactions forming the jth product
S_n	molar concentration of molecules pertaining to section n
$\Delta S^{\#}$	molar entropy variation in the transition state
T	absolute temperature
$\tilde{V}_s$	molar volume of liquid or polymer unit of flow
V_{coll}	collision rate per unit volume and time in liquid phase
X_i	molar concentration of the radical i
$\vec{X}$	vector of the radical concentration
y_n	mass fraction of component n

Z_i	concentration of the radical decomposed in the reaction ith
W_0	molar mass of the monomer unit
W_n	molecular weight of the n component
W_S	molar mass of the liquid or of the unit of flow
α	conversion in polymer degradation
ϕ_{AA}	orientation factor
$v_{n,j}$	stoichiometric coefficient of component n in reaction j
μ_S	viscosity of the liquid S
$^k\mu_{p,l}$	k moment of polymer distribution in the liquid phase
$^k\mu_{p,v}$	k moment of polymer distribution in the vapour phase
ρ	polymer density

Appendix 1. MAMA Program: Automatic Generation of Primary Lumped Reactions

The aim of this mechanism generator is to evaluate the set and weights of the end products obtained by the decomposition of a large, heavy radical. Starting with an initial radical, the propagation path includes isomerization and decomposition/dehydrogenation reactions with the formation of final products. The assumptions made in the generation of the primary propagation mechanism correspondingly define the structure of the algebraic system, which once solved, gives the distribution of the primary decomposition products. This subject will be dealt with first in the next paragraph. Due to the large number of components that can be involved in a propagation path, the concept of lumping of components as well as the classes of components which need to be lumped, will be then discussed.

The generation of a propagation path requires computing time which rapidly increases with the carbon number of the initial radical. However, the resulting generated pathway of a specific radical is independent from the formation mechanism of the radical itself, and therefore there is a clear need to save the path in a database. The description of the database structure will include details on the molecule representation as well as on the way the reaction path is stored and recorded. Finally, before the functioning of the MAMA Generator is summarized, the algorithm for molecule homomorphism will be described in brief.

A.1. EVALUATION OF END-PRODUCT COMPOSITION

The automatic generation of the primary product distributions is based on the assumption that the propagation reactions of radicals are described with an autonomous system, i.e. H-abstraction, radical addition and termination reactions and any interactions of the intermediate radicals with the reacting mixture are disregarded. Consequently, the end products of this propagation mechanism are small radicals and unsaturated components (UC). The total production rate P_j of the jth product is simply expressed in the form:

$$\dot{P}_j = \sum_{i \in R_j} k^D_{i,j} Z_i + \dot{P}_{j0} \tag{A.1}$$

where P_{j0} is the initial production rate of the jth component j (normally $= 0$ for end products), $k^D_{i,j}$ is the production rate in the ith reaction, Z_i is the concentration of the radical decomposed in the reaction i, and R_j is the set of decomposition reactions forming the jth product. For all the intermediate radical components, assuming quasi-steady conditions, mass balance equations are written in the form:

$$X_j \left(\sum_{i \in I_j} k^I_{j,i} + \sum_{i \in D_j} k^D_{j,i} \right) = \sum_{i \in I_j} k^I_{i,j} \cdot X_i + \dot{P}_j \tag{A.2}$$

where $X_j =$ concentration of the jth component; $I_j =$ set of isomerization reactions involving the jth radical component; $D_j =$ set of decomposition reactions involving the jth radical component; $k^I_{j,i} =$ isomerization rate constant of the jth radical component to form the ith radical component; $k^I_{i,j} =$ isomerization rate constant of the ith radical component to form the jth radical component; $k^D_{j,i} =$ decomposition rate constant of the jth radical component in the reaction i; $\dot{P}_j =$ total production rate of the jth radical component as described in Eq. (A.1), with $Z \in X$.

The left side of Eq. (A.2) represents the depletion of jth component due to its isomerization and decomposition. The right side is the formation contribution to jth component coming both from the isomerization of ith radical and from the decomposition of larger radicals.

At fixed temperature, the material balances turn into a system of linear equations in radical concentrations.

$$\overline{\overline{A}} \vec{X} - \vec{P}_0 = 0 \tag{A.3}$$

where $\overline{\overline{A}}$ is the coefficient matrix (size n,n), $\vec{X}$ the vector of radical concentrations (size n), $\vec{P}_0$ the vector of the initial production rates (size n), n the number of the intermediate radical components.

It is worth to notice the following properties of this linear system:

- The size of the system rapidly increases with the carbon number of the initial radical, reaching values of several hundred thousands when studying the primary propagation path of unsaturated components, such as alkenes and *cyclo*-alkanes with more than 14–15 carbon atoms,

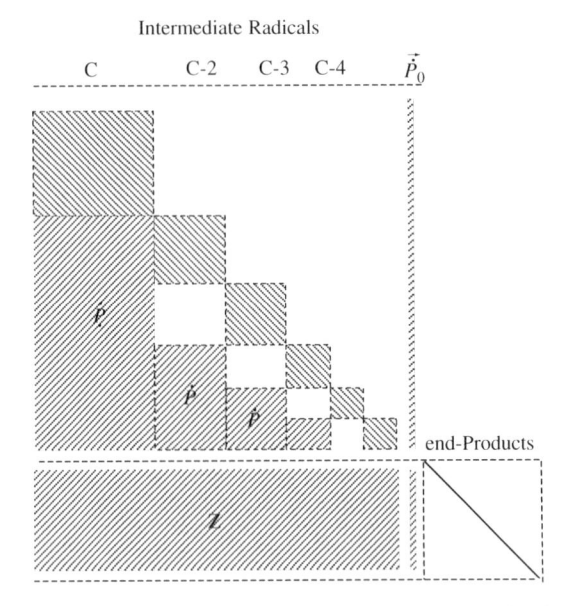

FIG. A1. Structure of the system (3) with end-product production rate.

- The system is dominantly sparse with a few tens of unknowns in each Eq. (A.2),
- The overall system may be decomposed into subsystems of reduced size as shown in Fig. A1.

Once the linear system is solved and the intermediate radical concentrations are known, the end product distribution (Eq. (A.1)) is directly obtained.

The upper part of Fig. A1 shows the structure of the system (3) while the set of Eq. (A.1) giving the total production of end products, is in the lower part. As previously mentioned, the two portions are solved disjointed, in sequence with the upper portion followed by the lower one.

In the system (3), the unknowns are ordered according by decreasing carbon number C. It should be stressed that a radical with C_n carbon atoms can only decompose generating radicals with C_{n-2}, C_{n-3}, ... carbon atoms. In fact, β-decomposition does not allow the formation of a radical with C_{n-1} carbon atoms. This explains the presence of empty blocks in the structure of the system. The equations are ordered by decreasing carbon number too. The resulting structure of the system in Fig. A1 suggests the most effective solution. By its nature, the block including the equations with the highest carbon number does not contain any production term (see Eq. (A.1)) except the initial values $\dot{P}_{j0}$, i.e. the set of decomposition reactions R_j is empty. This block can be solved and the calculated values of the unknowns factorize the vertical band "$\dot{P}$". "$\dot{P}$" contains the production terms of smaller radicals. The factorization renders as known all the production terms of the successive block with the highest C number. This block is solved and the results allow the factorization of the related vertical band. This procedure is then repeated until the end of the system. Finally, the production of the end products is easily calculated as all the related unknowns are available courtesy of the previous system solutions.

One important aspect derives from the initial assumption made: each radical evolves without interacting with the surrounding environment. In other words, the decomposition reactions are considered irreversible which results in the linearity of the algebraic system: the end product composition of an initial radical is not dependent on the initial composition of any other intermediate radical.

In fact if $X_j(..,\dot{P}_{j0},.)$ and $X_i(..,\dot{P}_{i0},.)$ stand for the solution vectors of the above system (3) for a null vector $\dot{P}_0$ except, respectively, for the initial production rates of components j and i, a trivial property of linear systems gives

$$X_{j,i}(...\dot{P}_{i0},..\dot{P}_{j0}..) = X_i(..,\dot{P}_{i0},.) + X_j(..,\dot{P}_{j0},.)$$

This property applied to the problem at hand turns into the statement that, given a mixture of initial radicals at $\dot{P}_{10}, \dot{P}_{20}, \ldots \dot{P}_{R0}$, their product distribution is equal to the sum of the distributions of each single radical. The advantage of this result that the product distributions of each radical can be considered to be independent of the mixture it belongs to, such as, for example, the initiation reactions which generated it. This fact anticipates an item relevant to the invariance of the propagation paths in the face of the problem in which they are included. The propagation paths are built once for all and stored in a database and they are ready to be analysed for any class of problems that requires them.

A.2. LUMPING OF COMPONENTS

The evaluation of the end product composition in a propagation chain may require the solution of linear systems involving a large number of unknowns referring to the concentrations of intermediate radicals. The number of end products may be so large that they are almost of the same magnitude.

Figure A2 shows the number of components involved in the primary propagation reactions of normal alkenes. These data clearly indicate that this number rapidly becomes very large, increasing the carbon number of the initial alkene. This fact justifies the need to turn to the component lumping when dealing with detailed and mechanistic models describing the hydrocarbons pyrolysis of such heavy species.

However, the following conclusions may help to make the solution of this complex problem more viable:

– The amount of alkenes and unsaturated components in the end products is large and strongly suggests the introduction of lumping of components, as it will be described below.

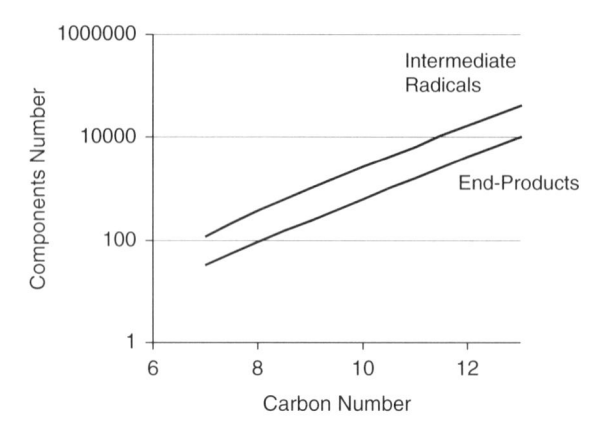

FIG. A2. Components number in the primary propagation reactions of normal-alkenes.

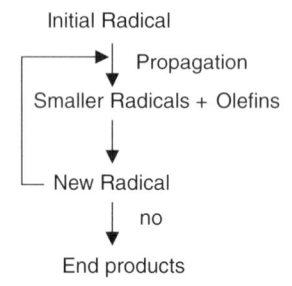

FIG. A3. Schematic propagation reaction chain of radical decomposition and formation of end products.

- As the end products consist of an initial radical not dependent on the initial composition of any other intermediate radical, the simplification and lumping procedures will not refer to such components.
- The amount of small and stabilized radicals in the end products is rather limited compared to the amount of the remaining components so that the lumping might be avoided for this class or applied to a small portion of it.

If just a single radical is considered (see Fig. A3), propagation reactions produce a smaller and/or stabilized radical and an olefin and/or UC. Each new radical is then processed according to its primary propagation reactions (isomerization and decomposition) which results in the production of new olefin or UC and a smaller radical. The olefins produced define the whole set of components involved in this chain so that the lumping technique conveniently applies only to them.

The grouping or lumping of components does not refer to unique rules, but it is only dependent on the successive uses of the kinetic scheme. Consequently, the Automatic Generator does not contain any lumping rule. Nevertheless, the resulting final products may be conveniently lumped according to pre-assigned rules.

A.3. DATABASE STRUCTURE

As already anticipated, a key feature of the automatic generator is a database storing the propagation paths of initial radicals. The choice of which radical has to be considered is dependent, as mentioned above, on the mathematical model (say steam cracker simulator) which will include the generator results. Provided that all the possible isomers of hydrocarbons up to six carbon number are limited to few hundred, it can be assumed that the database already stores the propagation paths of all the radicals included in this set. This means that the propagation path of a C7 hydrocarbon will produce new isomerization reactions and products with seven carbon atoms while the remaining reactions and products with a lower number of C-atoms are already present in the existing database (so that there is no need to further describe them). Thus the database is enlarged to contain only the new portion created by this component, as it shown in Fig. A4.

It will take some effort to enlarge the database to include all the possible isomers with seven carbon number as the isomers are still acceptably limited in their number. However, this becomes rather difficult and practically impossible with higher carbon numbers due to the rapid increase of the involved isomers set. A compromise is therefore required between the comprehensiveness and the size of the whole database.

Within the context of a steam cracker simulator, experience has shown that the accumulation of propagation paths in the database by starting with mixture of radicals obtained by H-abstraction reactions of normal paraffins and normal olefins (with progressively increased carbon number), produces a whole

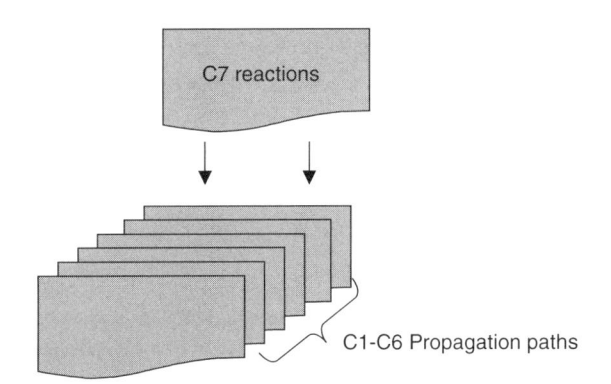

FIG. A4. Sketch of the accumulation of new portions in the database.

information set which covers the description of a good percentage of components present in the kinetic scheme. The rationale for this comes from the fact that isomerization reactions produce (via successive ring formation) and then break a set of isomers of alkenyl and *cyclo*-alkyl radicals which is close to the complete theoretical one. Dehydrogenation reactions therefore increase the total number of end products with a definite finger print of all the intermediate radicals.

The computing time required to create reaction paths is similar to that reported in Fig. A2: only few seconds are required for radicals with carbon atoms lower than five to six while almost a week is needed for olefins with carbon number larger than 16.

Because of the rapid increase in components and the consequent practical problem of storing all the data in an efficient way, it also worth mentioning the technique adopted for this purpose.

As the propagation path is a sequence of reactions steps, the basic elements for the description of the whole path are those ones which can describe a single reaction:

– Description of the components molecular structure
– Indication of the products and related reactions associated with the component.

For instance, the decomposition reaction of 6-hepten-2-yl radical

is characterized by the description of the molecular structures of the three components, the type of the reaction and the distinction between the reactant and the products.

As the components are normally involved in several different reactions, it is more convenient to collect all the component descriptions in a list, and to address their position in that list.

Similarly, the reactions too may be grouped in a proper list.

Figure A5 provides an example of a possible presentation of the cited reaction as a sequence of addresses:

k (reactant), **i** (olefin product if any), **j** (radical) and **n** (reaction type).

The molecule description is based on a sequence of integers. In fact, in dealing with hydrocarbons the molecule is univocally characterized by the carbon–carbon bonds and the type of the carbons. For example, Table A1 reports the assumed correspondence between integer indexes and carbon types: they are used to represent a molecule, in a plane, as a sequence of integers.

KINETIC MODELLING OF PYROLYSIS PROCESSES 159

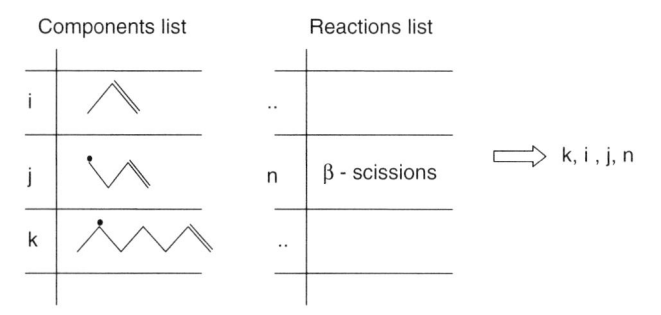

FIG. A5. Schematic structure of elements describing a reaction step.

TABLE A1
MOLECULE DESCRIPTION. CORRESPONDENCE BETWEEN INTEGER NUMBERS AND CARBON TYPES

| | → 21 22 23 22 2 1 |
	1
1 → C —	21 → C=	31 → C≡
2 → —C—	22 → —C=	32 → —C≡
3 → —C—	23 → —C=	42 → =C=
4 → —C—		

Similarly, *cyclo*-alkane with five- or six- membered rings too may be put into correspondence with integers which contain information about the ring members and their connections to the molecule chain.

It is worth to noting that in this type of molecule representation, there is no need to specify the connections between C- and H-atoms any further (Wauters and Marin 2001).

A.4. MOLECULE HOMOMORPHISM ALGORITHM

Molecule homomorphism is a non-ambiguous and effective statement whether two molecule descriptions refer to the same product or not. The importance of, or rather, the necessity for such a statement is obvious. In a certain sense, it is the key statement on which the Automatic Generator is based. Inability to recognize identical structure may cause unnecessary calculations at the best of times but more commonly will lead to infinite loops of calculations.

The molecule homomorphism is based on a "branch and chain" algorithm. A branch is a linear portion (including rings), which is connected to the molecule

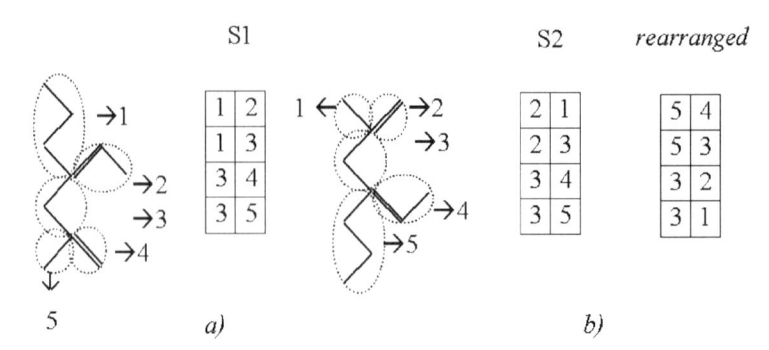

FIG. A6. (a) Molecule divided into branches and substitution matrix; (b) initial substitution matrix S2 of an equal molecule, and its content after rearrangement.

chain at its ends. A molecule is then divided into branches (see Fig. A6a). Due to its linearity, it is assumed that the branch homomorphism, i.e. the statement about two branches are identical, is a trivial problem. In fact, the two branches are checked for their length and position of possible unsaturated bonds. The molecule is represented as a graph whose nodes are the branches. The molecule homomorphism is then transformed into a graph homomorphism. Despite many algorithms present in literature, the one used by Steward (1965) for the definition of an admissible output set of an algebraic system of equations may be slightly modified to solve the problem at hand. For low interconnected graphs, substitution matrixes are an efficient method of representation (Fig. A6a). The homomorphism of two graphs analyses their representation with the matrixes S1 and S2 which start at the first row and column with a branch-edge, i.e. one of the two extreme points of a branch which is not connected to any other branch. While S1 description remains fixed, S2 is rearranged from top to bottom with the aim of creating a chain equal to S2 (Fig. A6b). The rearrangement involves an exhaustive analysis of all the possible chains obtained by connecting each branch of S2 with admissible branches, i.e. those which are equal to the corresponding one in S1.

A.5. SCHEMATIC FUNCTIONALITY OF THE KINETIC GENERATOR

The kinetic generator's functionality is based on three elements with three different functions (see Fig. A7).

– A pre-processor which prepares the kinetic data and the initial radical mixture according to specified user requirements (Initiation reactions, addition, ... etc.)
– The main processor that analyses each single radical of the mixture with the aim of calculating the composition of the end products. In the case of a

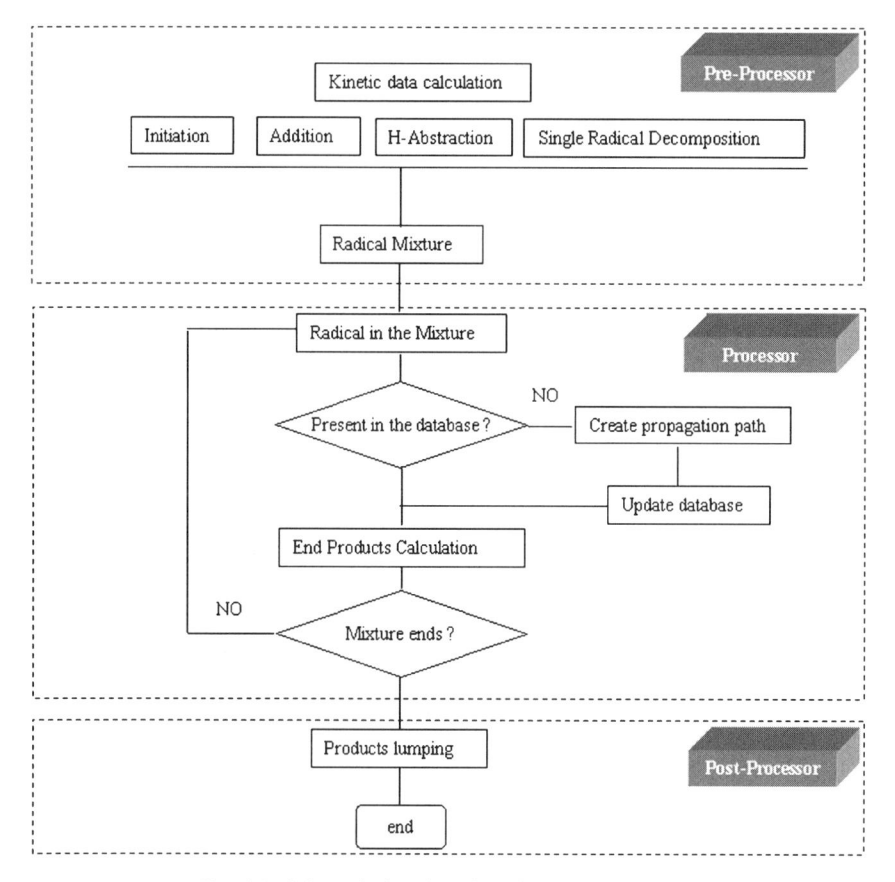

FIG. A7. Schematic functionality of the kinetic generator.

radical not present in the database, the propagation path of the radical is created for the portion not present in the database which is then updated with these new data.
– Lastly, a post-processor lumps the end products according with pre-assigned rules specified by the user.

REFERENCES

Aalund, L. *Oil Gas J.* **29**(March), 98–122 (1976a).
Aalund, L. *Oil Gas J.* **26**(April), 112–126 (1976b).
Abramowitz, M., and Stegun, I. A., "Handbook of Mathematical Functions with Formulas, Graphs, and Mathematical Tables". Dover Publications, New York, Chapter 26 (1968).
Albright, L. F. *Ind. Eng. Chem. Res.* **41**, 6210–6212 (2002).

Albright, L. F., and Marek, J. C. *Ind. Eng. Chem. Res.* **27**, 751–755 (1988).
Albright, L. F., and Marek, J. C. *Ind. Eng. Chem. Res.* **31**, 14–20 (1992).
Albright, L. F., McConnell, C. F., and Welther, K., *in* "Thermal Hydrocarbon Chemistry ACS 183" (A. G. Oblad, H. G. Davis, and R. T. Eddinger, Eds.), Chap. 10, pp. 175–191 (1979).
Alfè, M., Apicella, B., Barbella, R., Ciajolo, A., Kennedy, L., Merchan-Merchan, W., Rouzaud, J. N., and Tregrossi, A. Proceedings of Italian Combustion Meeting, Napoli III-7, (2005).
Ali, M. F., Hasan, M. U., Bukharl, A. M., Saleem, M. *Hydrocarb. Process.* February, 83–86 (1985).
Altgelt, K. H., and Boduszynski, M. M., "Composition and Analysis of Heavy Petroleum Fractions". Marcel Dekker, Inc, New York (1994).
Anderson, D. A., and Freeman, E. S. *J. Polymer Sci.* **54**, 253 (1961).
Appel, J., Bockhorn, H., and Frenklach, M. *Combust. Flame* **21**, 122–136 (2000).
Bach, G., Zimmermann, G., Kopinke, F. D., Barendregt, D., Oosterkamp, P. V. D., and Woerde, H. *Ind. Eng. Chem. Res.* **34**, 1132 (1995).
Balthasar, M., and Frenklach, M. *Combust. Flame* **140**, 130–145 (2005).
Barendregt, S., Dente, M., Ranzi, E., and Duim, F. *Oil Gas J.* **6**(April) (1981).
Battin-Leclerc, F., Fournet, R., Glaude, P. A., Judenherc, B., Warth, V., Come, G. M., and Scacchi, G. P. *Combust. Inst.* **28**, 1597–1605 (2000).
Baulch, D. L., Cobos, C. J., Cox, R. A., Esser, C., Frank, P., Just, T., Kerr, J. A., Pilling, M. J., Troe, J., Walker, R. W., and Warnatz, J. *J. Phys. Chem. Ref. Data* **21**, 411–429 (1992).
Baulch, D. L., Cobos, C. J., Cox, R. A., Frank, P., Hayman, G., Just, T., Kerr, J. A., Murrells, T., Pilling, M. J., Troe, J., Walker, R. W., and Warnatz, J. *J. Phys. Chem. Ref. Data* **23**, 847–1033 (1994).
Benson, S. W., "The Foundations of Chemical Kinetics". McGraw-Hill, New York (1960).
Benson, S. W., "Thermochemical Kinetics". 2nd ed. Wiley, New York (1976).
Blaikley, D. C. W., and Jorgensen, N. *Catal. Today* **7**, 227 (1990).
Blurock, E. S. *J. Chem. Inf. Comput. Sci.* **44**, 1336–1347 (2004).
Bockhorn, H., "Soot Formation in Combustion". Springer-Verlag, Berlin (1994).
Bockhorn, H., Hornung, A., and Hornung, U. *J. Anal. Appl. Pyrolysis* **50**, 77–101 (1999a).
Bockhorn, H., Hornung, A., Hornung, U., and Jakobstroer, P. *J. Anal. Appl. Pyrolysis* **49**, 53–74 (1999b).
Borchardt, H. J., and Daniels, F. *J. Am. Chem. Soc.* **79**, 41 (1957).
Bowman, C. T. *Combust. Flame* **25**, 343–354 (1975).
Bozzano, G., and Dente, M. ESCAPE 15-Barcelona (2005).
Bozzano, G., Dente, M., and Carlucci, F. *Comput. Chem. Eng.* **29**, 1439–1446 (2005).
Bozzano, G., Dente, M., Faravelli, T., and Ranzi, E. *Appl. Therm. Eng.* **22**(8), 919–927 (2002).
Bozzano, G., Dente, M., and Pirovano, C. *AIDIC Conference Series* **1**, 173–180 (1995).
Bozzano, G., Dente, M., Sugaya, M., and Mcgreavy, C. "ACS-National Meeting—Boston" 43, p. 653 (1998).
Bozzelli, J. W., and Dean, A. M. *J. Phys. Chem.* **94**, 3313–3317 (1990).
Braun-Unkhoff, M., Frank, P., and Just, T. *Ber. Bunsen. Phys. Chem.* **94**, 1417–1425 (1990).
Broadbelt, L. J., Stark, S. M., and Klein, M. T. *Ind. Eng. Chem. Res.* **33**, 790–799 (1994).
Broadbelt, L. J., Stark, S. M., and Klein, M. T. *Ind. Eng. Chem. Res.* **34**, 2566–2573 (1995).
Broadbelt, L. J., Stark, S. M., and Klein, M. T. *Comput. Chem. Eng.* **20**, 113–129 (1996).
Brown, T. C., and King, K. D. *Int. J. Chem. Kinet.* **21**, 251 (1989).
Cai, H., Krzywicki, A., and Oballa, M. C. *Chem. Eng. Proc.* **41**, 199 (2002).
Calcote, H. C. *Combust. Flame* **42**, 215 (1981).
Chan, K. Y. G., Inal, F., and Senkan, S. *Ind. Eng. Chem. Res.* **37**, 901 (1998).
Chang, P., Oka, M., and Hsia, Y. P. *Energy Sources* **6**, 67–83 (1982).
Chevalier, C., Warnatz, J., and Melenk, H. Proc. XII Task Leaders Meeting IEA (1988).
Clymans, P. J., and Froment, G. F. *Comput. Chem. Eng.* **8**, 137–142 (1984).
Coats, A. W., and Redfern, J. P. *Nature* **201**, 68 (1964).

CONCAWE, "Gas oils (diesel fuels/heating oils)", Report 95–107 (1995).
Curran, H. J., Gaffuri, P., Pitz, W. J., and Westbrook, C. K. *Combust. Flame* **114**, 149–177 (1998).
D'Anna, A., D'Alessio, A., and Kent, J. H. *Combust. Flame* **125**, 1196–1206 (2001a).
D'Anna, A., Violi, A., D'Alessio, A., and Sarofim, A. F. *Combust. Flame* **127**, 1995–2003 (2001b).
de Freitas Sugaya, M., "Analysis of stresses in reactors for pyrolysis of petroleum residues", Ph.D. Thesis, Leeds University, Department of Chemical Engineering (1999).
De Witt, M. J., Dooling, D. J., and Broadbelt, L. J. *Ind. Eng. Chem. Res.* **39**, 2228–2237 (2000).
Dean, A. M. *J. Phys. Chem.* **89**, 4600 (1985).
Del Bianco, A., Panariti, N., Anelli, M., Beltrame, P. L., and Carniti, P. *Fuel* **72**, 75–80 (1993).
Dente, M., Bozzano, G., and Bussani, G. *Comput. Chem. Eng.* **21**, 1125–1234 (1997).
Dente, M., Faravelli, T., Pierucci, S., Ranzi, E., Barendregt, S., and Valkenburg, P. AIChE Spring Meeting—Atlanta (2005).
Dente, M., Pierucci, S., Ranzi, E., and Bussani, G. *Chem. Eng. Sci.* **47**, 2629 (1992).
Dente, M., Ranzi, E., *in* "Pyrolysis: Theory and Industrial Practice" (L. F. Albright, B. L. Crines, and W. H. Corcoran Eds.), Academic Press, San Diego (1983).
Dente, M., Ranzi, E., Antolini, G., and Losco, F. 97th event of the EFCE, Florence (1970).
Dente, M., Ranzi, E., Barendregt, S., and Tsai, F. W. AIChE Meeting—Houston (1983).
Dente, M., Ranzi, E., and Goossens, A. G. *Comput. Chem. Eng.* **3**, 61 (1979).
Ebert, K. H., Ederer, H. J., Shroder, U. K. O., and Hamielec, A. W. *Makromol. Chem.* **183**, 1207 (1982).
Edwards, T., and Maurice, L. Q. *J. Propul. Power* **17**, 461–466 (2001).
Ellis, C., Scott, M. S., and Walker, R. W. *Combust. Flame* **132**, 291–304 (2003).
Evans, B., Dupon, E. J. P., Overwater, J. A. S., and Laghate, A. S. TECHNIP-COFLEXIP Ethylene Conference (2002).
Faravelli, T., Bozzano, G., Colombo, M., Ranzi, E., and Dente, M. *J. Anal. Appl. Pyrolysis* **70**, 761–777 (2003).
Faravelli, T., Bozzano, G., Scassa, C., Perego, M., Fabini, S., Ranzi, E., and Dente, M. *J. Anal. Appl. Pyrolysis* **52**, 87–103 (1999).
Faravelli, T., Goldaniga, A., and Ranzi, E. *P. Combust. Inst.* **27**, 1489–1495 (1998).
Faravelli, T., Pinciroli, M., Pisano, F., Bozzano, G., Dente, M., and Ranzi, E. *J. Anal. Appl. Pyrolysis* **60**, 103–121 (2001).
Filho, R. M., and Sugaya, M. *Comput. Chem. Eng.* **25**, 683–692 (2001).
Flynn, J. H., and Wall, L. A. *J. Polym. Sci. Pol. Lett.* **4**(5), 323 (1966).
Fournet, R., Battin-Leclerc, F., Glaude, P. A., Judenherc, B., Warth, V., Côme, G. M., Scacchi, G., Ristori, A., Pengloan, G., Dagaut, P., and Cathonnet, M. *Int. J. Chem. Kinet.* **33**, 574–586 (2001).
Frenklach, M. *Chem. Eng. Sci.* **40**, 1843 (1985).
Frenklach, M. *Phys. Chem. Chem. Phys.* **4**, 2028–2037 (2002).
Frenklach, M., Clary, D. W., Gardiner, W. C., and Stein, S. E. *P. Combust. Inst.* **20**, 887–901 (1984).
Frenklach, M., and Wang, H. *P. Combust. Inst.* **23**, 1559–1566 (1990).
Frenklach, M., and Wang, H., "Soot Formation in Combustion". Springer-Verlag, Berlin (1994).
Friedman, H. L. *J. Polym. Sci.* **6**, 183 (1963).
Frisch, S., Hippler, H., and Troe, J. *Zeitschrift fur Physikalische Chemie* **188**, 259–273 (1995).
Froment, G. F. *Rev. Chem. Eng.* **6**, 293 (1990).
Froment, G. F. *Chem. Eng. Sci.* **47**, 2163–2174 (1992).
Gambiroza-Jukic, M., and Cunko, R. *Acta Polym.* **43**, 258–260 (1992).
Gear, W., "Numerical Initial Value Problems in ODEs". Prentice-Hall, New Jersey (1968).
Gelbard, F., Tambour, Y., and Seinfeld, J. H. *J. Colloid Interf. Sci.* **76**, 541–556 (1980).
Glasman, I., "Combustion". Academic Press, San Diego (1996).
Goossens, A. G., Dente, M., and Ranzi, E. *Oil Gas J.*, Sept. 4, 89–94 (1978).
Granata, S., Cambianica, F., Zinesi, S., Faravelli, T., and Ranzi, E. ECM—Proceedings of European Combustion Meeting, Louvain-la-Neuve (2005).

Green, W. H., Barton, P. I., Bhattacharjee, B., Matheu, D. M., Schwer, D. A., Song, J., Sumathi, R., Carstensen, H.-H., Dean, A. M., and Grenda, J. M. *Ind. Eng. Chem. Res.* **40**, 5362–5370 (2001).

Grenda, J. M., Androulakis, I. P., Dean, A. M., and Green, W. H. *Ind. Eng. Chem. Res.* **42**, 1000–1010 (2003).

Grenda, J. M., Dean, A. M., Green, W. H. J., and Peczak, P. K. Work-in-Progress Poster (WIPP) Session, at the 27th International Symposium on Combustion, Boulder, CO, Aug 2–7 (1998).

Haynes, B. S., and Wagner, H. G. *Prog. Energ. Combust. Sci.* **7**, 229 (1981).

Helveg, S., and Hansen, P. L. *Catal. Today* **111**(1–2), 68–73 (2006).

Helveg, S., Lopez-Cartes, C., Sehested, J., Hansen, P. L., Clausen, B. S., Rostrup-Nielsen, J. R., Abild-Pedersen, F., and Norskov, J. K. *Nature* **427**(6973), 426–429 (2004).

Heynderickx, G. J., Cornelis, G. G., and Froment, G. F. *AIChE J.* **38**, 1905–1912 (1992).

Heynderickx, G. J., Oprins, A. J. M., Marin, G. B., and Dick, E. *AIChE J.* **47**, 388–400 (2001).

Hillewaert, L. P., Dierickx, J. L., and Froment, G. *AIChE J.* **34**, 17–24 (1988).

Homann, K. H., and Wagner, H. G. *P. Combust. Inst.* **11**, 371–379 (1967).

Hudgens, J. W. Workshop on combustion simulation databases for real transportation fuels NIST Gaithersburg, MD (2003).

Ikeda, E., Tranter, R. S., Kiefer, J. H., Kern, R. D., Singh, H. J., and Zhang, Q. *P. Combust. Inst.* **28**, 1725–1732 (2000).

Kee, R. J., and Rupley, F. M. Sandia Report SAND87-8215 (1990).

Knumann, R., and Bockhorn, H. *Combust. Sci. Technol.* **101**, 285–299 (1994).

Kopinke, F. D., Zimmermann, G., and Nowak, S. *Carbon* **26**, 117 (1988).

Krambeck, F. *J. Chemtech* **22**(5), 292–299 (1992).

Krestinin, A. V. *Combust. Flame* **121**, 513–524 (2000).

Kruse, T. M., Levine, S. E., Wong, H. W., Duoss, E., Lebovitz, A. H., Torkelson, J. M., and Broadbelt, L. J. *J. Anal. Appl. Pyrolysis* **73**, 342–354 (2005).

Kruse, T. M., Wong, H. W., and Broadbelt, L. J. *Macromolecules* **36**, 9594–9607 (2003).

Kruse, T. M., Woo, O. S., Wong, H. W., Khan, S. S., and Broadbelt, L. J. *Macromolecules* **35**, 7830–7844 (2002).

Lattimer, R. P., and Kroenke, W. J. *J. Appl. Polym. Sci.* **27**, 1355–1366 (1982).

Liguras, D. K., and Allen, D. T. *Ind. Eng. Chem. Res.* **28**, 665–673 (1989).

Liguras, D. K., and Allen, D. T. *Ind. Chem. Res.* **31**, 45–53 (1992).

Lobo, D. S., and Trimm, D. L. *J. Catal.* **29**, 15 (1973).

Lohr, B., and Dittman, H. *Oil Gas J.* July 18, 73–77 (1978).

Madorsky, S. L. *J. Polym. Sci.* **2**, 133 (1952).

Madras, G., and McCoy, B. J. *J. Colloid Interf. Sci.* **246**, 356–365 (2002).

Madras, G., Smith, J. M., and Mccoy, B. J. *Polym. Degrad. Stabil.* **52**, 349–358 (1996).

Magnan, J., Otsubo, K., Matsueda, S., and Fujiwara, M. ERTC Petrochemical Conference, Vienna (2004).

Marongiu, A. "Thermal Degradation of Polymers", Ph.D. Thesis, Politecnico di Milano, Department of Chemical Engineering (2006).

Marongiu, A., Faravelli, T., Bozzano, G., Dente, M., and Ranzi, E. *J. Anal. Appl. Pyrolysis* **70**, 519–553 (2003).

Matheu, D. M., Green, W. H., and Grenda, J. M. *Int. J. Chem. Kinet.* **35**, 95–119 (2003).

McCoy, B. J., and Wang, M. *Chem. Eng. Sci.* **49**, 3773–3785 (1994).

McGivervy, W. S., Manion, J. A., and Tsang, W. WIP 30th International Symposium on Combustion, Chicago, IL, USA July Paper 1F1-02, (2004).

Mehl, M., Marongiu, A., Faravelli, T., Bozzano, G., Dente, M., and Ranzi, E. *J. Anal. Appl. Pyrolysis* **72**, 253–272 (2004).

Merchan-Merchan, W., Saveliev, A. V., and Kennedy, L. A. *Combust. Sci. Technol.* **175**, 2217 (2003).

Migliavacca, G., Parodi, E., Bonfanti, L., Faravelli, T., Pierucci, S., and Ranzi, E. *Energy* **30**, 1453–1468 (2005).
Miller, J. A. *Faraday Discuss.* **119**, 461–475 (2001).
Miller, J. A., and Melius, C. F. *Combust. Flame* **91**, 21–39 (1992).
Miller, J. A., Pilling, M. J., and Troe, J. *P. Combust. Inst.* **30**, 43–88 (2005).
Miranda, R., Pakdel, H., Roy, C., and Vasile, C. *Polym. Degrad. Stabil.* **73**, 47–67 (2001).
Montaudo, G., and Puglisi, C. *Polym. Degrad. Stabil.* **33**, 229–262 (1991).
Moskaleva, L. V., Mebel, A. M., and Lin, M. C. *P. Combust. Inst.* **26**, 521–526 (1996).
Muller, C., Michel, V., Scacchi, G., and Côme, G. M. *J. Chim. Phys.* **92**, 1154 (1995).
Oberdorster, G., Sharp, Z., Atudorei, V., Elder, A., Gelein, R., Kreyling, W., and Cox, C. *Inhal. Toxicol.* **16**, 437–445 (2004).
Oka, M., Chang, P., and Gavales, G. R. *Fuel* **56**, 3–8 (1977).
Palmer, H. B., and Cullis, H. F., "The Chemistry and Physics of Carbon". Dekker, New York (1965).
Pierucci, S., Ranzi, E., and Barendregt, S. ESCAPE 15, Barcelona (2005).
Poutsma, M. L. *J. Anal. Appl. Pyrolysis* **54**, 5–35 (2000).
Poutsma, M. L. *Macromolecules* **36**, 8931–8957 (2003).
Quann, R. J., and Jaffe, S. B. *Ind. Eng. Chem. Res.* **31**, 2483–2497 (1992).
Quann, R. J., and Jaffe, S. B. *Chem. Eng. Sci.* **51**, 1615–1635 (1996).
Ranzi, E. *Energy Fuels* **20**, 1024–1032 (2006).
Ranzi, E., Dente, M., Faravelli, T., Bozzano, G., Fabini, S., Nava, R., Cozzani, V., and Tognotti, L. *J. Anal. Appl. Pyrolysis* **40/41**, 305–319 (1997a).
Ranzi, E., Dente, M., Faravelli, T., and Pennati, G. *Combust. Sci. Technol.* **95**, 1–50 (1994).
Ranzi, E., Dente, M., Goldaniga, A., Bozzano, G., and Faravelli, T. *Prog. Energ. Combust.* **27**, 99–139 (2001).
Ranzi, E., Dente, M., Pierucci, S., Barendregt, S., Cronin, P. *Oil Gas J.* Sept. 2, 49–52 (1985).
Ranzi, E., Faravelli, T., Gaffuri, P., Garavaglia, E., and Goldaniga, A. *Ind. Eng. Chem. Res.* **36**, 3336–3344 (1997b).
Ranzi, E., Faravelli, T., Gaffuri, P., and Sogaro, A. *Combust. Flame* **102**, 179–192 (1995).
Ranzi, E., Frassoldati, A., Granata, S., and Faravelli, T. *Ind. Eng. Chem. Res.* **44**, 5170–5183 (2005).
Ren, T., Patel, M., and Blok, K. *Energy* **31**, 425–451 (2006).
Rice, F. O., and Herzfeld, K. F. *J. Am. Chem. Soc.* **55**, 3035–3040 (1933).
Rice, F. O., and Herzfeld, K. F. *J. Am. Chem. Soc.* **56**, 284–289 (1934).
Richter, H., Granata, S., Green, W. H., and Howard, J. B. *P. Combust. Inst.* **30**, 1397–1405 (2005).
Ritter, E. R., and Bozelli, J. W. *Int. J. Chem. Kinet.* **23**, 767–778 (1991).
Robaugh, D., and Tsang, W. *J. Phys. Chem.* **90**, 4159 (1986).
Roth, W. R., Bauer, F., Beitat, A., Ebbrecht, T., and Wustefeld, M. *Chem. Ber.* **124**, 1453–1460 (1991).
Saeys, M., Reyniers, M. F., Marin, G. B., Van Speybroeck, V., and Waroquier, M. *J. Phys. Chem. A* **107**(43), 9147–9159 (2003).
Saeys, M., Reyniers, M. F., Marin, G. B., Van Speybroeck, V., and Waroquier, M. *AIChE J* **50**(2), 426–444 (2004).
Saeys, M., Reyniers, M. F., Van Speybroeck, V., Waroquier, M., and Marin, G. B. *ChemPhysChem* **7**(1), 188–199 (2006).
Sarofim, A. F., and Longwell, J. P., "Soot Formation in Combustion". Springer-Verlag, Berlin (1994).
Savage, P. E. *J. Anal. Appl. Pyrolysis* **54**, 109–126 (2000).
Seery, D. J., and Bowman, C. T. *Combust. Flame* **14**, 37–48 (1970).
Smith, H. M., "Qualitative and Quantitative Aspects of Crude Oil Composition". Bureau of Mines, Bartesville (1968).
Smoot, L. D., Hecker, W. C., and Williams, G. A. *Combust. Flame* **26**, 323–342 (1976).

Steward, D. V. *Soc. Ind. Appl. Math. J.* **2B**, 345 (1965).
Sundaram, K. M., and Froment, G. F. *Ind. Eng. Chem.* **17**, 174 (1978).
Susnow, R. G., Dean, A. M., Green, W. H., Peczak, P., and Broadbelt, L. J. *J. Phys. Chem. A* **101**, 3731–3740 (1997).
Touchard, S., Fournet, R., Glaude, P. A., Warth, V., Battin-Leclerc, F., Vanhove, G., Ribacour, M., and Minetti, R. *P. Combust. Inst.* **30** (2004).
Tsang, W. *Int. J. Chem. Kinet.* **10**, 1119 (1978a).
Tsang, W. *Int. J. Chem. Kinet.* **10**, 599 (1978b).
Tsang, W. *Int. J. Chem. Kinet.* **10**, 821 (1978c).
Tsang, W. *J. Phys. Chem. Ref. Data* **17**, 887 (1988).
Tsang, W. *Ind. Eng. Chem.* **31**, 3–8 (1992).
Tsang, W. *Data Sci. J.* **3**, 1–9 (2004).
Van Geem, K. M., Heynderickx, G. J., and Marin, G. B. *AIChE J.* **50**, 173–183 (2004).
Van Krevelen, D. W., "COAL, Typology, Chemistry-Physics-Constitution". Elsevier, Amsterdam (1961).
Van Krevelen, D. W., "Properties of Polymers, their Estimation and Correlation with Chemical Structure". Elsevier, Amsterdam (1976).
Van Nes, K., and Van Westen, H. A., "Aspects of the Constitution of Mineral Oils". Elsevier Publ. Company, New York (1951).
Vlasov, P. A., and Warnatz, J. *P. Combust. Inst.* **29**, 2335–2341 (2002).
Warnatz, J., Rate coefficients in the C/H/O system, *in* "Combustion Chemistry" (W.C. Gardiner Ed.), Springer-Verlag, New York (1984).
Warth, V., Battin-Leclerc, F., Fournet, R., Glaude, P. A., Côme, G. M., and Scacchi, G. *Comput. Chem.* **24**, 541–560 (2000).
Watson, K. M., Nelson, E. F., and Murphy, G. B. *Ind. Eng. Chem.* **27**(12), 1460–1466 (1935).
Wauters, S., and Marin, G. B. *Chem. Eng. J.* **82**(1–3), 267–279 (2001).
Wauters, S., and Marin, G. B. *Ind. Eng. Chem. Res.* **41**, 2379 (2002).
Weissman, M., and Benson, S. W. *Int. J. Chem. Kinet.* **16**, 307 (1984).
Westbrook, C. K., and Dryer, F. C. *Prog. Energ. Combust. Sci.* **10**, 1–57 (1984).
Westbrook, C. K., Mizobuchi, Y., Poinsot, T. J., Smith, P. J., and Warnatz, J. *P. Combust. Inst.* **30**, 125–157 (2004).
Westerhout, R. W. J., Waanders, J., Kuipers, J. A. M., and Van Swaaij, W. P. M. *Ind. Eng. Chem. Res.* **36**, 1955–1964 (1997).
Westmoreland, P. R., Dean, A. M., Howard, J. B., and Longwell, J. P. *J. Phys. Chem.* **93**, 8171–8180 (1989).
Wulkow, M. *Macromol. Theor. Simul.* **5**, 393 (1996).
Yoneda, Y. *Bull. Chem. Soc. Jpn.* **52**, 8–14 (1979).
Zhao, B., Yang, Z., Johnston, M. V., Wang, H., Wexler, A. S., Balthasar, M., and Kraft, M. *Combust. Flame* **133**, 173–188 (2003).
Zimmerman, G., Dente, M., and Van Leeuwen, C. AIChE Meeting, Orlando (1990).
Zou, R., Lou, Q., Liu, H., and Niu, F. *Ind. Eng. Chem. Res.* **26**, 2528 (1987).
Zychlinski, W., Wynns, K. A., and Ganser, B. *Mater. Corros.* **53**, 30–36 (2002).

KINETIC MODELS OF C_1–C_4 ALKANE OXIDATION AS APPLIED TO PROCESSING OF HYDROCARBON GASES: PRINCIPLES, APPROACHES AND DEVELOPMENTS

Mikhail Sinev[1],*, Vladimir Arutyunov[2] and Andrey Romanets[3]

[1]Laboratory of Heterogeneous Catalysis, Semenov Institute of Chemical Physics, Moscow, Russia
[2]Laboratory of Hydrocarbon Oxidation, Semenov Institute of Chemical Physics, Moscow, Russia
[3]Russian Research Center Kurchatov Institute, Moscow, Russia

I. Introduction: Problem Statement	169
A. Modeling of Complex Reacting Systems: Purposes and Expectations	172
B. Circumscription of Subject and Area of Parameters	176
C. Accounting of Heterogeneous Processes	179
D. Accuracy of Modeling and Influence of Macro-Kinetic Parameters	183
II. Alternative Approaches to Modeling	187
A. "Additive" Models	189
B. "Combinatorial" Models	193
C. Ruling Principles for Comprehensive Modeling	194
D. Reduction of Models	200
E. Modeling of Heterogeneous–Homogeneous Catalytic Reactions	201
III. "Elemental Base"	203
A. Limitations of Species Taken into Account	203
B. Elementary Reactions	206
C. Selection of Rate Constants	210
D. Pressure-Dependent Reactions	210
E. Heterogeneous–Homogeneous Catalytic Reactions on Oxide Catalysts	213
F. Heterogeneous–Homogeneous Catalytic Reactions on Metal Catalysts	227
IV. Modeling and Experimentation	231
A. Comparison with Experiment	231
B. Requirements of Experiment	233
V. Some Examples and Comments	237
A. C_1–C_2 Joint Description	237
B. Expansion on Higher Hydrocarbons	239

*Corresponding author. E-mail: sinev@center.chph.ras.ru

C. Reactions between Alkyl Radicals and Oxygen and
     Transformations of Alkylperoxy Radicals 243
  D. Capabilities of Process Influencing, Governing and
     Design 246
VI. Concluding Remarks 250
  Acknowledgments 251
  List of Symbols 251
  References 253

Abstract

The topical problems existing in the modeling of oxidation processes in gas chemistry, particularly in oxidative transformations of light (C_1–C_4) alkanes, are outlined. It is demonstrated that the requirements to adequate description of process kinetics in this case differ significantly from those which are satisfactory in combustion kinetics and, on the other hand, in near-equilibrium petrochemical processes. The principles that could be utilized in developing consistent kinetic models are determined by the goal of modeling, which appears to be the most important factor controlling all the approaches and procedures. The models under discussion are aimed at (i) the rationalization of the reaction network responsible for the main features of the overall process, (ii) the description of light alkane oxidation, and (iii) optimization of the yield of some reactive intermediate product, such as olefins, saturated and non-saturated alcohols, aldehydes, acids, synthesis gas, etc. The latter defines the ranges of parameters (temperature, pressure, initial composition of reaction mixture) and the main ruling principles of modeling. It is concluded that comprehensive kinetic models should be based on the following principles: thermodynamic consistency, fullness, independence of kinetic parameters, openness of the description. The possibility of accurate description of phenomena under consideration is determined by a series of factors, including fullness of the model, adequate description of the reaction space (reactor itself and connecting tubing), correct accounting of heat- and mass-transfer processes, precision of kinetic parameters and even the form of their representation, etc. Another complicated problem is model verification. It is demonstrated that the comparison with experimental data is associated with serious difficulties and its applicability for the model verification is limited. Special attention is paid to the modeling of heterogeneous processes. Possible approaches to the description of elementary steps on oxide and metal surfaces and building of combined heterogeneous–homogeneous models are discussed. Some examples are given, mainly related to the joint description of C_1–C_2 hydrocarbon oxidation, and the ways of its expansion on higher hydrocarbons are traced.

I. Introduction: Problem Statement

The increasing efficiency of chemical transformations is an ultimate goal of applied scientific research. Developments in hydrocarbon processing present numerous examples of the interplay between practical industrial needs and the advances in kinetic and mechanistic studies. For instance, the "oil shocks" of the mid- and late 1970s shifted interest from relatively high molecular weight hydrocarbons and light olefins to light alkanes as initial compounds for many industrial processes. Nowadays, this interest is increasing along with petroleum price. One of the most clearly revealed tendencies in both expert judgments and public opinion is a progressing understanding that oil is a finite resource and global peak in its production is imminent (see, for instance, web resources http://www.asponews.org, http://www.eia.doe.gov). According to some estimates, the decline in the world oil production and supply will become irreversible during the current decade.

Unlike oil, natural gas and its main components—light (C_1–C_4) alkanes—are much more abundant. The existing estimates of their assured resources in traditional gas fields are much more optimistic as compared to oil resources. Moreover, the Earth's mantle permanently releases additional amounts of methane, and its resources in the form of gas hydrates are estimated as more than a half of the organic carbon in the Earth's crust (Kvenvolden, 1988). The development of these giant resources becomes the strategic goal of the world economy and accordingly one of the priority directions in scientific research—both fundamental and applied (Melvin, 1988).

Nevertheless, the world economy remains oil-based and the production of almost the whole range of petrochemical and bulk organic chemical products (with a few exceptions, like methanol, urea, and cyanic acid) is still oriented toward the processing of liquid fossil hydrocarbons as the principal raw materials. Also, the same liquid hydrocarbons dominate in the production of motor fuels and lubricants. However, most if not all of the products traditionally produced from oil can be also obtained from gaseous hydrocarbons, including light alkanes. Unfortunately, up to now their utilization in the chemical industry is not satisfying.

One of the possible reasons for that is our insufficient understanding of light alkane chemistry and, consequently, our limited capability to develop efficient technologies. First of all, petrochemistry dominates not only in industry, but also in prevailing scientific and technical approaches. However, if we analyze the production of the same final products (for instance, liquid motor fuels) from light alkanes and higher hydrocarbons (oil), a distinct difference in the involved processes becomes clear. In the case of higher hydrocarbon processing we are dealing mainly with equilibrium reactions (cracking, isomerization, hydration, dehydration).

In contrast, in the case of low-molecular weight alkane processing, the main types of transformations relate to the cleavage of strong C–H bonds and building up the carbon skeleton. Such processes proceed most efficiently in the non-equilibrium oxidative (or redox) mode, which shifts the main emphasis to their kinetic aspects. As a result, if equilibrium petrochemical processes require first of all the use of active and durable catalysts, in light alkane processing (gas chemistry) the problem of selectivity with respect to reactive intermediate products is brought to the forefront.

Secondly, the problem of selectivity has been thoroughly scrutinized as applied to catalytic chemistry of olefins, aromatic hydrocarbons, oxygenates, etc. However, in the case of light alkanes we are facing a very different set of problems originating from relatively low reactivity of the starting molecules as compared to the molecules of the desired products. As a result, the applicability of existing approaches for the development of selective catalysts becomes doubtful and makes the role of kinetic factors crucial.

At last (not least!), oxidative transformations of light alkanes (both homogeneous and catalytic) as a rule proceed as high-temperature complex multi-step processes. In non-equilibrium conditions during alkane (e.g., methane) oxidation the complexity of the system is headily rising, as numerous intermediates participate in multiple elementary reactions. In such conditions the process is poorly accessible for direct mechanistic investigations. Even pathways of both formation and consecutive transformations of particular target products become difficult to trace. However, although processes taking place on the "elementary" level are undoubtedly responsible for the kinetically attainable yields of target product(s), the most successful and systematic descriptions of such limitations are still accessible in terms of comparative overall reactivity of products and initial hydrocarbon (Batiot and Hodnett, 1996; Isagulyants et al., 2000).

In the conditions of insufficient understanding of chemistry and kinetics of light alkane transformations, the family of industrial processes based on methane conversion to synthesis gas (syngas—$CO + H_2$ mixture) is dominating. In its turn, syngas production is based on near-equilibrium reactions: steam and "dry" (with CO_2) reforming of methane, water-gas shift reaction and to some extent on oxidative reforming, which proceeds also under conditions close to equilibrium. Methane conversion to syngas is a high-temperature energy-consuming and capital-intensive process, but nevertheless it remains the basis of large-scale chemical processing of natural gas.

Although several other processes allowing bypass of the stage of syngas formation have been extensively investigated (see Arutyunov et al., 1995a; Arutyunov and Krylov, 1998 and literature cited therein), now they are mostly considered as not proven and marginal technologies. There are just a few examples of their implementation on a small pilot-plant scale. The same can be said about oxidative processing of C_2–C_4 hydrocarbons with the exception of butane oxidation to maleic anhydride and (to some extent) propane oxidation to acrylic acid.

Obviously, some breakthrough in this field is required. All factors mentioned above make modeling the most promising (if not the only) approach to studying the details of kinetic behavior of such complex reaction systems and also the most valuable tool for process governing and optimization. Indeed, the number of publications in this area is rapidly increasing. However, it should be noted that up to now there is no unified and generally appreciated or accepted methodology for the modeling of complex processes. Moreover, in some cases the main purpose, basic statements and underlying principles of modeling are not clearly formulated, as well as the main procedures are not described. This makes the subject of study quite vague and the obtained results difficult to analyze.

In this work, the authors have attempted to outline the scope of topical problems existing in the modeling of oxidation processes in gas chemistry. We consider each section in this work as a formulation of the corresponding problem, but not as its solution, and, accordingly, the entire paper as an invitation to a broad discussion on the designated subject. Besides, we believe that the principles formulated below could be utilized in developing consistent kinetic models of the "open" type. The latter means that such models can be further built on as new reliable kinetic information becomes available.

Let us conclude the Introduction by a formulation of what cannot be found in the paper.

1. We tried to avoid any detailed or complete literature review concerning the subject under discussion; this would inevitably lead to vast and diffuse discussions and criticism. Instead, we attempted to point out the main problems and difficulties, to trace paths to possible solutions and to mention the achievements reached to date.
2. For the same reason we resisted the temptation to announce our own model, or even some core, which could be further built on. Moreover, all examples presented below are based on models already published by different authors, even if particular simulations are published here for the first time. We consider the substantiation and validation of basic principles as the ultimate goal of this paper. The core of the model based on such principles will be published separately.
3. The authors know from their own experience how attractive and fascinating, but at the same time meaningless can be "computer games" with equations, software, and model modifications. Based on this, we have almost completely refrained from consideration of any mathematical aspects and computational procedures, which are indeed very important in modeling and kinetic simulations. However, these topics require separate consideration, but only when

the main questions of physical, chemical, and kinetic matters are answered, or at least raised and discussed.

A. Modeling of Complex Reacting Systems: Purposes and Expectations

The modeling of any complex object is not trivial because there are no well-defined algorithms and recipes to perform it, so the procedures of model development in the general case cannot be formalized. The principal question in modeling that determines the entire sense of the work is its goal. There are several possible answers. First of all, as soon as a set of experiments is accomplished, the investigator may wish to build a continuous description of the process. In spite of the particular mathematical form of this description, it can very precisely fit the experimental data obtained in the given range of parameters. However, the width of a broader range to which the prediction of such a description can be extrapolated without a drastic decline in accuracy is always undefined.

The simplest mathematical description may be represented in the form of a table or as a set of curves demonstrating the process behavior (conversion, concentration of certain products, distribution of temperatures in the reacting system, etc.) at varying external parameters (initial composition and temperature, total pressure, flow rate, etc.). Such dependencies also can be expressed in the form of algebraic equations and used, for instance, for process optimization, almost exclusively by interpolation within the range of parameters where the initial set of experiments was carried out.

Compilation of a kinetic scheme containing a limited number of chemical equations and corresponding differential equations is another possible form of simple models. The chemical part of it cannot be considered as a detailed mechanism, but serves only as a phenomenological description of the process. As to the system of differential equations, on certain assumptions it can be solved and reduced to an algebraic form (for instance, in steady-state approximation). Models of this type are widely presented in the scientific literature.

When a detailed chemical description is not required, a limited set of a few stoichiometric equations can be included into the scheme just to describe the rate of heat evolution and change of total number of gas species in the system. Chemically oversimplified models of this kind are widely used, for instance, to describe heat-transfer and to optimize thermal regimes in reactors (see, e.g., Fukuhara and Igarashi, 2005; Kolios et al., 2001). A similar approach is used to describe the fuel combustion and corresponding dynamic phenomena in engines of different types: simplified equations describing "kinetic" features are solved together with complex equations of heat- and mass-transfer and fluid dynamics (Frolov et al., 1997; Williams, 1997).

More advanced kinetic models are developed for the interpretation and qualitative description of experimentally observed complex phenomena. The classical examples of such models are the first kinetic schemes proposed to describe oxidation of phosphorus, hydrogen, methane, and some other substances; they served as a basis for the creation and development of chain theory (Semenov, 1959). Of course, at that stage any detail in the mechanism for most of those processes was not and even could not be revealed. However, in spite of the "intuitive" character of those kinetic schemes, they laid a cornerstone for the theory, which in its turn has become the real basis of contemporary chemical kinetics of extraordinary descriptive and predictive power.

Let us finally dwell on one more type of model, which in principle does not require any preliminary experimental study of the process. Models of this kind themselves can be considered as tools for investigation of complex processes on a very intimate level. They constitute a hypothesis about the detailed chemical mechanism of the process (its micro-chemistry) that includes types of species present in the reacting system and chemical interactions between them. Being filled with kinetic parameters and treated in accordance with proper mathematical procedures, this type of model is supposed to reproduce the behavior (or evolution in time) of the system, which can be compared with some experiment in order to see if there are any matches or common features.

The obvious goal in this case could be an attempt to describe all phenomena ever observed in experiment within the framework of some broad, but uniform, mechanism. If this goal is reached, the model can be used to predict the system behavior even in the range of parameters far beyond those accessible in any realistic experiment. Such a range can be limited only by the physical and chemical sense of the kinetic scheme content and kinetic parameters. For instance, if we do not consider elementary reactions of excited species (e.g., carrying vibrational and/or electronic excitation), the scheme cannot be extrapolated to relatively high temperatures at which the effect of these factors becomes substantial.

In principle, there are no limitations for changing the chemical content of the model, as well as the values of kinetic parameters. Even some species and reactions, as well as the values of the kinetic parameters, which undoubtedly cannot exist in nature, can be included in order to answer the question such as: "What would happen if ...".

Another very important feature of such models is their "openness": they can be built on and on to any direction from any given level. On the one hand, this in principle allows a very detailed elaboration of a given reaction. On the other hand, such "openness" is very helpful if more and more complicated systems must be analyzed. For instance, any action of additives (promoters, inhibitors, etc.) on hydrocarbon oxidation can be analyzed. On the other hand, a kinetic scheme describing the reaction of a certain compound can be spread on similar reactions of related compounds with gradually increasing complexity (e.g., from methane oxidation to analogous reactions of higher alkanes).

The main content of this paper will be devoted to the models mainly of this type. This is why here, in the very beginning, we would like to highlight some serious difficulties. First of all, we must stress the difference between so-called direct and inverse kinetic tasks. In the first case, the system evolution is simulated based on the kinetic scheme and accepted values of parameters. Such a task always has a solution; the only problem is to have a proper "calculator" (most frequently—specialized computer software).

As to an inverse kinetic task, it consists of the determination (or correction) of the kinetic scheme and kinetic parameters based on repeated solution of the direct task and comparison with experimental data. In order not to run ahead, we would just mention here that in general an inverse kinetic task has no definite solution. The uncertainty increases with increasing complexity of the system for a very simple reason: if the system under consideration is complex enough, there are always several combinations of parameters giving satisfactory agreement of the simulated behavior with that observed experimentally. Another fundamental limitation is an uncertainty (qualitative and quantitative), which is always present in the experimental data. Some other aspects of the comparison between simulated and experimental data will be discussed below. However, even at this initial level of consideration we can state that the development of a model to be used as a tool for studying the system of certain (high enough) complexity requires a formulation of some ruling principles. In their turn, these principles are determined by the goal of modeling, which appears to be the most important factor controlling all other approaches and procedures.

Another principal statement can be formulated as follows: the model of any complex process is a compromise between the precision and thoroughness of the description and its applicability and comprehensibility. Any extreme is irrational. The model can be so bulky that any attempt to analyze it would become useless, or it may be so incomplete that substantial intrinsic structure and connections would distort or even disappear from the description.

The number of computer codes and software packages for kinetic simulations and published kinetic models keeps increasing. This creates an illusion of "simplicity" (as compared to "complexity" of modern experimental techniques) of these exercises and makes "kinetic modeling" very popular. As the main positive result of such "modeling" the good agreement between model predictions and some experimental results, selected by authors, is usually considered. At the same time the attention paid both to the structure of the model and to the thorough analysis of its consequences is not sufficient. On the other hand, the preliminary analysis of experimental results chosen for the comparison with simulated data is also required. We have to be sure that these results are definite, comprehensive, and reliable enough and also that the conditions in the experimental reactor are adequately reproduced during the simulations.

There is one more aspect to which we have to pay special attention. The newly developed model can be a "mechanical" combination (a "sum" or "enclosure") of models previously used for other purposes. In that case we are facing a risk to put together some principally incompatible matters—to combine the elements or subsystems from different levels based on controversial principles. In Sections III.B–III.F we will demonstrate several examples of how it is important to analyze the applicability of existing models if we wish to use them for solving new problems.

Besides the precise definition of the ultimate goal and the main purposes, an adequate phenomenological description of the system under consideration is required. This should include its qualitative composition, phase (aggregative) state, ranges of parameters, first of all temperature, pressure, concentrations of components, type of reactor, heat- and mass-transfer conditions, flow regimes, etc. In other words, we need to have a clear conceptual vision of the reacting system and corresponding understanding of the main factors determining its behavior as far as the proclaimed goal of modeling is concerned.

Now let us turn to one of the key questions: what goals and purposes are worthwhile to pose for the modeling of light alkane oxidation? As we already mentioned, these reacting systems are so complex that the experimental study of their detailed mechanism is practically impossible. But modeling is capable of providing such a possibility. On the other hand, because of the great practical importance of these processes, we also need a tool for their optimization, as far as the efficiency of certain product formation is concerned. These two aspects—mechanistic studies and process optimization—can be considered as the main purposes for developing adequate kinetic models of light alkane oxidation.

Now let us look at the same problem at an angle of realistic expectations. We have to confess that at the current stage of understanding micro- and macro-kinetic aspects of alkane oxidation it is practically impossible to anticipate a precise quantitative description and prediction of the system behavior, product yields, etc. This is not only due to our limited ability to build an adequate model, but also because of a limited accuracy and significant uncertainty in the parameters of any real experiment (see Sections IV.A and IV.B for a detailed analysis).

If we consider a better understanding as the main achievement of modeling, we should not fight for particular numbers or precise fit to certain experimental curve(s). The same model can describe very well one set of experimental data and be in a serious contradiction with the others. Also, models describing certain experimental curves well can be self-contradictory. Unfortunately, the examples of such "good" correlation between experiment and simulations are numerous.

A more realistic and useful result of modeling would be a qualitative description and interpretation of phenomena, which are of principal importance for the behavior of the system. Therefore, the predictive power of the model is the main criterion of its usefulness. In other words, the model should be considered as adequate if it provides a description and prediction of qualitative effects and phenomena that have been and/or can be experimentally observed. From this point of view, the development of the most efficient criteria for the comparison of simulated and experimental data is essential.

To summarize this section, we should state that a model cannot (and should not) be identical to the phenomenon or process (object) under consideration.

Even more so the object is not identical to its model. Consequently, we should not expect that the behavior of the real object is going to be necessarily equal to the model predictions and will be quantitatively described by the model. In many theoretically and practically important cases even a qualitative description and/or prediction of complex phenomena observed experimentally should be considered as a success of modeling and a serious breakthrough in understanding of the nature of the process under study.

Below we will focus mainly on the problems of compiling micro-chemical (micro-kinetic) schemes of alkane oxidation and populating it with kinetic parameters. Of course, we had to pay certain attention to some closely connected questions but it was impossible to analyze all of them in detail in the framework of one paper.

B. Circumscription of Subject and Area of Parameters

Oxidation of hydrocarbons is one of the oldest and traditional subjects for chemical kinetics. During several decades of widespread application of computers in chemistry, a vast number of publications concerning the modeling of these processes have appeared. As we noted above, the complexity of the chemical system designated as "hydrocarbon oxidation", even in the simplest case of light alkanes, practically excludes the possibility of the development of the sole and universal model suitable for all practically important applications. All models that have been extensively used have been developed for some particular processes over limited ranges of parameters, although by no means it is always clearly stated. Therefore, the first and the most important condition, which should be strongly executed when any kinetic model is announced, is the statement of the problem and the circumscription of the parameter space for which the model is to be developed.

Among the external parameters, temperature is the one that to the utmost determines the peculiar features of the oxidation process, to some extent even more than the number of carbon atoms in the alkane molecule. This is why all kinetic models for gas-phase alkane oxidation can be classified based on the temperature range in which the process is going to be described. In their turn, the temperature ranges could be delimited based on the main types of reactions taking place and typical products formed, as follows.

1. *Low-temperature oxidation*. It is characterized by very low rates, by the predominance of associative processes, and by the formation of complex peroxide compounds. Although chain oxidation can take place, the probability of branching is low. Characteristic times of homogeneous transformations in the low-temperature range are very long; they can exceed by far the times of diffusion to the walls of reactor from the flow core. On the other hand, even physically adsorbed species can form dense adsorbed layers on the surface.

This is why at such a temperature range the role of reactions taking place on the reactor walls is very important, even at elevated pressures. The surface can serve as the main "reaction zone" and, at the same time, as a "storage" for relatively reactive intermediates and heavy reaction products. The oxidation reactions of this type are very important for atmospheric chemistry (Gershenzon and Purmal, 1990). The upper temperature for them does not exceed 350–400 K. Nowadays, the interest to modeling low-temperature oxidation is increasing along with the environmental concerns (Gershenzon et al., 1990; Larin and Ugarov, 2002).

2. *Elevated (moderate) temperatures* for alkane oxidation range up to 1,200–1,300 K. At such temperatures the probability of chain-branching increases; moreover in many cases it determines the main kinetic features of the process. Accordingly, peroxide compounds are becoming the branching intermediates, and the most abundant reaction products are represented by non-peroxide oxygenates and olefins. Surface(s) present in the reactor still play an important role, but different from that for lower temperatures. Heterogeneous reactions are considered as initiation, termination, and even as chain propagation and branching steps.

3. *High-temperature oxidation*, including combustion, is characterized by very fast elementary transformations. As a result, product distributions usually reflect the thermodynamic equilibrium state to which the system can arrive from the given composition of the initial gas mixture. Correspondingly, carbon oxides, water, hydrogen, and carbonaceous solids are among the main products. Besides them, the most thermodynamically stable hydrocarbons, such as acetylene, benzene, and cracking products (hydrocarbons with lower number of carbon atoms in the molecule compared to the initial alkane), can also form. As to the reaction intermediates, the species formed due to deep dehydrogenation of hydrocarbon molecules (CH_2, CH, C_2H, etc.) and soot precursors (poly-cycles and carbonaceous agglomerates) play a significant role in this temperature range.

Interestingly, the reactions of the high-temperature type also have the upper temperature limit. It is restricted, on the one hand, by the maximum (adiabatic) heating, which in its turn is determined by the heat of reaction and by the heat capacity of the products. On the other hand, the reaction temperature is limited by the increasing probability of the formation of excited species (in higher vibrational and electronic states). The latter favors the intense additional removal of energy from the reacting system due to irradiation. So, there is a physical limit for practically accessible temperatures in oxidation processes. On the other hand, as far as the process kinetics is concerned, it cannot be described in terms of reactions involving species in the ground state at temperatures substantially exceeding 1,300–1,500 K.

Of course, the "border lines" for each of the ranges denoted above cannot be defined unambiguously; also, inside them some sub-ranges can be distinguished.

Let us consider in more detail the range of elevated (or moderate) temperatures. It is the most interesting from the standpoint of oxidative light alkane processing. On the one hand, the rates of processes in this temperature range are high enough. On the other hand, it is still possible to restrain the process and to optimize the yield of desired non-equilibrium products. However, both lower and upper limits for the range of "elevated" temperatures are the least definite. Inside this range, an additional partition can be made based on some "practical" or kinetic reasoning. For instance, in methane oxidation, below a certain threshold (around 900 K) a predominant formation of oxygenates is taking place, whereas above it the products of methyl radical recombination (C_{2+} hydrocarbons) are efficiently formed (see Section III.E for more detail). Similarly, in propane oxidation, the border above which cracking processes are pronounced or even dominating is ~700 K.

The range of temperatures between 600 and 1,300 K is most "practical" for the efficient production of several compounds via oxidation of C_1–C_4 alkanes. Oxidation at these temperatures is very extensively studied using experimental methods. However, numerous short-lived intermediates crucially important for the process are at the same time very difficult for direct observations and investigation. Some of them, such as complex organic radicals formed via the abstraction of hydrogen atom(s) from relatively bulky molecules, are mysterious, needless to say about the products of O_2 molecule addition to them. Therefore, modeling in many cases is the only real way to gain an understanding of such puzzling reaction systems. Although the modeling of these processes attracts growing attention, this area is not sufficiently well developed as yet.

Higher temperature oxidation (including combustion) has been much more attractive for researchers during the last several decades. This is due to the reason that gas combustion is the basis for many important technologies, including those used for transportation and power generation. Numerous types of engines and burners utilize the heat and the tractive forces produced during alkane oxidation. This generates a great deal of interest in modeling. Fortunately, many ideas developed in this area (see Gardiner, 1984) are applicable also for the "moderate" temperature range. Also, review articles and databases dealing mainly with the data obtained in neighboring temperature regions (see, for instance, Cohen and Westberg, 1991; Tsang and Hampson, 1986; Warnatz, 1984) contain information very valuable for modeling at 600–1,300 K.

To conclude this section, let us define the area of parameters the most suitable for light alkane oxidative processing and, for this reason, being of interest for the development of corresponding kinetic models. As we already mentioned, the temperature range can be restricted to 600–1,300 K (300–1,000°C). Concerning the total pressure, transformations of light alkanes to the products of partial oxidation proceed efficiently at 1–100 bar (0.1–10 MPa). Also, as far as selective partial oxidation is concerned, it requires as a rule an oxygen-deficient (hydrocarbon-rich) composition of the initial reaction mixture. For instance, in the case of methane oxidation, selectivity to any partial oxidation products substantially decreases if the methane-to-oxygen ratio does not exceed two. From a practical standpoint this means that total conversion of the starting alkane as a rule cannot be achieved in one pass through the reactor at high selectivity with respect to partial oxidation products.

C. ACCOUNTING OF HETEROGENEOUS PROCESSES

The adequate description of heterogeneous radical reactions is important both for gas phase kinetics (traditionally treated as "homogeneous") and for a broad and rapidly developing area of heterogeneous catalytic oxidation of alkanes in which homogeneous chemistry can play a very significant role. Until recently, researchers working in these two areas had applied very different methodologies to solve the problem of heterogeneous–homogeneous coupling. This had not allowed even the comparison of results obtained in these two areas, not to mention their superposition. Fortunately, now we can conclude that things are slowly, but surely, changing. We consider this paper as our contribution to bridging the gap between the two areas. At least we attempt to define the cases when it is possible and to which extent and to highlight the remaining problems.

The problem of interplay of heterogeneous and homogeneous processes during hydrocarbon oxidation has a long and complex history. On the one hand, the effect of heterogeneous factors upon homogeneous oxidation is commonly recognized as principally important. Suffice it to mention that heterogeneous termination is ranked as one of key factors in the theory of ignition (see, for instance, Semenov, 1959; Shtern, 1964). There is substantial experimental evidence for the participation of reactor walls in gas-phase oxidation. In some cases they control general kinetic features, degrees of conversion, and product distributions. It was repeatedly noted that initiation steps can proceed on the surfaces of reactors and/or "inert" fillings much faster than in the free gas phase. This may result in a substantial decrease of reaction temperatures and shortening (or even disappearance) of induction periods as compared to the predictions of "purely homogeneous" kinetic schemes.

There is substantial evidence (see, for instance, Garibyan, 1969, 1971) for the heterogeneous initiation in the homogeneous chain reaction of hydrogen and hydrocarbon oxidation in quartz reactors at relatively low temperatures. In particular, in the range between 723 and 988 K the initiation reaction

$$RH + O_2 \rightarrow R + HO_2$$

is characterized by two different values of apparent activation energy E_a. A lower value observed at lower temperatures ($E_a = 100 \, kJ/mole$) is attributed to a heterogeneous mechanism, whereas at higher temperatures ($E_a = 200-210 \, kJ/mole$) a homogeneous process is predominant.

The problem discussed in this section is caused to a considerable degree by the difference in approaches dominating two parts of chemical kinetics, namely gas-phase kinetics and in heterogeneous kinetics, including catalysis. Even the basic terminology and main principles of kinetic analyses formed in these two areas are different. This in turn causes serious difficulties in the description of borderline phenomena.

It is worth noticing here that during several decades so-called heterogeneous–homogeneous processes have been subject of debates. Interest in them had appeared in the bosom of "undivided" chemical kinetics when scientists studying gas reactions and heterogeneous processes in many cases were the same and at least spoke the same language (see Bogoyavlenskaya and Kovalskii, 1946; Polyakov, 1948). The main results obtained during this initial period were summarized in the monograph by Gorokhovatskii *et al.* (1972). In this book

- the processes including heterogeneous and homogeneous steps are classified;
- possible mechanisms of initial formation of free radicals are analyzed from a thermochemical point of view;
- experimental methods for the registration of active species in the gas phase and for the investigation of the overall heterogeneous–homogeneous processes are described;
- mechanistic and kinetic aspects of some particular processes are discussed.

Unfortunately, later the development of two areas—"homogeneous" kinetics and "heterogeneous" catalysis—occurred almost independently, which caused serious intrinsic discrepancies. For instance, the traditional chain theory implies the participation of surfaces also in chain termination, which determines the existence of the low-pressure ignition limit. In the framework of this approach, two regimes—diffusional and kinetic—are distinguished. In the latter case the parameter that describes the process is the probability of surface decay of chain carriers per one collision. It is worth noticing, however, that this assumes only a disappearance of active species from the gas phase, without any analysis of its mechanism and even stoichiometry. This is why the heterogeneous termination reactions are usually represented in kinetic models as a formal reaction:

$$C_X H_Y O_Z \to \text{decay}$$

usually breaking even a material (atomic) balance. Their kinetic parameters, as a rule, are taken from macro-kinetic considerations (from the rate of radical diffusion through the gas to the reactor surface). They also are used as adjustable parameters during the modeling (Vedeneev *et al.*, 1988a, b, c, d) without paying any attention even to the fate of atoms constituting the decayed radical species. Especially in the latter case it is very difficult to assign any physical sense to the parameters, and even to the modeling procedure as a whole.

There is a series of studies by Azatyan and co-workers that stands apart from the main stream in gas-phase kinetics (see Azatyan, 1985, 2005 and literature cited therein). Based on the analysis of experimental data and kinetic schemes these authors claim that heterogeneous reactions participate in all main types of reactions taking place during the chain process (besides initiation and termination, also chain-branching and propagation). Therefore, they must be included in the kinetic scheme on equal terms with homogeneous steps. In our opinion such an approach is very logical from kinetic and even general chemical standpoints, since in its framework an artificial gap and difference in the description of elementary processes taking place simultaneously in one reaction system can be overcome.

One more reason why heterogeneous reactions are usually taken outside of regular analysis in gas kinetics is related to the possibility of experimental study of micro-kinetics. Whereas in the case of homogeneous reactions elementary reactions can be investigated and their kinetic parameters can be measured directly, at least in some important cases, this is practically impossible for heterogeneous reactions for several reasons. First of all, in most cases even the reaction stoichiometry is not definite. Even if only the surface of the reaction vessel is considered, it is influenced simultaneously by the totality of species present in the gas phase. Some of them can physically adsorb and reside on the surface for some time keeping their chemical nature and reactivity. Some of the others being very reactive towards the reactor material can rapidly interact with it causing serious local chemical changes in the surface layer and transforming themselves into different gas species with different reactivity. Species of the third type can be inert with respect to the wall material, but rapidly react with ad-species of different types, etc. The possibility to isolate the particular reaction, to determine its stoichiometry, and to measure its kinetics against a background of other reactions seems very doubtful.

Secondly, any solid surface present in the reactor (reactor walls and parts, various solid fillings inserted into it for different reasons) cannot be unambiguously characterized from the standpoint of its reactivity and adsorption power with respect to different gas species. This is due to the fact that the reaction medium strongly modifies (physically and chemically) any surface, up to almost total change of its nature. It is known, for instance, that during hydrocarbon oxidation any surface influenced by the reaction mixture is being covered by carbonaceous residues (see, for instance, Evlanov and Lavrov, 1977; Suleimanov *et al.*, 1986). The chemical nature and morphology of the latter are not strictly definite and unchanging (neither in time, nor moreover at varied surface and initial gas composition). As a result, surface reactions of species present in the gas phase become even less accessible for direct and distinct study.

At last, to study such interactions it is required to utilize certain experimental methods capable of monitoring the state of the surface and its evolution with time. As far as any spectral methods are concerned, their use requires a sufficient density of states subjected for observation in the zone of measurements. This can be achieved in some cases for heterogeneous catalysts with relatively high surface areas: even in their thin layers at short length of optical path the probing irradiation can collect enough information for further analysis. This is by far more complicated as far as reactor walls and other flat surfaces are concerned. In rare cases—mainly for metal surfaces—some spectral methods can provide valuable information about the composition of adsorbed layers and even about the rates of chemical transformations of certain ad-species (see for instance Firsov et al., 1990; Shafranovsky, 1988). However, even in these rare cases we can speak only about model "*ex situ*" experiments, and the obtained information possesses a limited value for the analysis of real oxidation processes. In most cases, namely modeling provides the possibility to elucidate the nature of heterogeneous processes, which are thus far inaccessible to experimental investigation.

The revival of the interest in heterogeneous–homogeneous processes is associated with the discovery and extensive studies of oxidative coupling of methane (OCM) (Fang and Yeh, 1981; Keller and Bhasin, 1982). The direct observations of free radicals and preliminary kinetic analysis (Driscoll et al., 1987; Lunsford, 1989; Sinev et al., 1987, 1988) clearly demonstrated that the OCM process has a substantially heterogeneous–homogeneous nature. This implies that both heterogeneous and homogeneous steps are substantial for the main features of this process and—what is even more important—they cannot proceed and be analyzed separately. Particularly, activation of methane takes place at the catalyst surface and results in the formation of free methyl radicals. Further recombination of these radicals leads to the formation of the primary OCM product—ethane. As soon as such a scheme of the process became clear, it has become also obvious that its correct description requires a superposition of approaches accepted in heterogeneous catalysis with those developed in homogeneous kinetics of free radical reactions. Such an understanding has marked the new stage in modeling; this is why the OCM process remains a very important and valuable model reaction, without regard to the prospect of its practical implementation. The same can be said about oxidative dehydrogenation (ODH) of light alkanes over oxide catalysts, which in many aspects is very similar to OCM.

There is another important group of processes in which both heterogeneous and homogeneous processes play a substantial role, namely so-called stabilized combustion. Following Pfefferle and Pfefferle (1987), we assign this definition to the gas-phase combustion proceeding in the presence of the surface possessing a catalytic activity. This allows burning various fuels (hydrocarbons, hydrogen, CO) in their lean mixtures with air at substantially reduced temperatures and with almost negligible yield of pollutants (first of all, nitrogen oxides). Simultaneously the efficiency of the evolving heat is increasing (Pfefferle, 1974, 1975). Direct measurements of active species, such as OH radicals (Cattolica and Schefer, 1982; Pfefferle et al., 1988, 1989) and O-atoms (Dyer et al., 1990; Pfefferle et al., 1989), demonstrated that depending on the conditions and properties of the surface, concentrations of such species in the gas can steadily decrease or pass through a maximum. If in the first case the process is likely localized on the surface or in near-surface gas layer, in the latter case we have a clear evidence for the chain process initiated by surface reaction.

Finally, one more very important group of processes in which a joint analysis of heterogeneous and homogeneous reactions is required proceeds in the presence of Pt-group metal catalysts at very short (millisecond) contact times (Davis et al., 2000; Huff and Schmidt, 1993; Schmidt, 2001). It can be easily demonstrated that only some part of the gas molecules (in some cases as low as 20%) can access the active surface within the time of gas passing through the catalyst zone. If, nevertheless, complete conversion of the limiting reactant (as a rule, oxygen) is taking place, this strongly suggests that the reaction substantially proceeds in the gas phase. Another important indication for the conjugation between surface and gas reactions is product distributions. In some cases they are typical for gas-phase oxidation, whereas the reaction times are too short for the exclusively homogeneous reaction at a given temperature. However, in other cases product distributions vary substantially from catalyst to catalyst indicating that surface reactions strongly modify the process that occurs in the gas phase.

One can assume that most (if not all) high-temperature gas-phase catalytic reactions proceeding at temperatures above 800 K are substantially heterogeneous–homogeneous (in the meaning defined above). Consequently, their adequate kinetic description and modeling require a development of corresponding approaches and procedures.

D. Accuracy of Modeling and Influence of Macro-Kinetic Parameters

The possibility of more or less accurate description of a phenomenon based on hundreds of non-accurate parameters could be considered as really surprising. As follows from formal error analysis theory, the results of kinetic modeling should be regarded as completely inconsistent. Probably there exist some deep reasons, which lead to self-consistency of complex kinetic models and allow trusting the results of simulations.

It should be admitted that the question about kinetic parameters of elementary reactions that form complex process mechanisms is very complicated and even painful. Referring to the existing kinetic databases demonstrates that for some reactions the difference in kinetic rate constants reaches orders of magnitude. Each group of authors has particular arguments in support of "their" values. Obviously, among existing values only few (or even none!) are "correct". The reason why we put the word "correct" in quotation marks originates from the problem of representation of rate constants—it will be discussed in more detail in Section III.B. Here we just mention that for practical modeling the so-called "three-parametric" equation

$$k = A\left(\frac{T}{T_0}\right)^n \exp\left[\frac{-E}{RT}\right]$$

is generally accepted for representation of temperature dependence of rate constants.

Since the problem of severe discrepancy in kinetic parameters for one reaction does really exist, several very important questions arise, among which the following are crucial:

– Are there any rational criteria for selection of "true" parameters?
– What meaning should be assigned to the results of simulations if the uncertainty in the values of kinetic parameters of practically all elementary reactions cannot be eliminated?

The former question should probably be answered positively. Some approaches of such a kind are described in the literature and used for the compilation of "recommended" databases of kinetic parameters (Atkinson *et al.*, 1989; Baulch *et al.*, 1982, 1984, 1992, 1994; Cohen and Westberg, 1991; Kondratiev, 1970; Tsang and Hampson, 1986; and other). In brief, the rational approach should include at least the following essential criteria:

- The agreement of kinetic parameters with the physical sense of "three-parametric" equation. For example, if the value of a pre-exponential factor for a bi-molecular reaction in the gas phase is a few orders of magnitude higher than the possible frequency of collisions between certain species it should be considered as contradiction to the physical sense of the corresponding parameter. Another example of a similar kind is a negative activation energy, which most probably should be treated as evidence for the non-elementary character of the corresponding reaction.
- The agreement between the values of kinetic parameters and the thermodynamics of the process. On the one hand, the activation energy for an elementary endothermic reaction cannot be less than the heat of reaction. On the other hand, if the data about rate constants for both direct and reverse reactions are available, their ratio should be equal to the equilibrium constant calculated independently from thermodynamic data for species participating in the reaction.
- The conformity of given reaction parameters to those obtained from empirical correlations for reactions of the same type for one reactant with other partners and/or even with participation of other species. For instance, the Polanyi–Semenov equation is a good approximation for activation energy of hydrogen atom transfer between two species (radicals or molecules) (see, for instance, Kondratyev and Nikitin, 1974). If the parameters for some reaction seriously

contradict the values obtained from such estimations, this needs an explanation, or at least should be taken into account.

Let us come back to the problem of "accuracy" of kinetic simulations based on parameters, which in their turn have at least uncertain accuracy. On the one hand, all values of kinetic parameters contain both systematic and stochastic errors of measurement or calculation. On the other hand, it is known from modeling practice that variation of values for some parameters even within a narrow range allowed by criteria discussed above can lead to dramatic changes in the kinetics of the overall complex process. The simplest example of this type is the effect of chain-branching parameters. The commonly used approach to solve the problem is the "adjustment" of parameters for a number of stages, which allows fitting of the results of simulations to existing experimental data. Such an approach is acceptable, especially for interpolation-type models (Section I.A). Unlike that, the use of any adjustment or fitting procedures becomes doubtful, if the emphasis in modeling is placed on the mechanistic studies or on the analysis of the process over a wide range of parameters, especially beyond the range of basic experiments.

As a rational compromise, the comparison of modeling results with experiments may be considered as the way to select between alternative descriptions if other criteria cannot be applied. Otherwise it is hard to define the ambit beyond which an "optimization" of parameters for a limited set of important stages transforms into the total adjustment of the kinetic scheme to some particular experiment. This aspect will be further discussed in Section II.C.

In general, the problem of required accuracy of kinetic simulations strongly correlates with both the goal of modeling and the choice of criteria for comparison with experiments. We have to notice again that from the above formulated standpoint, the results of modeling can be considered successful if it allows one to describe some non-trivial effects observed in the processes under study, as well as qualitative trends in the yields of particular products upon some variation of external parameters. The latter means that the particular ranges of parameters (temperature, pressure, concentrations) where those effects are observed experimentally and found in modeling can significantly differ. One example of this kind is presented in Fig. 1: the experimentally observed negative temperature coefficient of reaction (NTC) in propane oxidation was reproduced in simulations (Vedeneev et al., 1997a, b). Although there is no quantitative coincidence between experimental and simulated curves, the model makes such an effect comprehensible. This should be considered as a success, especially taking into account both the imperfections of the model and the inaccuracy of the parameters used, which in principal cannot be eliminated for objective reasons.

One prevalent source of deviations between experimental data and simulated kinetics is an inadequate accounting of macro-kinetic factors (heat- and mass-transfer), which can strongly affect the process. A more or less adequate description of transfer processes and fluid dynamics in chemical reactions requires

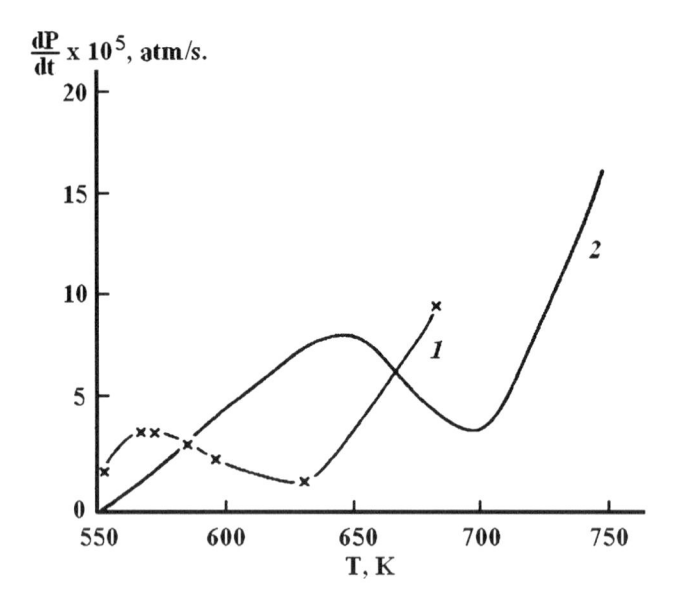

Fig. 1. Maximum pressure jump $d(\Delta p)/dt$ (reaction rate) during homogeneous propane oxidation vs. temperature (Vedeneev et al., 1997a, b). (1) experimental; (2) calculated. Reaction conditions: initial pressure $P_0 = 200$ Torr; propane-to-oxygen ratio = 23.

the utilization of very complicated mathematical and computational procedures; as a result, micro-chemical filling of the model is usually sacrificed (see, for instance, Frolov et al., 1997; Williams, 1997). On the other hand, an adequate description of macro-kinetic factors for each real experiment is hardly accessible. This makes their accounting in simulations even more complicated.

In any real reaction system, chemical processes generate temperature, pressure, and concentration gradients making the description more complex. In general, macro-kinetic factors play an important role in homogeneous partial oxidation, but probably in these processes they are not as important as in combustion. On the other hand, as far as heterogeneous catalytic reactions are concerned, transfer processes must be accounted for, at least in models aimed at practical application, since all these systems are spatially distributed. A vast literature is devoted to heat- and mass-transfer in both homogeneous kinetics and heterogeneous catalysis (see, for instance, Gardiner, 1984; Frank-Kamenetskii, 1969; Kiperman, 1979; Satterfield, 1970, 1980).

In the strict sense, the real reacting system could be mathematically described by chemical kinetic equations only taking into account the processes of diffusion and convection in self-generated non-uniform temperature fields. However, this problem, as we mentioned above, can be solved only at the expense of significant simplification of the chemical description. It is usually acceptable in studies of combustion and explosion processes where the chemical part may be reduced to a very schematic description, which defines the rate of heat release only. If we are particularly interested in routes of intermediate product formation and decay, a cardinal simplification of the chemical mechanism is impossible and we have to look for a trade-off between chemical and macro-kinetic precision.

One possible approach to estimate the potential significance of different macro-kinetic factors is a separate analysis of extreme cases. For example, isothermal and adiabatic regimes of chemical processes are two extreme cases characterized by extremely fast or extremely slow (in comparison with chemical transformation) heat-exchange between the reacting system and the external medium. All real heat-transfer regimes including both external heating of the reaction zone and auto-thermal reaction are intermediate cases between these two regimes. It is possible to compare the modeling of these two outermost cases in order to estimate how (qualitatively) and to what degree a heat-exchange regime can affect product yields. Figure 2 presents an example of such a comparison. Homogeneous methane oxidation was simulated using the kinetic model developed previously for methanol production at high pressures (Vedeneev *et al.*, 1995). In this case a comparison of certain product (methanol and formaldehyde) yields is plotted vs. conversion of the limiting reactant (oxygen). Without going into detail, let us conclude that the thermal regime of reaction has a strong effect on the behavior of the reacting system and heat-transfer should be optimized, as far as maximum yields of certain products are concerned. However, different products are subjected to this effect in somewhat different ways. The difference in the behavior of formaldehyde and methanol reflects particular differences in their formation and transformation pathways.

An accurate accounting of mass-transfer is especially important in heterogeneous systems, and, first of all, in catalytic oxidation. In this case the simultaneous presence of active surfaces and an "active" gas phase and occurrence of different processes involving different species in these two reaction zones makes the system puzzling. Some additional reasoning and examples illustrating the complexity of such systems and possible solutions will be discussed in Sections III.E and III.F. However, we must confess that a thorough elucidation of this extremely important and complicated problem is far beyond the scope of this paper.

II. Alternative Approaches to Modeling

As we stated in the Introduction, a complete review and thorough analysis of publications devoted to the subject overstep the bounds of this work. Below in this section we only present an overview illustrating some typical approaches to modeling of hydrocarbon oxidation. Among them, we distinguish two groups of models stipulatively defined as "additive" and "combinatorial". The former

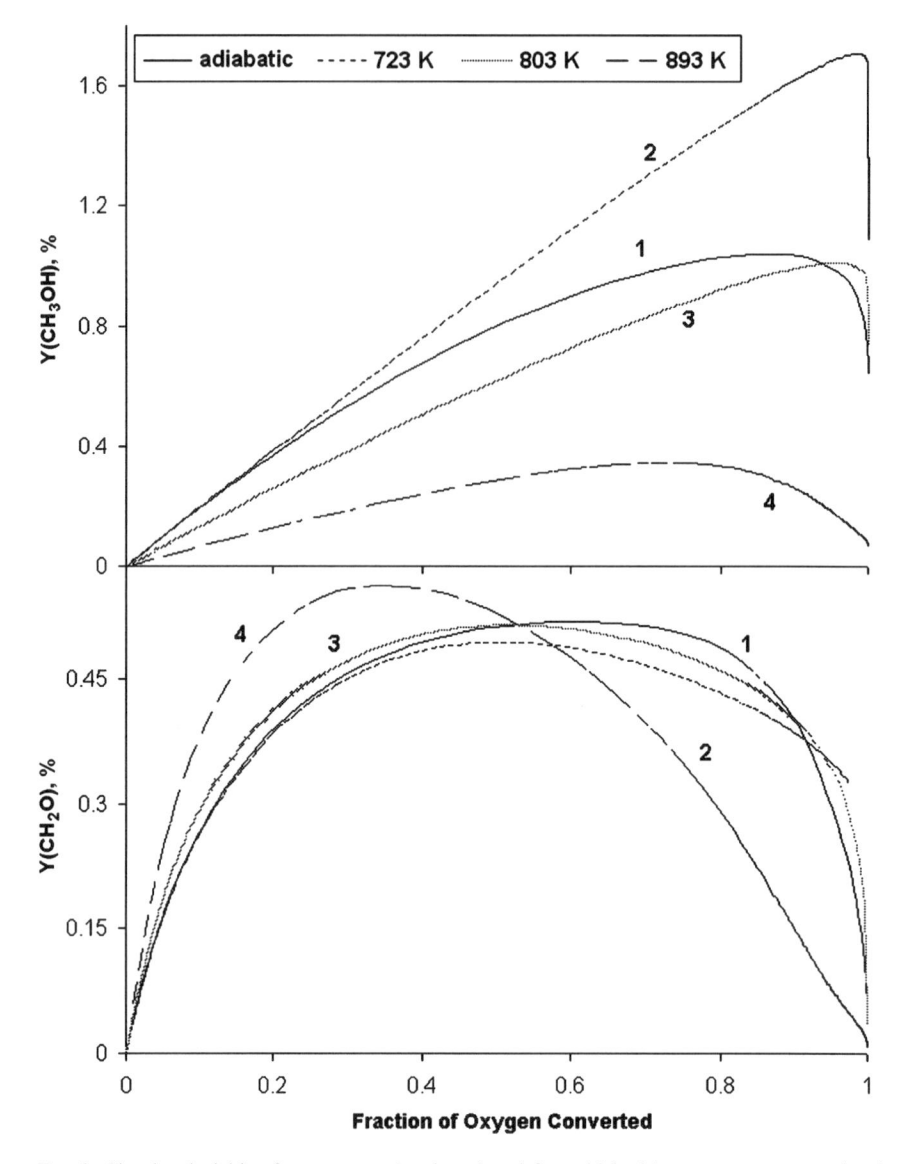

FIG. 2. Simulated yields of oxygenates (methanol and formaldehyde) vs. oxygen conversion in different heat-transfer regimes. (1) Adiabatic, $T_0 = 723$ K; (2) isothermal, $T = 723$ K; (3) isothermal, $T = 803$ K; (4) isothermal, $T = 893$ K.

group includes the models compiled by the addition of selected elementary steps, or groups of steps, to some initial core. The criteria for the selection of reactions to be included into the core, as well as for the further addition to it, are both defined by the investigator based on his (her) understanding of the process and simulation procedure. Alternatively, if the model of "combinatorial" type is developed, the investigator first defines the list of species taking part in the complex reaction network and then the kinetic scheme is formed as a combination of all chemically relevant elementary reactions of selected species.

A. "ADDITIVE" MODELS

The model aimed at the optimization of methanol production by methane partial oxidation at high pressures has been developed by Vedeneev and co-workers since 1988. Up to now it is one of the most successful examples of specialized models for moderate-temperature oxidative conversion of light alkanes. It is based on a deep understanding of the main ruling factors responsible for particular kinetic features of the process. Correspondingly, the blocks of elementary reactions for such features, and for the formation of methanol at high pressures in particular, were included into the kinetic scheme.

Next, it was declared that the comparison of calculations with selected experiments cannot be used as a universal criterion for model development and selection of rate constants, or should be restricted and at least performed very carefully. The values of the kinetic parameters were taken from independent measurements, theoretical calculations, or semi-empirical estimations. So, the invariability of rate constants for the description of any particular experiments was accepted as the one of the main principle.

Two possibilities were accepted for the further improvement of the model. The first one is a step-by-step addition of new elementary reactions in the case of the necessity for an adequate description or in the case if new reliable kinetic data are obtained. The second was a more accurate evaluation of kinetic parameters for elementary reactions already included into the model, but only on the basis of new direct experimental data or new calculations that are more advanced and precise. In one of the latter versions of this model (Vedeneev *et al.*, 1995), the thermodynamic consistency principle (see Section II.C for details) was realized.

As a result, a relatively compact model capable of predicting and capturing some features of the process over a given range of parameters (relatively low-temperature, high-pressure, low oxygen-to-methane ratios) was developed. As applied to the direct oxidation of methane-to-methanol (DOMM), it allowed the determination of several principally important qualitative features of the process. In particular, it was found that methane partial oxidation at high-pressure is a degenerative chain-branched process with very short chains and a long induction period. The chain-branching occurs via several parallel reactions, none of them which dominates.

It was shown that there are three main phases of the process (see Fig. 3). A very short initial phase is an auto-accelerating chain-branched reaction; the total concentration of chain carriers vs. time follows an S-shape curve, which is

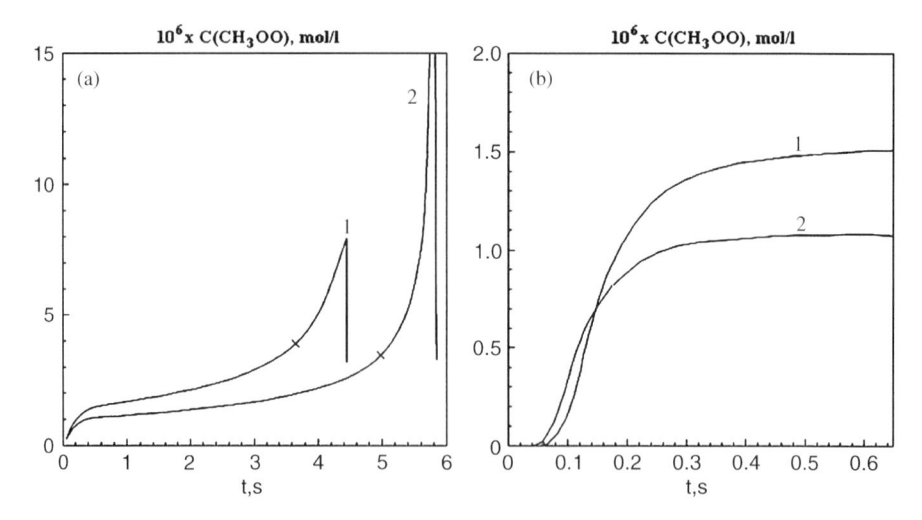

FIG. 3. Simulated concentration of CH$_3$OO radicals as a function of time during the DOMM reaction (Vedeneev et al., 1988a, b, c, d; Arutyunov, 2004). P$_0$(CH$_4$) = 80.9 bar, T = 680 K. (1) P(O$_2$) = 1.8 bar, (2) P(O$_2$) = 8.4 bar; (a) for the first phase, (b) for the initial part of the second phase of the overall process.

very similar to hydrogen oxidation. This phase had not been observed experimentally before the modeling was performed.

The second phase is a quasi-stationary chain-branched process with a quadratic radical termination, which proceeds mainly via the recombination of methylperoxy radicals

$$CH_3O_2 + CH_3O_2 \rightarrow CH_2O + CH_3OH + O_2 \qquad (1)$$

at approximate equality of the rates of branching and radical recombination. During this phase, a slow accumulation of intermediate products and a slow increase of temperature are taking place. This leads to the fast self-acceleration of the reaction in the third phase, which is due to the chain-branching in the reactions involving intermediate products and self-heating. It was clearly demonstrated that the quasi-stationary behavior is the most striking feature of the process (Arutyunov, 2002, 2004; Arutyunov and Krylov, 1998).

The model also revealed the importance of non-linear interactions of chains, which determine the existence of two quasi-stationary oxidation modes. The difference in reaction rate between them is about four orders of magnitude. A critical transition between these modes is taking place due to small changes in reaction parameters, e.g., pressure. This critical transition leads to a very fast increase in the oxidation rate and an abrupt transition from the low-pressure slow oxidation to the fast high-pressure stationary chain-branching process. In the latter mode, radical generation is no more determined by a slow initiation reaction, but by very fast chain-branching reactions. It is the phenomenon that allows accomplishing this non-catalytic gas phase process at moderate temperatures.

The analysis of this model revealed some interesting qualitative effects, which were experimentally verified later. The competition of reactions (2) and (3) of active methoxy radicals CH_3O with methane and oxygen

$$CH_3O + CH_4 \rightarrow CH_3OH + CH_3 \tag{2}$$

$$CH_3O + O_2 \rightarrow CH_2O + HO_2 \tag{3}$$

leads in the first case to the formation of very active chain propagating CH_3 radicals and in the second case to much less active HO_2 radicals that mainly recombine to form hydrogen peroxide

$$HO_2 + HO_2 \rightarrow HOOH + O_2 \tag{4}$$

The latter reaction is equal to chain termination in this temperature range. This is the reason why in this reaction oxygen acts as an inhibitor if its initial concentration does not exceed 10 vol.% (see Fig. 4). Later this effect was experimentally proved (Arutyunov *et al.*, 1996). Thus, the model demonstrated its capability of finding new intimate information about the intrinsic reaction mechanism.

The approach called above as "additive" can also be identified as "pragmatic". Indeed, if we wish to describe some combination of kinetic features of the system, we include only a limited number of elementary reactions, which are responsible

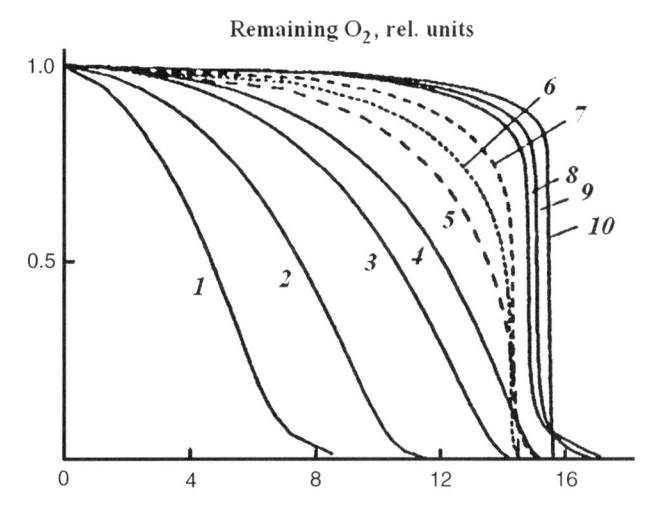

FIG. 4. Calculated kinetics of oxygen conversion (fraction of remaining oxygen in relative units) during methane oxidation at different initial oxygen concentrations in the mixture (in vol.%): 0.5 (1), 1 (2), 2 (3), 3 (4), 5 (5), 7 (6), 10 (7), 15 (8), 17 (9), 20 (10); $T_0 = 693$ K, $P = 10$ bar.

for the behavior of our interest. Any other reaction if it does not interfere with the phenomena to be described is not included into the scheme. It is clear, however, that such additional reactions can be very important for some other kinetic features, which for some reason remained out of consideration. Consequently, if one is intended to transfer such a "pragmatic" model to the description of related, but different processes, or to solve some very similar problem, but involving slightly different kinetic peculiarities, this attempt may fail or give just misleading results.

In this sense, the attempt to extend the use of the above model for the description of the OCM process (Vedeneev *et al.*, 1997a, b) is very illustrative. A simple addition of a limited set of new reactions to the initial kinetic scheme is not successful: the process proceeds in different range of temperatures, pressures, and methane-to-oxygen ratios. As a result, in order to be adapted to the new application, the initial model has to be totally revised. The required changes must be deep to an extent that it is possible to speak about the compilation of a different model.

The applicability of an "additive" approach is very doubtful in many other cases too. For instance, a typical task for modeling is a description of the behavior of small additives to the reacting system. Such additives may serve as reaction rate modifiers (promoters or inhibitors). Another important area is a detoxification of pollutants in the conditions of hydrocarbon oxidation (combustion). In both cases the detailed mechanism of the process in the presence of additives (especially, if they contain heteroatoms, such as nitrogen, halogens, sulfur, or phosphorus) can undergo very serious changes, which can hardly be accounted for by the addition of a limited number of new elementary steps.

The above discussion can be illustrated by numerous attempts to describe oxidation of fuel-rich mixtures at 900–1,200 K in the framework of the kinetic scheme known as GRI-Mech (see Frenklach *et al.*, 1995). These attempts consisted of the replenishment of the initial scheme by new blocks containing the reactions of complex radicals that are characteristic for such temperatures. Then the whole scheme was subjected to the procedure routinely accepted for the optimization in the framework of GRI-Mech approach (see Section II.B below). However, no reliable proof of the effectiveness of such an "extended" description has been obtained.

Borisov *et al.* (2000) mentioned that some widely used kinetic models used even for combustion are self-contradictory. For instance, the scheme applicable for the developed high-temperature combustion process is also used to describe the values of self-ignition delays (SIDs). The latter usually take place at temperatures several hundred degrees lower than the combustion itself and may obey very different rules, at least if the combustion process is described in terms of the "pragmatic" approach. Authors of the cited work suggested the approach to modeling both ignition and combustion of C_1–C_3 hydrocarbons consisting of a combination of two blocks, one of which describes combustion, and the second—low-temperature oxidation. As a result, the authors arrived at a typical "additive" model and verified it by comparing measured and calculated values of SIDs.

B. "Combinatorial" Models

At present, the most known and widely used kinetic model for light hydrocarbon oxidation and combustion is GRI-Mech (Smith *et al.*), which has been developed since the 1990s by the Gas Research Institute. The project was aimed at the development of the detailed micro-chemical mechanism that can be used for simulations of natural gas ignition and combustion. Now the model is optimized for two fuels, methane and natural gas. Recommended temperature and pressure ranges are 1,000–2,500 K and 1.3×10^{-3}–1 MPa, respectively; the fuel-to-oxidant equivalence ratio in premixed compositions is 0.1–5. Although the scheme includes some reactions of C_2–C_3 hydrocarbons and oxygenates participating in methane (natural gas) combustion, it is not recommended to use GRI-Mech for simulations of their oxidation (combustion) as the main initial fuels, since the model is not optimized for these purposes.

Initially the model was compiled using both experimentally measured and theoretically calculated kinetic parameters. Then, the results of simulations were compared with the data of multiple experiments and sensitivity analysis was employed to select the parameters, which should be corrected for the better agreement between experimentally observed and simulated kinetic behavior. The computation routine can perform the modification of each kinetic parameter within the range of its initial uncertainty. Such an approach gives a serious cause for criticism, since the discrepancies with experimental data are eliminated (or minimized) by changing the values of multiple parameters. First, this makes all of them correlated. Next, an independent correction of just one parameter in the model, or just a slight modification of the micro-chemical scheme leads to the readjustment of the whole system of kinetic parameters. This is in a certain sense equal to the solution of the inverse kinetic task, which, as we mentioned above, is an ill-conditioned problem.

Nevertheless, in spite of some difficulties, GRI-Mech signifies a new era in modeling of combustion and oxidation of light alkanes and other substances. The main principles of GRI-Mech serve as a basis for further widening the area of its application for related processes and substances (Curran, 2004).

It is appropriate mentioning here again the "accuracy" problem with respect to kinetic parameters and its relation with general modeling concepts. A distinct uncertainty in the selection of parameters and a wide "tolerance box" for their variation are among the main reasons for discrepancies between different models, which in turn stimulates the creation of new ones. This is the reason why the idea of using exclusively the kinetic parameters obtained by quantum chemical calculations instead of experimentally determined values seems attractive. Indeed, contemporary theoretical methods can give results, which in some simple cases are more reliable than mutually contradictory experimental data. One may even assume than in the case of serious discrepancies between experimental and theoretical data, the former should be called in question and revised first.

We cannot exclude that in the course of time this approach will become dominant and "consistent theoretical" models will help us to avoid the temptation to adjust the whole set of kinetic parameters as soon as some new experimental curve appears. However, the remaining problems are still tremendous and require serious effort; some of them will be discussed more thoroughly in Section III.B. Nevertheless, several publications reflect the results already obtained via this way (see, e.g., Green, 2001; Zabarnick, 2005).

The use of a computer not just for simulations as such, but also for the compilation of the model, is also stimulated by the sharply rising complexity with an increasing number of carbon atoms in the reacting hydrocarbon molecule. Some authors advance an opinion (Buda et al., 2005; Green, 2001) that any comprehensive model for oxidation of alkanes higher than propane can be developed exclusively based on automated (computer) generation methods, which enable one to avoid serious errors in the course of its compilation and further processing. Green (2001) suggested such a combined computational procedure that includes quantum chemical and group-additivity methods to calculate most of the molecular properties and kinetic parameters. These authors claim that they for the first time succeeded to implement $k(T, P)$ calculations in a general way that is suitable for use in an automatic reaction generation algorithm.

A similar combinatorial approach can be also applied for complex catalytic reactions (Shalgunov et al., 1999; Temkin, 2000; Temkin et al., 1996; Zeigarnik and Valdes-Perez, 1998). It is based on the notion that the reaction network involves the formation of certain products (final or intermediate) from one precursor via different routes. Only a complete accounting of all junctions and connecting reactions in such a network treated together with the appropriate kinetic parameters can give a realistic representation of the overall process. Network compilation is a subject for a formal computerized procedure.

C. Ruling Principles for Comprehensive Modeling

Special analysis that was performed by request of Gas Technology Institute in 2001–2002 (private communication) had clearly demonstrated that all kinetic models developed to date for high-temperature oxidation of methane and its homologues can describe well some selected experiments but contradict others. The groups of experiments described well by different models cannot be clearly classified by their types, methods, or conditions. It was concluded that different models are initially connected with some groups of arbitrary selected experiments.

Nevertheless, we have to confess that any experimental data requires some "theoretical" comprehension and rationalization. This is an obvious driving force for the development of new models for "local use". If their authors need to describe a relatively narrow group of experiments, this creates certain problems for co-ordination with existing models describing the general behavior of a reacting system. On the one hand, the latter may be quite far from the particular needs of the given experimental study. But, on the other hand, such basic models are usually very bulky, and the temptation to skip "the excess details" in favor of those, which seem more important, can be very strong. Besides, expanded models may account for the reactions of very complex species (such as aromatic poly-cycles), but include them into the scheme in an arbitrary manner. Thus, although the description includes numerous elementary reactions, it bears the stamp of certain randomness. The latter is aggravated by the use of adjustment procedures often employed to compensate for the lack of reliable kinetic parameters. As a result, even if a good fit of experimental data is reached, such a model strongly suffers from the disturbance of inner ties.

We would not discount the utility of such models for certain practical needs. However, one should be very cautious while applying them for the analysis of intimate details of the reaction mechanism. Also, their capability to predict the system behavior outside of the given parameter range is at least suspect.

To proceed to the development of comprehensive models capable of working over a wide range of parameters and for diverse applications, we have to formulate first of all the basic requirements of such a model, which will be further called "ruling principles". A rigorous observance of them would impart to the model an intrinsic consistency and predictive power.

The starting-point and basic-level principles of models, which can be defined as comprehensive or inclusive, must be the following: the combinatorial approach for the compilation of a kinetic scheme and the use of independent values of kinetic parameters. The latter means the flat refusal to use any "parameter optimization" algorithms based on adjustment of the whole model or its blocks to some selected experimental data. These two basic principles are already discussed above in Sections II.A and II.B with reference to the GRI-Mech (combinatorial approach) and the methane-to-methanol oxidation model developed by Vedeneev and co-authors (use of independent kinetic parameters). However, we could not find any example of consistent employment of both principles in conjunction.

Let us mention again, however, that in the course of real modeling multiple compromises are inevitable. This first of all relates to the compromise between the thoroughness and rigorousness of the description and, on the other hand, the accessibility of primary information and "usefulness" of the model. This should be also addressed to both formulation of ruling principles and their execution. Accordingly, it is necessary to formulate them in order to realize their inherent limitations, i.e., the area of phenomena and ranges of parameters where the model can (and cannot) be applied.

The following principles have been formulated based on the stated goals and tasks of modeling and on the analysis of the existing approaches in this area.

Thermodynamic consistency. This is one of the most fundamental principles for the compiling of any model purporting to have a wide applicability. It can be considered as a particular manifestation of the more general kinetic principle of microscopic reversibility that insists the possibility for any elementary process to proceed in both the forward and backward directions. Accordingly, in a complex kinetic scheme any chemical step must be written together with a corresponding reverse reaction. Their rate constants are strongly correlated: their ratio must be equal to the equilibrium constant, which in its turn is determined by thermodynamic functions of substances being in equilibrium. No one elementary reaction consisting of a pair of reactants and products should be ignored in favor of another one, no matter what is the value of corresponding rate constant under given conditions. In other words, the model must assure the asymptotic approach to equilibrium for a given atomic composition at infinite reaction time.

Unfortunately this principle is very often disregarded, or is not taken into account in modeling. We have to stress, however, that only thermodynamically consistent kinetic schemes correctly vector the overall process simulation. Moreover, the faster the reaction in the "forward" direction, the sooner is the equilibration. That means that disregarding of reverse reactions can distort the whole reaction network.

It should be noted that the execution of this principle can meet formal difficulties related to the representation of the temperature dependence of rate and equilibrium constants. Because of the strong dependence of thermodynamic functions on temperature, it is difficult to describe rate constants for both forward and backward reactions in terms of Arrhenius or "three-parametric" equations (see Section III.B). It is possible that the form generally accepted for the representation of temperature dependence of rate constants requires modification. Anyway, it is evident that such formal difficulties should not put obstacles in the way of more adequate modeling of reaction kinetics.

Model fullness. Any kinetic model aimed at the solution of a wide range of tasks over a given range of parameters should be based on the full list of elementary steps. When we state such "fullness" as one of the main ruling principles, it is assumed that the scheme must reflect the whole totality of our knowledge about micro-chemical relationships in the reacting system under consideration. In other words, the kinetic scheme must be compiled in accordance with the combinatorial approach as thoroughly as possible. In addition, some new elementary reactions or even groups of reactions are to be added to the scheme, as new data pointing to such a necessity becomes available. And vice versa, some reactions can be removed from the scheme for the same reason.

Even though the model is supposed to be "full", it should be "rationally full". This means that species, which are believed to be not important for a given range of parameters, can be excluded from consideration. For instance, as far as alkane oxidation at moderate temperatures is concerned, reactions of excited molecules and radicals, which are important for high-temperature combustion, can be excluded. On the other hand, over the same temperature range one can also ignore the reactions of complex peroxides containing chains of more than two oxygen atoms, which are considered sometimes as important for low-temperature oxidation.

In Section V.A we will provide arguments for the joint kinetic description of oxidative transformations of methane and C_2 hydrocarbons. Regarding molecules containing more than two carbon atoms, their influence on the overall kinetics and on the formation of many important products is below the anticipated accuracy of simulations (Arutyunov et al., 2005). This is why their formation and transformations can be not accounted for in methane and ethane oxidation models for many applications. At least it would not compensate the excessive complication of the model accounting for reactions of C_{3+} species.

Of course, in special cases when the modeling is aimed at the formation of higher hydrocarbons (Marinov et al., 1996; Mims et al., 1994), their reactions must be included. However, due to a relatively low concentration of these compounds, the requirements to the complete accounting of their reactions can be not very strict due to their relatively low importance. A typical example of such kind could be a chain-branching at the expense of hydroperoxides containing higher alkyl-groups.

The same can be related to oxidation of any other C_{2+} hydrocarbon. The kinetic scheme must include the reactions of lower hydrocarbons as its substantial part, but the reactions of species containing more C–C bonds than the initial hydrocarbon can be excluded for many applications without a substantial loss of accuracy.

In brief, the fullness is not a quantitative measure of the model, but more a qualitative one. The scheme containing fewer elementary steps may better meet this criterion than a much vaster one, if in the former case the reaction blocks related to species of a certain type (for instance, containing the same number of carbon atoms as the initial hydrocarbon and less) are more thoroughly taken into account.

Independence of kinetic parameters. Let us mention again that there are two alternative approaches to the relationships between the model and the kinetic parameters. In the first case, the data obtained in some experiment(s) are considered as the basis. The values of kinetic parameters for all (or some selected) elementary reactions included in the kinetic scheme are optimized in the framework of a certain fitting procedure, and the range of their variation can be determined in advance based on some considerations. As a result, the set of parameters assuring the best adjustment of the kinetic scheme to the basic experiment(s) is defined.

This approach provokes strong objections. First, the basic definition of "rate constant" becomes blurry. Indeed, none of the coefficients in the mass action law in the whole kinetic scheme can be considered as an independent physical value with a clear sense. If, however, selected parameters receive such privilege, this breaks the general principle and requires additional criteria.

Second, it is evident that the selection of different "basic" experiments should lead to different sets of "optimized" kinetic parameters. Since, as it was already mentioned, the inverse kinetic task is in general an ill-conditioned problem, such sets can differ significantly and it is difficult to define the criteria for the selection of the one (only one!) that should be used for modeling.

Lastly, practically in any kinetic experiment there are some uncertainties in its conditions and results. This concerns such factors as mixing of reactants, regimes of gas flow in different zones of the reaction system, wall effects, axial and radial temperature gradients, spatial distribution of reaction rates, concentrations, pressure, etc. A lot of effort must be applied even to measure the reactant conversion and the main product distributions at the exit of the reactor, needless to say anything about concentrations of active intermediates and concentration distributions inside the reactor. Moreover, up to now there is no clear understanding about which measured parameters are the most informative and should be preferred.

An alternative approach implies that all kinetic parameters used in the model are determined independently from the modeling itself. The probability of serious errors is still high in this case too, although for many parameters important for alkane oxidation and combustion such errors are evaluated (see, for instance, Tsang and Hampson, 1986; Warnatz, 1984). Despite that, it seems that in this case we are dealing with the lesser of two evils. Moreover,—and this is of principal importance—every value included into the scheme can be independently corrected as soon as newer and more reliable data appear. This does not require any modification of any other kinetic parameter in the model.

Therefore, we are inclined to use in modeling only the reliable recommended rate constants with the possibility of their correction only on the basis of independent experimental or theoretical results without any adjustment or fitting to experimental data.

The possibility to significantly improve the agreement between calculations and very reliable experimental data by variation of not more than one or two parameters may be considered as the maximum allowed adjustment procedure and as a cause for a more thorough analysis of accepted sources of kinetic parameters. Concerning the choice of the most reliable values of kinetic parameters, it is very difficult to make any general recommendation equally good for all possible cases. Every multi-step kinetic scheme consists of elementary reactions, which have been previously studied with very different thoroughness. Some reactions attract very rapt attention and are studied by various methods (direct and indirect). In this case such experimental data can be analyzed and some "weighted" values of parameters can be chosen. Another extreme can be represented by the reactions very poorly studied experimentally, or even hypothetical. In such case theoretical evaluation or even a semi-quantitative estimation using *ab initio*, semi-empirical, and/or empirical methods can be considered as the only way to obtain the primary information for modeling. But in any case it is possible to trace the sequence of sources with different levels of reliability. As far as the homogeneous elementary reactions in the gas phase are concerned, we no doubt believe that the review articles and databases, in which the analysis of the data is performed, can be considered as the most reliable ones. For instance, in the case of methane and ethane oxidation at moderate-temperature and pressures (600–1,300 K and 1–100 atm) the following sequence of sources (with decreasing universality and applicability) could be suggested.

- The recommended values presented in (Tsang and Hampson, 1986) and added to those recommended by Baulch *et al.* (1982, 1984, 1992, 1994) have a priority against any other.
- If the data are absent in these sources, the recommendations of (Warnatz, 1984) can be accepted.
- If none of the above reviews contains the recommendations about the reaction under consideration, but original experimental works containing the temperature dependence of rate constant(s) are available, the value for the closest temperature interval can be recommended.
- If no reliable data are available at all, the kinetic parameters can be evaluated by analogy with reactions for which the similar data are available. For instance, to evaluate rate constants for some bi-molecular reactions of C_2-radicals (C_2H_5O, $C_2H_5O_2$), the data for CH_3O and CH_3O_2 radicals can be used as a rough estimation.

Openness of the description. If we accept the independence of kinetic parameters as one of the main ruling principles and totally deny any adjustment of the kinetic scheme to particular experimental data, this gives the model an additional advantage making it fundamentally open. This means that the arrangement and tuning of the model can be done by the selection of more accurate values of kinetic parameters from independent sources and by the addition of elementary reactions required for the more detailed description of the reacting system. Moreover, once elaborated the core model can be further developed for a wider area of application by the addition of new elementary steps or their blocks. One example of such widening will be given in Section V.B.

To conclude this section, it is worth mentioning that any model, no matter how deeply it is developed, cannot be identical to the described phenomenon. This makes even more important the precise formulation of the goal of modeling and the right choice of corresponding instruments and tools. If we state that our goal is to develop the model, which in turn is going to be a tool for mechanistic investigations and to carry a certain predictive power, it must have no internal ties and connections with experimentation it should describe. We believe that such a goal can be reached in the framework of the united approach based on the principles formulated above.

D. REDUCTION OF MODELS

The development of comprehensive quantitative models is the main goal and the "high road" in modeling. However, compact "illustrative" models for more clear understanding of the essential phenomena and their qualitative description will remain in demand. In the ideal case, such an "illustrative" model should result from the reduction of a more comprehensive one and keep the whole totality of features important for the behavior or for the phenomenon we wish to illustrate, as well as the relation to the overall process. Whereas the inverse procedure—the initial formulation of relatively "simple" schemes—can be acceptable only as a starting hypothesis that requires further development.

Several methods and mathematical tools have been suggested and elaborated to reduce complex kinetic models to be more concise. We believe that it would be important to mention briefly the main guiding principles and criteria for such reduction. First of all, again we must stress that the tools and ways are caused by the goal—in this case the goal of reduction.

The "importance" criterion implies the selection of reactions (which will remain in a "concise" scheme) based on sensitivity analysis. Note that this widely used approach may cause very serious faults even with very small changes of external parameters. Moreover, one may lose sight of the fact that reactions which do not substantially contribute in the oxidation processes in a given regime may become of crucial importance if the conditions are changed. Below we will give an example of such a type when the reactions, which do not participate over a certain temperature range, allow the process proceed into the particular direction (namely—the OCM products formation during methane oxidation, and the appearance of the "OCM Window"). Thus, even to describe qualitative effects, "rational" description may be based on the "reverse" importance criterion.

The partial equilibrium criterion goes back to the steady-state concentration method, which was widely used before the computer age. In the framework of this approach, only species participating in the slowest reaction are considered, while a steady-state equilibrium for all other components and fast reactions is assumed. If two or more species are connected by fast chemical reactions and the rate of their reciprocal conversion in these reactions is much higher than into other compounds, they form a steady-state subsystem. Their ratio in such a subsystem is determined by the rate constants of their reciprocal conversion. In the case of two fast reciprocally transforming compounds, their ratio is determined only by the equilibrium constant and current concentration. The particular values of the rate constants determine only the time of equilibration, and other reactions of these compounds only slightly disturb their ratio. In such a case the use of the partial equilibrium criterion can be fruitful.

The minimization (even neglect) of chemical differences criterion admits the unified description of isomers (and even chemically related substances), if they can undergo the same types of chemical transformations. This criterion also should be applied with caution: even the comparison of two propyl radicals—of normal and *iso*-structure—demonstrates their serious difference, especially when relative importance of different reaction channels, including dehydrogenation and cracking (oxidative and non-oxidative), is analyzed. On the other hand, if the target products are olefins, any difference in the behavior of multiple oxygenates can be neglected, provided their further transformations do not lead to the target molecules and to chain-branching intermediates. In this case their formation with acceptable accuracy may be treated merely as the withdrawal of a sufficient amount of species from the main process.

The similarity criterion is analogous to the one just mentioned above. In this case, the same kinetic parameters are ascribed to different species participating in the reaction of particular type. Evidently, similar caution also should be taken. However, this approach becomes of principal importance for the modeling of higher hydrocarbon oxidation (e.g., transfer from C_1–C_2 to propane and butanes), or for processes in which some products with increasing complexity (like soot species) are formed.

The utility criterion supposes the selection (flow analysis) of the reaction set the most important for the particular behavior of the system, or for the particular product formation and transformations. This approach is very useful for the segregation of reactions, in which a particular substance (target product, leading radical, or key branching intermediate) forms or transforms. As a result, the analysis of the process under different conditions, or at different stages (when the ratio of various elementary reactions is changing) becomes available.

One should keep in mind that any reduction breaks the stringency of the description. The "reduced" description is only an illustration that should emphasize the chemical essence of the phenomenon under consideration, but cannot replace the "full-scale" modeling. Like in many other cases, we can consider reduced models as a compromise—on this occasion as a concession to clarity at the expense of thoroughness. Anyway, simulations using the comprehensive model should be employed to control the validity of thus obtained qualitative conclusions.

E. Modeling of Heterogeneous–Homogeneous Catalytic Reactions

Modeling of processes belonging to this group is discussed separately because of their significant specificity. As we already mentioned in Section I.C, systematic studies in this area started about 20 years ago, but they are still in the development stage.

Even from the very beginning it became clear that kinetic methods generally accepted in heterogeneous catalysis, which are based on the analysis of adsorption–desorption equilibria and surface reactions in the steady-state approximation, cannot be applied to the studies of heterogeneous–homogeneous processes. However, on the eve of 1980s no developed methodology applicable for joint analysis of catalytic and gas-phase reactions was at a researcher's disposal. For the description of the former group, models of the Langmuir–Hinshelwood, Eley–Rideal, and Mars–van Krevelen types were traditionally applied. Unlike them, the analysis of gas-phase reactions usually is based on kinetic schemes, which include multiple elementary steps and sets of corresponding kinetic parameters. The latter—in principle—can be determined (measured experimentally or theoretically calculated) independently.

The strategy previously applied for homogeneous reactions has been also accepted for modeling heterogeneous–homogeneous processes. It consists of

- Compiling a set of elementary steps (kinetic scheme) assumed to proceed in the reacting system with a corresponding kinetic description of each step (kinetic equation and parameters).
- Certain notions about heat- and mass-transfer in the reactor.
- Mathematical (computational) procedures enable the prediction of the development of the process in time (and space—in the case of spatially distributed systems).

Even if the influence of macro-kinetic factors (heat- and mass-transfer) is taken into account, the models of this type are usually called "micro-kinetic". This is due to the principal importance of the set of elementary reactions. In all cases these models are based on a guess (more or less experimentally and/or theoretically grounded) about the factors—chemical and physical—important for the behavior of the systems, as well as about the factors which can be ignored for certain purposes.

At least two circumstances stimulated a growing attention to modeling of heterogeneous–homogeneous processes. First is a great practical importance of natural gas processing reactions. For some catalytic processes belonging to this group (first of all OCM) their radical nature was proven; for others (methane partial oxidation to formaldehyde and synthesis gas, ODH of C_2–C_4 alkanes) the probability of such mechanisms was evaluated as being high. Scaling-up of such processes requires reliable kinetic models.

On the other hand, because of the serious complexity of the direct investigation of heterogeneous free radical reactions, as well as processes in adsorbed layers, modeling can serve as a powerful tool for mechanistic studies. However, poor knowledge about possible mechanisms and the lack of basic concepts still restrain progress in elaboration of kinetic schemes in which all kinetic parameters would be measured or evaluated independently. Moreover, as compared to elementary gas reactions, rate constants for heterogeneous processes, especially for reactions between adsorbed species, are hardly amenable to theoretical calculations. Even required thermochemical data are not accessible in most cases. Besides that, energetic characteristics and reactivity of surface sites and adsorbed species present a puzzling function of the state of surface and bulk of the solid and can be adequately accounted for only in rare cases.

Nevertheless, recently several approaches have been suggested and certain progress in theoretical methods applicable for heterogeneous kinetics has been achieved (see, for instance, Baranek et al., 2000; Hansen and Neurock, 1999; Lichanot et al., 1999; Neurock, 2003; Neurock et al., 2004; Palmer et al., 2002a, b; Shustorovich, 1990). Although the detailed analysis of this subject exceeds the bounds of our work, it is worth mentioning the most promising of them, namely Density Functional Theory (DFT) and the Unity Bond Index-Quadratic Exponential Potential (UBI-QEP) method. Whereas DFT is quantum-mechanics-based, UBI-QEP can be characterized as a phenomenological or semi-empirical approach. According to Shustorovich and Zeigarnik (2006), these two approaches are complimentary, and good agreement between their predictions testifies to a good understanding of the system under study. These are some examples of fruitful use of these two methods in the framework of one study (see, for instance, Hansen and Neurock, 1999; Mhadeshwar and Vlachos, 2005). The details of these two methods and examples of their application can be found in corresponding reviews (DFT: Bengaard et al., 2002; Hammer and Nørskov, 2000; Van Santen and Neurock, 1995; UBI-QEP: Sellers and Shustorovich, 2002; Shustorovich and Sellers, 1998; Shustorovich and Zeigarnik, 2006; Zeigarnik and Shustorovich, in press).

Bypassing the detailed analysis of some achievements and remaining problems, let us just mention that there is a distinct difference between reactions catalyzed by oxides and metals. Here we provide preliminary characteristics of these two groups (see Table I). They will be further treated separately in Sections III.E and III.F.

III. "Elemental Base"

As was already mentioned, the core of each model is presented by the set of elementary reactions included. Their selection is based on the concept of the process under study and is determined by the goal of modeling. We assume that in the framework of the ultimate goal announced in this paper, maximum possible fullness is one of the main ruling principles for model compilation. A grounded selection of kinetic parameters is another crucial aspect of modeling. These two topics form a micro-kinetic basis of each model, which is the main subject of this paper. Below we present our vision of the sequence of stages needed to compile a working micro-kinetic scheme. Also, some difficulties existing along this path are highlighted.

A. Limitations of Species Taken into Account

As we mentioned in the beginning, the most important aspect of modeling is the stated objective. In the present case we attempt to compile a kinetic scheme,

TABLE I
Some Characteristics of Heterogeneous–Homogeneous Catalytic Reactions of Light Alkanes

Characteristics	Reactions over oxides	Reactions over metals
Examples	Oxidative coupling of methane (OCM), oxidative dehydrogenation of C_1–C_4 alkanes, partial oxidation of methane to synthesis gas[a], combined oxidative coupling of methane and toluene to styrene	Surface-stabilized combustion, partial oxidation of methane to synthesis gas[a], synthesis of cyanic acid from methane, ammonia, and oxygen[a]
Availability of direct experimental kinetic studies of elementary reactions	Limited; techniques—matrix isolation ESR (MIESR) Driscoll et al. (1987), Lunsford (1989), Garibyan and Margolis (1989–90), Tong and Lunsford (1991), Politenkova et al. (2001); ESR, optic spectroscopy Radzig (1993), Firsov et al. (1990)	Limited; techniques—molecular beams (including those with angular resolution) Smudde et al. (1993), Fairbrother et al. (1994), Hall et al. (1994), Kislyuk et al. (2000)
Availability of direct experimental studies of thermochemistry of surface centers/adsorbates	Limited; technique—in situ differential scanning calorimetry Sinev and Bychkov (1999)	(+/−)
Availability of theoretical (computation) methods for:		
a. Structure and energetics;	(+)[b]	(+)[a]
b. Kinetic parameters	(+)[b]	(+/−)[b]
Degree of surface (active sites) coverage by adsorbed species	Low[c]	Can be high
The main accepted method of kinetic scheme compilation	Addition of heterogeneous steps to existing homogeneous scheme	Compilation of assumed scheme of transformations is adsorbed layer and addition of homogeneous steps (or combination with homogeneous schemes); accounting of heat- and mass-transfer processes/fluid dynamics
Approaches to evaluation of kinetic parameters	1. By analogy with corresponding homogeneous reactions; 2. Use of thermochemical correlations	1. Evaluation using thermochemical data; 2. Quantum chemistry; 3. Optimization against experimental data

[a]Heterogeneous–homogeneous mechanism is assumed, but not proven for all cases;
[b]See text for comments;
[c]Hydroxylation of surface sites can be very high.

which would be applicable for the reactions of light alkane processing and satisfy the requirement of fullness. Stating this, we arrive at the necessity to identify first the range of species participating in the reactions to be further analyzed in the framework of the model.

This task is not trivial even in the case of the simplest hydrocarbons (C_1–C_4). Indeed, there are "obvious" participants in the process—initial reactants (hydrocarbon and oxidant), total oxidation products (water, CO_2), intermediates (target and side partial oxidation products). But already here we are facing a dilemma. The broader the list of species included (composition and structure of which do not contradict the chemical sense and which can be at least in principle present in the reaction mixture), the less definite is the available information about the kinetics of their transformation.

Further progress in building the model leads to increasing difficulties, since while deciding which species (and their reactions) should be included into the scheme, serious doubts about the reliability of even their thermochemical parameters appear.

Nevertheless, let us try to define which species should be included into the scheme, for instance, of methane transformations. They are supposed to be

1. Initial reactants (methane, oxygen).
2. Final products of total oxidation (water, carbon dioxide).
3. Incomplete oxidation products (methanol, formaldehyde, formic acid, CO, H_2).
4. Molecular products of radical transformations (first of all, C_2-hydrocarbons), molecular peroxides (H_2O_2, CH_3OOH).
5. Free radicals—products of sequential abstraction of H-atoms from the above molecules (CH_3, H, HO_2, OH, CH_3OO, CH_2OH, CH_3O, CHO, COOH, C_2H_5, C_2H_3).

In most cases in the range of conditions of interest to us (moderate temperatures and elevated—not less than ambient atmospheric—pressure), reactions of excited molecules can be excluded. The only exception will be discussed in detail in Section III.D—the group of total pressure-dependent reactions, which proceed via the formation and deactivation of excited molecules (compounds).

As we go further, some new questions arise about the necessity to account for additional species, for instance, products of deeper dehydrogenation of methane and ethane molecules (acetylene, CH_2, CH, C_2H radicals) and products of CH_3OO, CH_3O, and CH_2OH dehydration. Not all of these questions can be answered before the real analysis of the model begins. Some of those species should be kept in mind and added to the scheme (of course, together with complete blocks of their elementary reactions) as required based on the accepted logic of modeling.

B. ELEMENTARY REACTIONS

The elementary (or simple) reaction is one of the key concepts in the modeling of complex processes. An elementary reaction can be defined as a totality of all chemically identical elementary acts in which reordering of chemical substances and/or change of their state are taking place (see, for instance, Emanuel and Knorre, 1984). Since the following discussion deals with the development of kinetic schemes consisting of multiple reactions to which such "elementary" sense is attributed, a clear definition of this term is required. We will further call the reaction elementary if it passes through not more than one potential barrier in both directions (forward and reverse).

While modeling complex process kinetics, almost generally accepted is the rate equation for elementary reactions in the form of mass action law

$$W = k\Pi(C_i^m) \tag{3.1}$$

where k is the rate constant, C_i the concentration of reactant (i), m the reaction order with respect to reactant (i).

For many cases Eq. (3.1) serves as a good approximation. However, one must keep in mind that the more correct form of mass action equation must contain activity values, which in the general case differ from concentrations (or, in other words, activity coefficients are not equal to 1). Deviations of activities from concentrations are most pronounced at relatively low temperatures and high pressures, i.e., when properties of the reaction system display a pronounced difference from those of an ideal gas. Uncertainty of kinetic simulations can therefore increase if values of kinetic parameters obtained at low pressures are used to model high-pressure processes in the framework of Eq. (3.1). Among processes of interest announced in this work, at least one—oxidation of methane-to-methanol—severely needs high pressures, at which the non-ideality of the reaction system can in principle manifest itself.

The rate constant, k, in turn, as a rule, is presented in the form of the Arrhenius equation

$$k = Z \exp\left(\frac{-E_a}{RT}\right) \tag{3.2}$$

where Z and E_a are constants, to which somewhat different meanings are ascribed depending on the accepted interpretation of this exponential dependence. The activation energy E_a is usually treated as a barrier to be crossed by the reacting system to transform from reactant(s) into product(s). As to the pre-exponential factor Z, different theories interpret its physical sense differently and, correspondingly, give different mathematical expressions for it (see Alekseev, 1988; Emanuel and Knorre, 1984; Eyring et al., 1980; Glasstone et al., 1941). In particular, in the framework of active collision theory the pre-exponential factor for a bi-molecular reaction is considered as the frequency of collisions of a certain type. It is counted as

$$Z = \sigma \left(\frac{8RT}{\pi \mu}\right)^{1/2} \tag{3.3}$$

where σ is the collision cross-section, R gas constant, and μ the reduced molecular weight of colliding molecules.

On the other hand, following the activated complex theory, one can derive the value of the pre-exponential factor from the thermodynamic expression of the rate constant

$$k = \chi \frac{RT}{hN_A} \exp\left[\frac{\Delta S^{\#}}{R}\right] \exp\left[-\left(\frac{\Delta H^{\#}}{RT}\right)\right] \tag{3.4}$$

where h is Planck's constant, N_A Avogadro number, χ the transmission coefficient, $\Delta S^{\neq}$ and $\Delta H^{\neq}$ are standard entropy and enthalpy of formation of activated complex.

In spite of the difference in the underlying concepts and the forms of equations, Eqs. (3.3) and (3.4), both descriptions reflect the statistical sense of the rate constant. The latter statement is crucially important for better understanding of the problem existing in heterogeneous kinetics. Indeed, the above-mentioned theories are based on gas statistics and the given equations assume an equilibrium Maxwell–Boltzman distribution for gas species, which in the absence of reaction interact only via elastic collisions. If this can be considered as a satisfactory approximation for gas reactions at moderate temperatures and pressures discussed here (with some exceptions—see Section III.D), its applicability to the processes involving surface sites (i.e., elements of solid lattice) or adsorbed species is not so obvious.

Moreover, when rate constants are measured or derived somehow from experimental data, one can assume that the expression for the rate constant in the Arrhenius form can be used as an approximation, even if the meanings of Z and E_a are different from those followed from Eqs. (3.3) or (3.4). However, the situation is substantially different when we evaluate or calculate rate constants basing on some theoretical considerations. The discussion about applicability of such simple equations to reactions in condensed phase and/or interfaces has repeatedly ignited and extinguished during the last few decades. To our opinion, nowadays, when various theoretical methods are widely used for evaluation and calculation of rate constants, it is time to revert to the question.

A substantially macroscopic (in the thermodynamic sense) character of the parameter called the "rate constant" is its fundamental feature. Rigorously speaking, it is a statistical value (as well as temperature is) and can be applied only if the equilibrium distribution of Maxwell–Boltzman type is kept (see Knyazev and Tsang, 1999). Unfortunately, the requirement of using the more precise concept, namely reaction cross-section, in everyday practice would not be realistic even for homogeneous gas reactions. This is both because of the absence of corresponding data for a number of reactions, as well as due to the immense size of such a database if it would exist.

Taking a realistic stand, we nevertheless should keep in mind that not only any acceptable set of parameters, but also the shape of the mathematical expression, must be treated as approximations and may vary, for instance, with temperature. Evidently, we have to agree with the impossibility to derive an "absolutely-precise" expression of the rate constant, no matter how deep we go into the nature of elementary reactions.

One of the most striking examples of compromise between physical reality and a need for a "compact" and uniform mathematical description is the use of the so-called three-parameter form of the Arrhenius equation for the representation of the temperature dependence of rate constants

$$k = A\left(\frac{T}{T_0}\right)^n \exp\left(\frac{-E_a}{RT}\right) \tag{3.5}$$

Although non-Arrhenius behavior over a wide temperature range is more frequently a rule than an exception, the physical sense of the parameters A, n, and even E_a is not always clear. Moreover, parameters optimized for one temperature range may be poorly acceptable for extrapolation to another.

It is worth mentioning here that some other forms for representation of reaction rate and rate constants exist. Some of them are presented in Sandia National Laboratories document (Kee et al., 1996). Nevertheless, the three-parameter form now is almost generally accepted in modeling practice, and the corresponding parameters (A, n, and E_a) can be found for the largest number of elementary reactions in original publications, databases, and review articles (see, for instance, http://www.nist.gov/srd/nist17.htm; Tsang and Hampson, 1986; Warnatz, 1984).

As we mentioned in Section II.C, one of the most important requirements, which is applicable to the overall structure of reaction schemes, is its *thermodynamic consistency*. In other words, the scheme, as written, must allow the process to proceed asymptotically to its equilibrium state at infinite time. This can be reached only if any elementary step is included into the overall scheme together with its reverse reaction. Let us consider the consequences for the description of the process kinetics. For the sake of simplicity, we assume that the reaction proceeds in the ideal gas mixture. The value of rate constants $k_{(+)}$ for forward and $k_{(-)}$ for reverse reactions must satisfy the connecting equation

$$\frac{k_{(+)}}{k_{(-)}} = K \tag{3.6}$$

where K is the equilibrium constant.

Equilibrium constants, in their turn, can be calculated from available thermodynamic parameters of reactants and products

$$K = \exp\left[\frac{-(\Delta H - T\Delta S)}{RT}\right] \quad (3.7)$$

where ΔH and ΔS are the differences between corresponding thermodynamic functions (enthalpy of formation and standard entropy) of products and reactants. These functions are tabulated nearly for all molecular compounds involved in the processes discussed here, as well as for many radical species. However, the form of their representation—especially fractional polynomial temperature dependence—may contradict the above equations for rate constants. What could be recommended for the calculation of rate constants for reverse reactions if $k_{(+)}$ and thermodynamic functions are available is to use the averaged values of ΔH and ΔS for the temperature range of interest. If this range is too wide, or better precision is required, one can calculate K and $k_{(-)}$ values for several temperatures using Eqs. (3.6 and 3.7), and then approximate them by the optimal set of parameters A, n, and E_a. In some software packages developed for kinetic simulations these or analogous procedures can be performed automatically.

The analysis of particular kinetic schemes and the results of simulations indicate that models which meet the requirement of thermodynamic consistency possess additional benefits, such as possibilities to describe multiple processes within one kinetic scheme:

– Oxidation of one hydrocarbon to different intermediate partial oxidation products;
– Catalytic reactions of different starting compounds.

Unfortunately, up to now we possess no generally accepted standards of selection, treatment, and representation of kinetic data obtained by one means or another. Although the attempts undertaken by the CHEMKIN package developers and other similar institutions and research groups must be appreciated, the analysis of publications and discussions taking place on several Internet forums demonstrates the absence of any consensus or even co-ordination in this crucial aspect. This makes the comparison of results obtained by different research groups almost impossible. The most complete and popular kinetic database nowadays is the one developed by the National Institute of Standards (see web resource http://www.nist.gov/srd/nist17.htm), but this is, in fact, a collection of all published experimental and calculated data, which require further critical evaluation before being utilized in simulation practice. On the other hand, widely known databases of recommended kinetic parameters (Baulch et al., 1982, 1984, 1992, 1994; Kondratiev, 1970; Tsang and Hampson, 1986; Warnatz, 1984 and others) seriously suffer from incompleteness.

C. SELECTION OF RATE CONSTANTS

The most consistent way to develop a theoretical kinetic model of a complex process can be described as follows. First, we determine the list of species participating in the process, and then compile a set of elementary reactions based on the fullness principle. The most logical step after this would consist of *ab initio* calculations of kinetic parameters for elementary reactions included into the model (kinetic scheme). If the parameters calculated in this way are in significant contradiction with values obtained from independent experiments, this should cause a re-consideration of the underlying principles of both the calculations and the experimental measurements. However, this must not influence the core of the model and values of other kinetic parameters.

The above statement reflects an ideal, or extreme, view. It is clear that such approach is not realistic. Moreover, taking into account our limited knowledge about dynamics of elementary processes (even homogeneous) and other limitations discussed here, we must conclude that strict adherence to such "rules" cannot lead to a satisfactory result. In other words, the experimentally measured values of kinetic parameters must be taken into account.

Unfortunately, even in this case we are facing several difficulties. First, different reactions included into a complex kinetic scheme cannot be preliminarily studied to the same extent using independent methods. So, if for some reactions numerous data are available in corresponding literature and databases, in other cases not a single reliable value could be found. This means that in the framework of one kinetic model we have to use parameters obtained using very different experimental methods (direct and indirect) and also by more or less well-grounded evaluations.

Second, if for some reactions multiple experimental data are available, kinetic parameters obtained by different authors using different methods as a rule are not in a close agreement, and in some cases are even contradictory. In this case we have to employ some reference argumentation in order to choose the one to be used. Some criteria for such selection are already discussed in Sections I.D and III.B. In any case, the final decision about each value is going to be taken by a particular investigator, no matter if he/she gives credence to the authors of preceding studies or makes his/her own evaluations.

D. PRESSURE-DEPENDENT REACTIONS

Due to a broad width of temperature and pressure ranges to which the models of the processes under discussion should be applied, additional difficulties with the description of certain reactions arise. Such reactions proceed as multi-step processes with intermediate formation of excited species and their further deactivation via decay (dissociation) or via stabilization in collisions with other molecules ("third bodies"). Because of the latter, this group of processes is called sometimes "three-body reactions". This group includes recombination of atoms and relatively small radicals, formation of RO_2 (and HO_2) radicals from light alkyl radicals (or H-atoms) and O_2 molecules. Also, the processes which formally proceed as a mono-molecular decay in fact require an excitation in molecular collisions to gain the excess energy for crossing the dissociation barrier. Reactions of these types (e.g., recombination and dissociation) can and must be treated simultaneously within the same approach. The latter requirement follows from the stated thermodynamic consistency principle.

These reactions play an extremely important role in hydrocarbon oxidation over the temperature range discussed here, and in some cases even small variations in their parameters can lead to serious deviations in the simulated kinetics.

For all reactions of this group, at a "low-pressure limit" the rate increases proportionally (as a rough approximation) to the total pressure, or to the sum of the concentration of all species present in the system, since any collision can lead to stabilization of excited species. Also, all gas species contribute to the excitation of stable molecules for further dissociation. As the total pressure increases, another limit can be reached, in which the reaction rate depends only on the concentration(s) of the reactant(s), but not on the total pressure, and the overall kinetic order decreases. Corresponding rate constants at the "low" and "high" pressure limits are usually denoted as k_0 and k_∞.

Several approaches can be realized in kinetic models to describe the rate of "three-body" reactions in the transition (fall-off) regions (at "moderate pressures"). One of them represents the rate of reaction in terms of the "high pressure limit", but with a rate constant somewhat lower than k_∞. Corresponding decreasing coefficient asymptotically approaches 1 at increasing pressure and also decreases with temperature at a given total pressure. This approach requires knowledge about actual temperature and pressure dependencies; also additional correction of rate constants during the course of reaction must be performed. This substantially limits its adaptability and applicability.

Another approach proposed by Lindemann (1922) suggests the extended description of such reactions, which includes an explicit accounting of the formation of the excited species (marked with an asterisk) and two channels of its degradation

$$A + B \rightarrow AB^* \tag{L1}$$

$$AB^* + M \rightarrow AB + M \tag{L2}$$

$$AB^* \rightarrow A + B \tag{L3}$$

$$AB + M \rightarrow AB^* + M \tag{L4}$$

One can see that in the framework of this approach both directions (coupling and dissociation) are treated within one description. The main problem here consists of the absence of direct data about kinetic parameters of reactions (L1)–(L4). Their values should meet the following requirement: the overall description must approach the certain limits at high and low pressures and fit the data for fall-off region (if any available). Also, it is very desirable that these parameters are evaluated based on more or less grounded considerations.

The analysis of the scheme (L1)–(L4) shows that a steady-state concentration of excited species AB^* can be described as follows:

$$[AB^*] = \frac{k_{L1}[A][B] + k_{L4}[AB][M]}{k_{L2}[M] + k_{L3}} \quad (3.8)$$

If we ignore the contribution of reaction (L4), the rate of the product formation is equal to

$$W_{AB} \approx \frac{k_{L1}k_{L2}[A][B][M]}{k_{L2}[M] + k_{L3}} \quad (3.9)$$

The latter equation can be used to describe two extreme cases. If decomposition by far prevails over deactivation a "low-pressure limit" approximation can be applied

$$k_{L2}[M] \ll k_{L3} \quad \text{and} \quad k_0 = \frac{k_{L1}k_{L2}}{k_{L3}} \quad (3.10)$$

In an opposite case with deactivation prevailing, the kinetics typical for the "high-pressure regime" are observed

$$k_{L2}[M] \gg k_{L3} \quad \text{and} \quad k_\infty = k_{L1} \quad (3.11)$$

Thus, in the scheme (L1)–(L4) one rate constant (k_{L1}) can be equated to the "high-pressure" constant k_∞. Constant k_{L4} can also be easily estimated: with high probability its activation energy can be equated to the enthalpy of step (L1), or to the energy of A–B bond dissociation. As to the pre-exponential factor, it can be derived from the active collision theory.

One has to resort to additional assumptions to evaluate the constants k_{L2} and k_{L3} (such as lifetimes of excited species, deactivation probability in reaction (L2)) and the connecting relation subsequent upon the Eqs. (L1–L4)

$$\frac{k_{L3}}{k_{L2}} = \frac{k_\infty}{k_0} \quad (3.12)$$

In some cases an adjustment of both constants (k_{L2} and k_{L3}) must be performed in Eq. (3.12) in order to impart a physical sense upon them. A substantial simplification of this procedure is taking place if direct data about deactivation and/or about the pressure dependence of the apparent "high limit" constant are available.

As we already mentioned, similar considerations can be suggested and analogous expressions can be obtained for reverse processes, i.e., for reactions, which are usually considered as dissociation under the conditions of hydrocarbon oxidation. Typical examples are hydrogen peroxide and alkyl hydroperoxide decomposition. Despite existing difficulties, the Lindemann approach can be recommended as it allows:

– To use a uniform representation of all rate constants included into the kinetic scheme;
– To describe direct and reverse processes within one scheme;
– To account for new data about the kinetics of excited species (e.g., about efficiency of different "colliders" in reactions (L2) and (L4)) without changing the structure of the overall scheme.

There are also some other methods that provide a more accurate description of the fall-off region than does the simple Lindemann form. The details can be found in corresponding publications (see, for example, Allison, 2005; Gilbert et al., 1983; Stewart et al., 1989; Wagner and Wardlaw, 1988). Some kinetic simulation packages (e.g., CHEMKIN) optionally realize these approaches for representation of "three-body reaction" rates (Kee et al., 1996).

Finally, we mention that there are certain reactions, which to our opinion must be with no doubt described as "three-body processes". Among them, first of all, are

– Recombination of methyl radicals and H-atoms;
– Formation and dissociation of hydro-, methyl- and ethyl-peroxy radicals;
– Dissociation of hydrogen peroxide and alkylhydroperoxides.

E. Heterogeneous–Homogeneous Catalytic Reactions on Oxide Catalysts

Evidently, OCM was the first process the heterogeneous–homogeneous nature of which has been studied in detail. In this example many substantial features of this class of processes were rationalized. This includes the elaboration of the principles of modeling starting from the micro-chemical level. It is worth noticing that OCM is a striking example of the impossibility to adequately describe the behavior of complex systems if some "insignificant" elementary reactions are not included into the kinetic scheme. In this process the key step leading to the formation of higher hydrocarbons—the recombination of methyl radicals—can take place only if gas reactions between CH_3 and O_2 are not taking place. This is exactly what happens in the temperature range of 873–1,223 K at atmospheric pressure and oxygen concentrations below 20 vol.%. In these conditions the equilibrium constant of reaction

$$CH_3 + O_2 \rightleftharpoons CH_3O_2 \qquad (5)$$

is already small enough, but the rate constants of methyl radical oxidation

$$CH_3 + O_2 \rightarrow CH_3O + O \qquad (6)$$

$$CH_3 + O_2 \rightarrow CH_2O + OH \qquad (7)$$

are not yet high enough to compete with recombination, despite the fact that the oxygen concentration in the reaction mixture is much higher than that of CH_3 radicals (see Fig. 5). As a result, the "OCM window" appears. It is important that in the same temperature range reactions with $E_a \approx 200$ kJ/mole (roughly one-half of the C–H bond strength in methane) can proceed with a relatively high rate. This is an additional, but not the main, factor providing for the possibility of the formation of higher hydrocarbons during methane oxidation. Although any standard procedure can demonstrate "insignificance" of reactions (5)–(7) under typical OCM conditions, their removal from the kinetic scheme of the reaction leads to the impossibility to describe the existence of the "OCM window". This would also lead to the loss of predictive power of the model, as far as variation of parameters (total pressure, concentration of oxygen) is concerned.

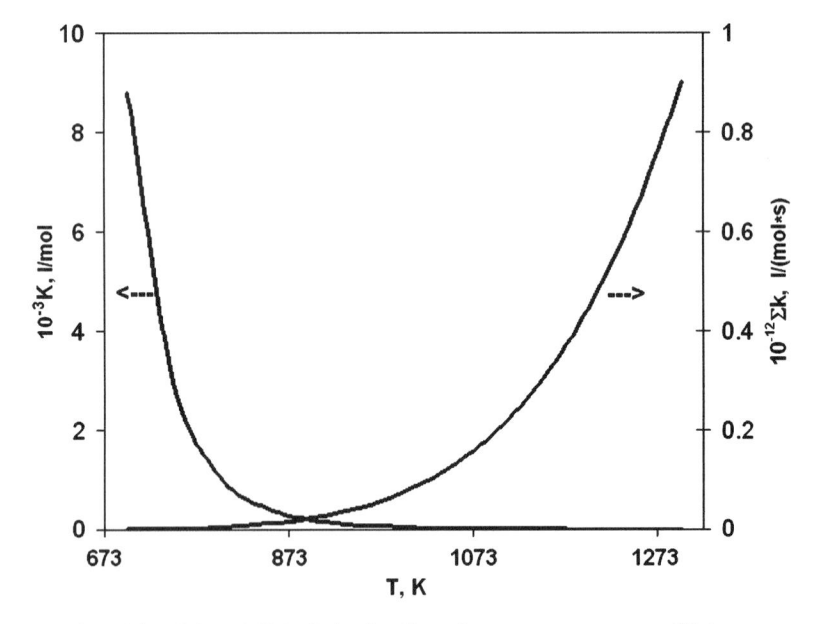

FIG. 5. The origin of the "OCM window": effect of temperature upon equilibrium constant of CH_3O_2 formation (K) and CH_3 oxidation rate constant (Σk).

At the initial stage of studying the mechanism of OCM, the analysis of the reciprocal influence of heterogeneous and homogeneous factors was confined to explanations of some important, but selected, peculiarities of the process using a limited number of CH_3 reactions in the gas phase. In particular, Lunsford and co-workers (Ito and Lunsford, 1985; Lin *et al.*, 1986) first pointed out that the increase of C_2-selectivity at rising temperatures can be explained by the competition of recombination with other reactions of CH_3 radicals: interaction with active surface sites and with oxygen in the gas phase. Soon after that, a "short" kinetic scheme consisting of five elementary reactions was suggested (Sinev *et al.*, 1987) that reflected some substantial channels of CH_3 transformation

$$CH_3 + O_2 \rightleftharpoons CH_3O_2 \quad (5)$$

$$CH_3 + CH_3 \rightarrow C_2H_6 \quad (8)$$

$$CH_3O_2 + CH_3O_2 \rightarrow 2\,CH_3O + O_2 \quad (9)$$

$$CH_3O_2 + CH_3O_2 \rightarrow CH_3OH + CH_2O + O_2 \quad (10)$$

$$CH_3 + CH_3O_2 \rightarrow 2\,CH_3O \quad (11)$$

If one assumes that

– reaction (5) is in equilibrium,
– only reaction (8) leads to the formation of OCM products,
– reactions (9)–(11) lead to O-containing products and to the removal of methyl radicals from the main OCM path,

the following expression for ethane selectivity at low conversions was derived:

$$S_{Lim} = \left\{ 1 + K_5 C_{O_2} \left[\frac{K_5 C_{O_2}(k_9 + k_{10})}{k_8} \right] + \frac{k_{11}}{k_8} \right\}^{-1} \quad (3.13)$$

The S_{LIM} value was named "limiting" as in real system in addition to reactions (5), (8)–(11) many processes (homogeneous and heterogeneous, parallel and sequential) leading to the formation of products other than C_2-hydrocarbons can proceed. Figure 6 shows "S_{LIM} vs. temperature" curves simulated assuming temperature independence of rate constants for reactions (8)–(11). Experimental data obtained by different authors prove that at low conversions it is impossible to obtain an OCM selectivity higher than that predicted by Eq. (3.13). At the early stage of investigations in this area, such a kinetic argument was strong evidence for the exclusive formation of ethane via CH_3 recombination in the gas phase. Later on, clear experimental evidence for this conclusion has been obtained using TAP (Temporal Analysis of Products)

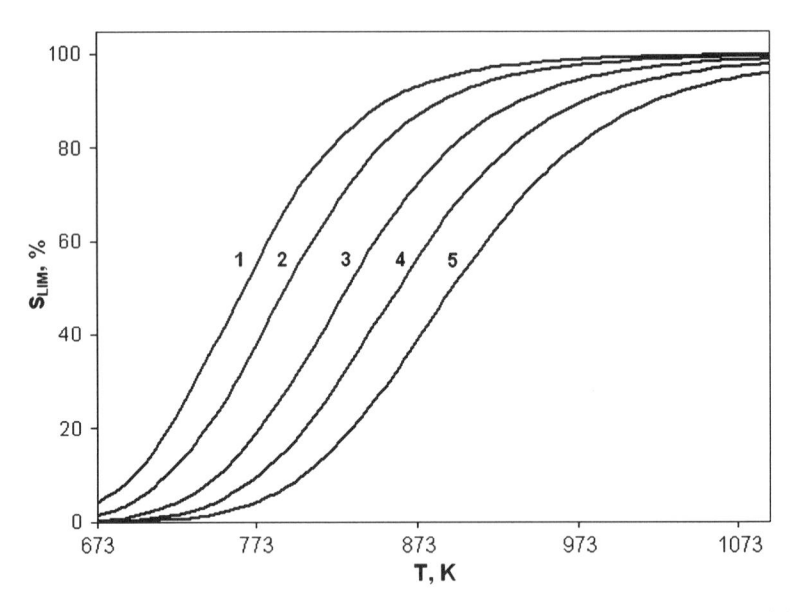

FIG. 6. Limiting selectivity (S_{LIM}) as a function of temperature at different oxygen partial pressures: (1) 0.01 bar; (2) 0.02 bar; (3) 0.05 bar; (4) 0.1 bar; (5) 0.2 bar.

(Buyevskaya et al., 1994; Mallens et al., 1994) and SSITKA (Steady-State Isotopic Transient Kinetic Analysis) (Nibbelke et al., 1995) techniques. It was demonstrated that ethane—the primary OCM product—is leaving the reactor with the same characteristic time as an inert tracer. This surely indicates that no intermediates noticeably residing on the surface participate in its formation.

Further development of kinetic models for the OCM process followed the path of addition of a limited number of heterogeneous steps (first of all—initiation or generation of primary methyl radicals) to homogeneous schemes of methane oxidation (Aparicio et al., 1991; Hatano et al., 1990; McCarty et al., 1990; Shi et al., 1992; Vedeneev et al., 1995; Zanthoff and Baerns, 1990). There was certain logic in such an approach: since the most efficient OCM catalysts are almost exclusively oxides with no transition metal ions (some Mn-containing oxide systems are the only exception), any reactions in adsorbed layers at such temperatures can be neglected. In the framework of such models some substantial features of the process could be described. For instance, they predicted the limit in the C_2-hydrocarbon yield close to that reliably observed experimentally over the most efficient catalysts (20–25%).

Such relatively simple kinetic models also allowed one to analyze simultaneously the development of radical processes and mass-transfer on different levels (inside catalyst pores, in inter-particle space, in reactor bulk—see for instance Bristolfi et al., 1992; Couwenberg, 1994; Hoebink et al., 1994; Reyes et al., 1993). This can be considered as another important achievement.

However, the more careful analysis indicates that all the schemes discussed above suffer from one intrinsic contradiction. Indeed, if the main process of activation of the initial hydrocarbon (e.g., methane) is a homolytic C–H bond cleavage on an oxidative surface center

$$[O]_S + CH_4 \rightarrow [OH]_S + CH_3 \tag{12}$$

it is logical to assume that any other species from the gas phase must undergo reactions of the same type (H-atom transfer). Moreover, for most of them corresponding rate constants should be higher than that for methane (due to weaker H-atom bonding).

The same can be said about re-oxidation of reduced surface sites (e.g., oxygen vacancies $[..]_S$): if the addition of an oxygen molecule to vacancies

$$2[\,]_S + O_2 \rightarrow 2[O]_S \tag{13}$$

is considered as the main re-oxidation path, any other O-containing species, such as radicals (peroxy, alkoxy, hydroxy) and even more oxygen atoms, should actively interact with the same vacancies, for instance via reactions

$$[\,]_S + CH_3O_2 \rightarrow [O]_S + CH_3O \tag{14}$$

$$[\,]_S + CH_3O \rightarrow [OCH_3]_S \tag{15}$$

Moreover, since most oxides known as efficient OCM and ODH catalysts are characterized by high values of the oxygen binding energy, dehydroxylation, which is usually considered as an evident step in catalytic cycle, is strongly reversible. This means that in the presence of high concentrations of water in the reaction mixture one should anticipate high degrees of surface coverage by OH-groups. Consequently, it is logical also to include into the consideration their interactions with gaseous species having high H-atom affinity—practically with all atoms and radicals present in the gas phase during hydrocarbon oxidation, as well as with molecular oxygen

$$[OH]_S + X \rightarrow [O]_S + XH \tag{16}$$

It is also important to notice that reactions similar to (12)–(16) must be taken into account from another point of view. Even in the presence of low-surface area catalysts, the total number of species in the gas phase inside the catalyst bed can be less than the number of surface sites. For instance, at 1 bar and 1,000 K the total number of species in 1 cm^3 of gas is 8×10^{18}. It is less than the total number of sites on an MgO surface at 1 m^2/g specific surface area and 1 g/cm^3 bulk density ($\sim 2 \times 10^{19}$ sites/cm^3). This means that even statistically gas species have a higher probability to collide with surface sites than with any other gas particle.

Such extension of a kinetic scheme can very seriously influence the results of simulations qualitatively changing our notions about pathways of processes under consideration. Indeed, rapid interaction of the most active gas species with surface sites and an increasing number of transformation channels would inevitably cause corresponding changes in the state of the working catalyst surface, as well as in the gas composition predicted by the model. As a result, the interpretation of experimental data of the overall rate and product distributions will change correspondingly. However, if this approach is valid, the reliability of the description and the predictive power of the model should increase significantly.

There is a serious obstacle on a path leading to building such advanced models, namely the absence of a generally accepted concept of how free radicals react with surfaces of different nature. Does this interaction always proceed as a direct collision or in some cases it is proceeded by more or less stable adsorbed precursor? What properties of the solid surface—local or collective—are responsible for directions and rates of such reactions to a greater extent? Which approaches are more fruitful for evaluation of kinetic parameters in this case? What experimental information could be relevant and helpful for building more adequate models and for a more precise evaluation of kinetic parameters? What type of experimental data should be employed to examine the efficiency of such models?

Before we expound upon possible alternative to answer the above set of questions, it should be noted once again that no model of complex phenomena can be identical to the object under study. Proceeding to the modeling we should say to ourselves: "Let us imagine that some totality of properties of the system can be described in such-and-such terms in such-and-such way (chemically, physically, mathematically). Let us look how the model based on these principles behaves in the range of parameters of our interest. Let us check if this behavior has any common features with the behavior of the real system under study. If so, there is a chance that we have succeeded in reflecting some substantial properties of the system in the model. Then let us check if there is any difference in the behavior of the model and the real system—this would suggest the direction of the further development of the concept and help to estimate whether or not any further improvement can be achieved at the current level of knowledge".

Such target setting allowed one of us to suggest the following approach to the analysis of heterogeneous–homogeneous OCM over oxide catalysts (Sinev, 1992, 1994, 1995; Sinev et al., 1995), which later has been extended over some other processes of lower alkane oxidation (Sinev et al., 1997a, b; Sinev, 2003, 2006).

1. Substantial features of processes under consideration can be described in the framework of a multi-step scheme, analogous to those used in homogeneous gas-phase kinetics.

2. All heterogeneous reactions of gas species are considered as their stoichiometric elementary reactions with surface sites of distinct chemical nature.
3. Active surface sites participating in the main catalytic route are strongly-bonded oxygen species (surface lattice oxygen). They can exist in two main forms—oxidized and reduced (in the form of surface OH-groups) further denoted as $[O]_S$ and $[OH]_S$, respectively. Subsequent elimination of surface OH-groups (dehydroxylation) can lead to the formation of surface oxygen vacancies.
4. Reactions of gas species with surface sites in oxidized and reduced forms, as well as with oxygen vacancies, proceed in the same way as their bi-molecular homogeneous reactions, i.e., as direct collisions without any preliminary equilibrium adsorption. The same groups of reactions as in the gas phase are considered:
   – H-atom transfer;
   – O-atom transfer;
   – Capture of particle by a surface site (analogous to recombination in the gas).
5. All paired interactions leading to any reaction of above type should be considered, if possible.
6. Since a direct measurement of rate constants for the heterogeneous reactions included into the scheme is practically impossible, all of them must be evaluated basing on uniform principles following the same procedure.
7. No adjustment of kinetic parameters for "improvement" or "fitting" to any experimental data is allowed.
8. The main regularities of heterogeneous reactions are analogous to those for elementary gas processes of the same type.

Let us consider the last statement in more detail. Keeping in mind serious doubts stated in Section III.B, we nevertheless assume that pre-exponential factors for rate constants of collision-type reactions between gas species and surface sites can be evaluated as a gas-surface collision frequency (v). The dimension (or cross-section) of the surface sites (σ) and their concentration per unit surface area (ϑ), as well as a steric factor (ξ) should also be taken into account. The latter value shows the probability of reciprocal orientation of two reactants optimal for the reaction to proceed during the collision. As a result, the pre-exponential factor can be calculated as follows:

$$Z = \sigma \vartheta \xi \left[\frac{RT}{2\pi\mu_g}\right]^{1/2} \qquad (3.14)$$

where μ_g is the molecular weight of gas particle.

Data on ionic radii can be utilized for the evaluation of σ. For instance, for lattice oxygen ions $\sigma \approx 6 \times 10^{-20}\,m^2$ can be accepted. Concentrations of surface sites ϑ can be obtained, for instance, from gas titration experiments (Sinev *et al.*, 1990), if bulk diffusion of oxygen does not play a significant role. As to the steric factors ξ, their values can be taken the same as for analogous homogeneous reactions of the same particle with spatially non-constrained partners (e.g., O- or H-atoms).

Thus, what we still need for description of heterogeneous rate constants is a method for evaluation of activation energies. One important observation helped to solve this problem. While studying kinetics and thermochemistry of redox processes over typical OCM catalysts, it was found (Bychkov *et al.*, 1989; Sinev *et al.*, 1990) that the activation energy of methane interaction with [O]$_S$ sites can be sufficiently well described in terms of the well-known Polanyi–Semenov correlation (see Fig. 7)

$$E_a = a + b|\Delta H| \tag{3.15}$$

where ΔH is enthalpy change in the process, a and b are constants.

Simultaneously the same conclusion was derived from methane interactions with active centers on activated silica surface (Bobyshev and Radzig, 1988, 1990).

The analysis of existing data for numerous gas processes shows that the values of a and b vary for reactions of different types. For H-transfer reactions we can set $a = 0$, if the transfer proceeds between two electronegative atoms (e.g., O and F, or O and O), and $a \approx 30$ kJ/mole in all other cases. As to the b value, it depends on the sign of ΔH: in the case of endothermic reactions it can be taken with acceptable precision as 1; for exothermic H-transfer processes

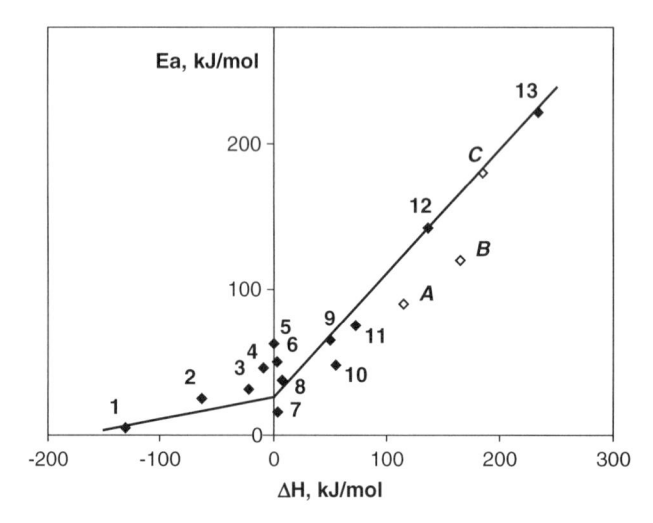

FIG. 7. Polanyi–Semenov-type "E_a vs. ΔH" correlation for $CH_4 + X \rightarrow CH_3 + XH$ reactions. X = F (1), OH (2), C_6H_5 (3), CF_3 (4), CH_3 (5), H (6), Cl (7), O (8), SH (9), CH_3O (10), Br (11), I (12), O_2 (13), Li/MgO (A), K/Al$_2$O$_3$ (B), Pb/Al$_2$O$_3$ (C).

$b \approx 0.25$–0.3. In the case of O-transfer reactions, $a \approx 0$ and $b \approx 1$. For capture processes, $E_a = 0$ was accepted.

Thus, thermochemistry of catalyst surface sites is a key factor determining their reactivity and, consequently, kinetic parameters of elementary heterogeneous reactions. For all of them the values of ΔH can be evaluated based on two thermochemical values—H-atom affinity of surface active sites (or O–H bond dissociation energy in surface OH-groups, $E_{[O-H]}$) and oxygen binding energy, $E_{[O]s}$. These values were measured experimentally for three different OCM catalysts—Li/MgO, Pb/Al$_2$O$_3$, and K/Al$_2$O$_3$ (Bychkov et al., 1989; Sinev et al., 1990); they are presented in Table II.

The above thermochemical values were used to fill the "heterogeneous module" of the kinetic scheme for the OCM reaction over a model Li/MgO catalyst with corresponding kinetic parameters (see Table III). In combination with a scheme of homogeneous methane oxidation, this set of reactions forms the desired micro-kinetic description. It allowed us to re-consider specific features of the OCM process and to obtain some unexpected results.

Let us first discuss the simulated data obtained with increasing concentration of catalytic active sites [O]$_S$ in a reacting methane–oxygen mixture. The results presented in Fig. 8 show corresponding changes of the development of the process in time. Whereas the kinetic curve for the "purely homogeneous" process is typical for radical chain scenario with a pronounced induction period and almost exponential growth of the reaction rate, at gradually increasing concentration of active sites the initial reaction rate linearly increases. Simultaneously the acceleration of the process caused by development of chains becomes less and less significant. Finally, at [O]$_S$ concentrations comparable with those present in real catalytic systems, the process becomes practically "linear", i.e., it starts with a maximum rate which gradually decreases along with the current concentrations of initial reactants. This indicates that catalytic sites play a combined role: they actively generate primary radicals, but at the same time they terminate development of chains. As a result, the catalyst totally changes the character of the process: remaining free radical, it becomes non-chain due to a very rapid quadratic termination via recombination of heterogeneously-generated CH$_3$ radicals and to a high probability of secondary reactions of any radical existing in the gas mixture.

This substantially extended model allows one to simulate more subtle effects, which are, however, extremely important for a better understanding of both

TABLE II
Measured Thermochemical Characteristics (Parameters) of Three OCM Catalysts

Parameter	Li/MgO	Pb/Al$_2$O$_3$	K/Al$_2$O$_3$
$E_{[O-H]}$, kJ/mole	320	250	270
$E_{[O]s}$, kJ/mole	535	407	450

TABLE III
ACCOUNTED HETEROGENEOUS REACTIONS AND THEIR ESTIMATED KINETIC PARAMETERS (Li/MgO)

No.	Reaction	K^a	N	E_a (kJ/mole)
1	$LO + CH_4 = LOH + CH_3$	1.225×10^8	0.50	90
2	$LOH + O_2 = LO + HO_2$	8.7×10^7	0.50	115
3	$LOH + LOH = LO + L + H_2O$	5.6×10^{13}	-0.50	250
4	$LO + L + H_2O = LOH + LOH$	1.16×10^5	0.50	0
5	$L + L + O_2 = LO + LO$	8.7×10^4	0.50	0
6	$LOH + H = LO + H_2$	4.9×10^8	0.50	15
7	$LO + H = LOH$	4.9×10^8	0.50	0
8	$LOH + HO_2 = LO + H_2O_2$	8.4×10^5	0.50	0
9	$L + HO_2 = LO + OH$	8.4×10^5	0.50	10
10	$LOH + OH = LO + H_2O$	1.20×10^8	0.50	0
11	$LO + H_2O_2 = LOH + HO_2$	8.4×10^7	0.50	44.5
12	$L + OH = LOH$	1.20×10^8	0.50	0
13	$LO + HO_2 = LOH + O_2$	8.4×10^6	0.50	0
14	$L + O = LO$	1.20×10^8	0.50	0
15	$LOH + O = LO + OH$	1.20×10^8	0.50	0
16	$LO + CH_3 = LOCH_3$	1.26×10^6	0.50	0
17	$L + OCH_3 = LOCH_3$	1.0×10^7	0.50	0
18	$LOCH_3 + O_2 = LOCH_2 + HO_2$	6.0×10^7	0.50	188
19	$LOCH_3 + HO_2 = LOCH_2 + H_2O_2$	4.0×10^5	0.50	52.5
20	$LOCH_3 + OH = LOCH_2 + H_2O$	1.8×10^8	0.50	19
21	$LOCH_3 + CH_3 = LOCH_2 + CH_4$	1.5×10^5	0.50	50
22	$LOCH_3 + CH_3O = LOCH_2 + CH_3OH$	1.20×10^6	0.50	16.5
23	$LOCH_3 + CH_3O_2 = LOCH_2 + CH_3OOH$	7.0×10^4	0.50	56.5
24	$LOCH_3 + CH_2 = LOCH_2 + CH_3$	1.5×10^7	0.50	48
25	$LOCH_3 + C_2H_5 = LOCH_2 + C_2H_6$	1.5×10^7	0.50	56.5
26	$LOCH_3 + C_2H_3 = LOCH_2 + C_2H_4$	1.5×10^7	0.50	48
27	$LOCH_3 + C_2H = LOCH_2 + C_2H_2$	2.5×10^7	0.50	0
28	$LOCH_3 + O = LOCH_2 + OH$	8.0×10^7	0.50	21
29	$LOCH_3 + H = LOCH_2 + H_2$	1.00×10^8	0.50	27
30	$LOCH_3 + LO = LOCH_2 + LOH$	1.5×10^{14}	-0.50	42
31	$LOCH_2 + O_2 = LOCHO + OH$	1.4×10^6	0.50	21
32	$LOCHO + O_2 = LO + CO_2 + OH$	1.0×10^7	0.50	0
33	$LOH + CH_3 = LO + CH_4$	1.26×10^6	0.50	15
34	$LOH + CH_3O_2 = LO + CH_3OOH$	7.0×10^5	0.50	0
35	$L + CH_3O_2 = LO + CH_3O$	7.0×10^7	0.50	10
36	$LOH + CH_3O = LO + CH_3OH$	3.0×10^7	0.50	0
37	$L + CH_3O = LO + CH_3$	1.0×10^7	0.50	10
38	$L + CH_3O = LOCH_3$	1.0×10^7	0.50	0
39	$LO + CH_3O = LOH + CH_2O$	6.0×10^7	0.50	0
40	$LOH + CH_2OH = LO + CH_3OH$	6.0×10^7	0.50	21
41	$LO + CHO = LOH + CO$	6.0×10^7	0.50	0
42	$LO + CH_2O = LOH + CHO$	6.0×10^7	0.50	12.5
43	$LO + CH_3OH = LOH + CH_2OH$	6.0×10^7	0.50	41.5
44	$LO + CH_3OH = LOH + CH_3O$	3.0×10^7	0.50	83.5
45	$L + CH_2O = LOCH_2$	1.0×10^7	0.50	0
46	$LOCH_2 + CH_3 = LOH + C_2H_4$	1.26×10^6	0.50	0

Table III (*continued*)

No.	Reaction	K^a	N	E_a (kJ/mole)
47	$LO + C_2H_5 = LOH + C_2H_4$	6.0×10^7	0.50	0
48	$LO + C_2H_6 = LOH + C_2H_5$	9.0×10^7	0.50	75.5
49	$LO + C_2H_4 = LOH + C_2H_3$	9.5×10^7	0.50	103.5
50	$LO + H_2 = LOH + H$	3.5×10^8	0.50	95

$^a k$—in mole/l, s.

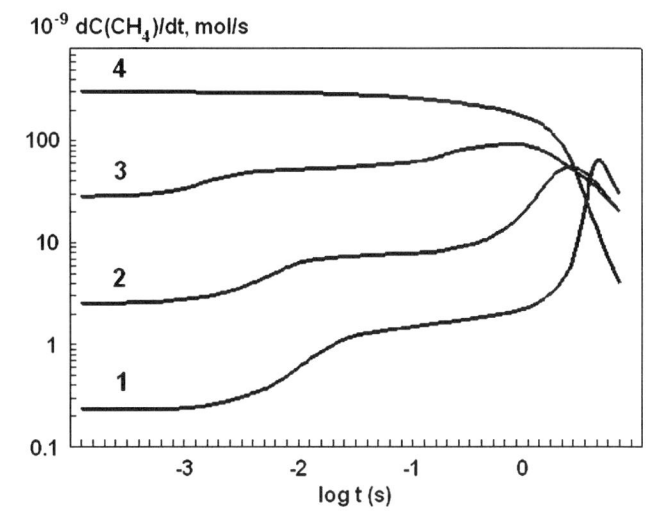

FIG. 8. Methane conversion rate vs. time curves at different concentrations of catalytic active sites; quasi-homogeneous simulation; 1,000 K, 1 bar, $CH_4:O_2 = 10:1$. C(LX), mol/l = 0 (1); 5×10^{14} (2); 5×10^{15} (3); 5×10^{16} (4).

surface and free radical chemistries. Let us consider as an example in more detail the regularities of catalyst re-oxidation. The following experimental results were obtained (Sinev *et al.*, 1990) during Li/MgO catalyst sequential reduction by hydrogen or methane and re-oxidation with molecular oxygen and N_2O:

- Reduction by methane pulses results in the formation of ethane, but not water;
- Reduction with hydrogen leads to its consumption without water evolution;
- If N_2O pulses are supplied onto the sample pre-reduced in methane or H_2, only some insignificant amounts of nitrogen and oxygen are formed; only purging in dry helium for 1 h at 700°C leads to some additional N_2 evolution and oxygen consumption in the few first few pulses of nitrous oxide indicating the appearance of some oxygen vacancies;

- Re-oxidation of the catalyst by O_2 pulses leads to the immediate formation of water, i.e., the characteristic time of this process is less than the time during which the oxygen pulse resides in the layer of catalyst (<1 s);
- Apparent activation energies of reduction and re-oxidation evaluated from initial rates of consumption of methane over pre-oxidized and oxygen over pre-reduced catalyst are equal to 63 and 105 kJ/mole, respectively, whereas under steady-state conditions the apparent activation energy of the process at low conversions exceeds 210 kJ/mole.

These facts are indicative of a complex mechanism of the dehydroxylation process, which is not surprising if we keep in mind a high oxygen binding energy for this catalyst (see Table II). Also, they are suggestive of the possibility of re-oxidation without intermediate formation of oxygen vacancies.

Simulations based on the kinetic scheme described above (Sinev and Bychkov, 1993) indeed show the possibility of an alternative re-oxidation path via the sequence of steps:

$$[OH]_S + O_2 \rightarrow [O]_S + HO_2 \tag{17}$$

$$[OH]_S + HO_2 \rightarrow [O]_S + H_2O_2 \tag{18}$$

$$H_2O_2 \rightarrow 2OH \tag{19}$$

$$[OH]_S + OH \rightarrow [O]_S + H_2O \tag{20}$$

Among them, steps (18) and (20) are exothermic and their activation energies should be insignificant. Step (17) must have a relatively low E_a close to the value obtained from the re-oxidation experiments (with oxygen as on oxidant). As to step (19), its activation energy is high (about 190 kJ/mole), which determines the apparent value obtained from experimental data. Thus, the following conclusions were formulated:

- Re-oxidation of the oxide catalyst can proceed via a heterogeneous–homogeneous free radical path;
- The rate of steady-state catalytic reaction is determined not by activation of methane and formation of ethane, but by the most activated step in the sequence that leads to the formation of the second main product, namely water; the latter is required for closing the catalytic cycle.

It is also important to notice that the nature of the prevailing re-oxidation mechanism—via dehydroxylation or via the reaction set (17)–(20)—may differ from one catalyst to another depending on the values of the thermochemical parameters of surface oxygen species (Sinev and Bychkov, 1993). In this case an elegant experimental way to validate the model predictions can be suggested. If the catalyst was pre-treated by oxygen of normal isotopic composition, and methane (or any other hydrogen-containing reactant) is oxidized in the presence of $^{18}O_2$, the isotopic composition of forming water (correctly taking into account independently measured rates of oxygen hetero-exchange) should unambiguously indicate the proportion of two routes.

It is necessary to acknowledge that some existing experimental data indicates that oxygen can be removed from Li/MgO directly during the reduction in hydrogen at temperatures as low as 873 K (see, for instance, Leveles, 2002). Such discrepancies with the data described above might be due to some difference in catalyst preparation/pretreatment procedures, which leads to the formation of active sites with somewhat different thermochemical characteristics. What is important is to attribute the evaluated kinetic parameters to the catalysts of particular thermochemistry.

In the framework of this description an attempt to model an effect of spatial non-uniformity of real catalytic systems was made (Bychkov *et al.*, 1997). It was assumed that reaction proceeds in a heterogeneous system represented by two active infinite plane surfaces and in the gas gap between them. Surface chemistry was treated as for the Li/MgO catalyst (see Table III). Because of substantial complexity of the kinetic scheme consisting of several hundred elementary steps, the mass-transfer was described in this case as follows. The whole gas gap was divided into several (up to 10) layers of the same thickness, and each of them was treated as a well-stirred reactor. The rate of particle exchange between two layers was described in terms of the first-order chemical reaction with a rate constant:

$$k_{Di} = \frac{D_i}{(d)^2} \qquad (3.16)$$

where k_{Di} is the "diffusional" rate constant for ith particle, D_i the coefficient of gas-phase molecular diffusion for ith particle (which for simplicity was estimated as diffusion coefficient in air), and d the characteristic dimension (thickness) of one layer.

Two edge layers included surface species chemically and kinetically described as discussed above and considered together with all other species within one quasi-homogeneous mixture. In other words, in edge layers the catalytic active sites were equally accessible to all gas species. Surface sites were not allowed to react with each other and to migrate from the edge layers.

Evidently, such an approach represents a significant simplification. In particular, the use of molecular diffusion coefficients at small distances from reaction surfaces contradicts its physical sense. Nevertheless, even this estimation reveals some essential features of a spatially distributed reaction system. First, as shown in Fig. 9, at low gas gap thickness the total rate of conversion is entirely determined by the generation of primary CH_3 radicals in catalytic cycles on the surface. Second, even when the thickness of gas gaps reaches macroscopic values, heterogeneous reactions have a strong effect on the process that

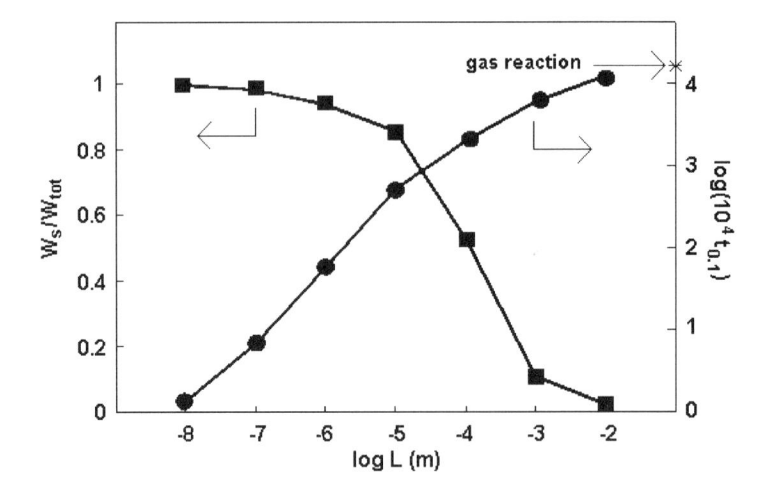

FIG. 9. Methane oxidation in spatially distributed system: effect of gas gap thickness on $t_{0.1}$ (time of reaching 10%-conversion) and relative efficiency of heterogeneous and homogeneous methane activation (W_{hom}/W_{tot}) (Sinev et al., 1997a, b).

proceeds predominantly in the gas phase. This conclusion is also important for evaluation of the significance of surface ("wall") reactions in gas-phase oxidation usually considered as "homogeneous".

At last, when the gas gap thickness is within some "critical" area (~0,1 ÷ 1 mm), a sharp change in the probability of chain propagation in the gas phase is predicted. It is worth noticing that the variation of catalyst particle size within the same range leads to sharp changes in the contribution of homogenous reactions to total conversion also during the ODH of C_3–C_4 alkanes in the presence of V-containing oxide catalysts (Vislovskiy et al., 2000). This is very likely due to a sharp increase of chain length in homogenous reactions in gaps between catalyst species.

Of course, such accounting for mass-transfer is an oversimplification of real processes taking place during alkane oxidation over real catalysts. Additional studies are required to estimate the possibility to integrate a detailed microchemical (and micro-kinetic) description with methods capable of advanced accounting of mass-transfer on the inter- and intra-particle level and in the bulk of reactor (see, for instance, Couwenberg, 1994; Hoebink et al., 1994).

To conclude this section, we should mention again that the phenomena taking place during alkane oxidation over oxide catalysts are much more complex than any description which can be expressed in the form of micro-kinetic schemes of the type given in Table III. For instance, such schemes cannot reflect deep changes in the catalyst, which can occur under the influence of gas mixtures containing water and CO_2 as reaction products. Such changes can lead to a very substantial modification of catalyst chemistry ranging from poisoning (reversible and irreversible) of active sites with carbon oxides

$$[O]_S + CO_2 \rightleftharpoons [OCO_2]_S \qquad (21)$$

$$[O]_S + CO \rightarrow [OCO]_S \qquad (22)$$

to total changes in catalyst chemical and phase compositions, morphology, surface area, and density of active sites. Of course, reactions similar to (21) can be included into the overall scheme in order to describe the modification of catalyst activity by carbon oxides, especially if the data on adsorbed CO_X lifetimes under different conditions are accessible (Leveles, 2002). Their influence can also be accounted for in analytical form, as it is done by Xu *et al.* (1992). However, at the moment we do not see any rational way of accounting for and including into models of the discussed type the whole totality of factors, which can be called "effect of reaction mixture upon catalyst chemistry and morphology". Consequently, any quantitative coincidence of simulated and experimental data should not be expected. And vice versa, this is an additional reason to resist the adjustment of kinetic parameters or their "optimization" against experimental data.

F. Heterogeneous–Homogeneous Catalytic Reactions on Metal Catalysts

Another very important group of light alkane oxidation processes is represented by reactions catalyzed by metals, and by Pt-group metals in particular. Pioneering publications by Schmidt and co-workers (Hickman and Schmidt, 1992, 1993; Huff and Schmidt, 1993) had opened the area of catalysis at short (millisecond) contact times. After that, several groups have contributed to methane (Bui *et al.*, 1997; Deutschmann and Schmidt, 1998; Deutschmann *et al.*, 1994; Heitnes *et al.*, 1994; Mallens *et al.*, 1995) and higher alkane (Beretta and Forzatti, 2001; Beretta *et al.*, 1999; Reyes *et al.*, 2001; Zerkle *et al.*, 2000) oxidation over noble metal catalysts (monoliths, foils, gauzes, in annular reactors). Both experimental attempts and modeling have been employed in order to elucidate the main features of reaction mechanisms and also to find the conditions for optimal yields of particular products (oxygenates, olefins, synthesis gas).

In such systems modeling meets very serious difficulties, since the problem of formulation of a kinetic description of reactions in the adsorbed layer on the active metal is added and interferes with other problems stated above. One of the first attempts to suggest such a description was done by Hickman and Schmidt (1992, 1993). Analyzing a nearly 10-year period of development in the area, Schmidt (2001) concluded that "… these apparently simple processes are in fact far more complicated than the usual packed bed catalytic reactor assumptions used for typical modeling. First, the temperatures are sufficiently high that some homogeneous reaction may be expected to occur, even at very short reaction times. Second, the gradients in all properties are so large that all conventional assumptions may be inaccurate".

Indeed, a great discrepancy between different authors exists concerning the role and contribution of surface and gas-phase reactions in these very complex processes. Evaluations range from the recognition that the metal surface plays a dominant (or even sole) role in the formation of products (de Smet *et al.*, 2000; Deutschmann *et al.*, 2001; Mhadeshwar and Vlachos, 2005) to the conclusion about almost complete formation of target products in the gas phase ignited by the oxidation reaction on the metal surface (Beretta and Forzatti, 2001; Beretta *et al.*, 1999; Reyes *et al.*, 2001). Even for practically the same reactions there are very different opinions. Liebmann and Schmidt (1999) claim on the purely heterogeneous reaction systems for C_1–C_3 partial oxidation, heterogeneously assisted homogeneous reactions of C_{5+} reaction systems, and a mixed mechanism for C_4. However, other authors (e.g., Beretta and Forzatti, 2001; Beretta *et al.*, 1999; Reyes *et al.*, 2001; Silberova *et al.*, 2003; Tulenin *et al.*, 2001) provide arguments for the formation of olefins from light alkanes via catalyst-assisted gas-phase process.

By contrast, there is no doubt or any contradictions in literature concerning the importance of macro-kinetic factors (heat- and mass-transfer processes) in the reactions discussed in this section. It is generally realized that these processes proceed in spatially distributed systems and generate sharp gradients of parameters (temperature, pressure, density, concentrations, etc.). It could be noted here a distinct similarity of alkane oxidation over Pt-group metals at short contact times and well studied and practically implemented ammonia oxidation and cyanic acid synthesis reactions, in which transfer processes play a dominant role (see Satterfield, 1970).

Let us, however, focus on the chemical and micro-kinetic sense of modeling. As we mentioned several times, the success in kinetic simulations that is sparking further progress in gas processing is associated with the development of models meeting certain requirements. Among the latter, the requirement of "fullness" is crucial.

Unfortunately, our knowledge about reactions in the adsorbed layer is far from complete. In such conditions, it is very difficult to meet the requirement of "fullness" of the reaction scheme. As a result, we fall into an exclusive circle. On the one hand, the necessary information about certain reaction intermediates and elementary steps, which has to be accounted for in the model, is missing. But, on the other hand, if some intermediates are not accounted for, reaction pathways related to them become omitted in simulations. For instance, if the escape of active species (atoms, radicals, excited molecules) from the metal surface to the gas phase and their capture by surface active sites are not accounted for, no coupling between gas-phase and surface chemistries can be derived from the modeling.

Another important aspect is a detailed accounting of reverse elementary reactions, i.e., the thermodynamic consistency of the reaction scheme. If this requirement is not executed, even the possibility of achieving correct final state in equilibrium reactions (or in limited quasi-equilibrium blocks inside a complex reaction network) is doubtful. We have to mention here that it is very difficult to say *a priori* which reverse reactions can be neglected in particular conditions. Unfortunately, in the case of reactions in adsorbed layers any reliable information about thermodynamic properties of surface intermediates, especially in mixed ad-layers, is scarce. Although recent developments in quantum chemistry and complimentary semi-empirical and phenomenological methods are encouraging (as applied to methane oxidation—see, for instance, Mhadeshwar and Vlachos, 2005, and literature cited therein), there is still a lack of information required for modeling of alkane oxidation kinetics over metal catalysts.

We would like also to mention some specific problems as far as the modeling of reactions on active metals is concerned.

1. Rate constants of heterogeneous reactions are usually represented in the "three-parameter" form discussed above. However, this form has a clear physical sense only for reactions in ideal gases at a strictly kept equilibrium Maxwell–Boltzman distribution. Despite several attempts to adopt transition-state theory to heterogeneous reactions, and to those proceeding in adsorbed layers in particular (e.g., Krylov *et al.*, 1972; Zhdanov *et al.*, 1988), its applicability in these cases is doubtful.
2. The most efficient metal catalysts consist of relatively small clusters with tens or hundreds of atoms exposed to the gas phase and accessible as adsorption sites. On the other hand, a detailed description of reaction in the adsorbed layer, even in the case of methane, can include a comparable number of *types* of surface intermediates: for instance, Mhadeshwar and Vlachos (2005) take into consideration 18 different surface species. In the latter case the authors modeled the reaction on a "macroscopic" surface, but anyway, certain caution is required when we apply "normal" kinetic equations to a reacting system where the conditions for statistical behavior are not assured.
3. Reactions on metals, including many oxidation processes, are known to proceed in a way very different from stochastic collision types, which can be described by mass action (or "acting surfaces"). The number of systems in which collective effects or topochemical type processes (via nucleation and growth of nuclei) are proved to determine the kinetic behavior is increasing. Despite the extensive literature on reactions in oscillatory regimes and spatially-structured reactions on surfaces (Gorodetskii *et al.*, 2005; Latkin *et al.*, 2003; Peskov *et al.*, 2003), such facts have not yet found an adequate reflection in the area under consideration.
4. In some practically important cases, e.g., partial oxidation of methane over Pd, Ni, and Co catalysts (Deng and Nevell, 1996; Hu and Ruckenstein, 1998; König *et al.*, 1994; Ozkan *et al.*, 1977; Tulenin *et al.*, 1996; Zhang *et al.*, 2003) complex kinetic behavior is caused by oxygen storage in the catalyst, which leads to the formation of bulk or sub-surface oxides. This process is

accompanied (or alternated) by the periodic formation of oxide layer(s) and carbonaceous deposits (see, for instance, Tulenin et al., 1999, 2004; Zhang et al., 2005). The latter form in thermodynamically non-equilibrium state and their thermochemistry and reactivity are changing with time (Bychkov et al., 2002a, b, 2003). Although these factors make the process extremely complicated, they also must be accounted for inadequate kinetic models.

In spite of the above listed complications, there is distinct progress in the kinetic description of alkane (first of all—methane) partial oxidation over metal catalysts in some particular cases. One bright example of this kind is represented by a recent publication by Mhadeshwar and Vlachos (2005) summarizing their achievements in the modeling of methane transformations and related reactions of oxygenates (methanol and formaldehyde) over a Rh surface. Among them:

1. Compiling a kinetic scheme that meets the requirement of thermodynamic consistency;
2. Accounting of homogeneous reactions and analysis of heterogeneous–homogeneous coupling;
3. Utilization of kinetic parameters independently measured or calculated using UBI-QEP and DFT;
4. Accounting of adsorbate–adsorbate self- and cross-interactions for a more accurate evaluation of kinetic parameters in the adsorbed layer.

Even though the authors could not avoid some adjustment of selected kinetic parameters, what is explicable taking into account the extraordinary complexity of the system. As a result, they succeeded in reproducing in their simulations some important features of the real system and validated their micro-kinetic model against high-pressure spatially resolved experimental data for catalytic partial oxidation of methane.

To conclude this section, we state that the area is very actively developing. The main achievements by now are reached in experimental techniques, in developing computational codes and software, and in accounting for the role of heat- and mass-transfer. These results and developments are worth to be used in related areas, first of all in catalytic oxidation of alkanes over oxides. Corresponding information can be found (and in some cases even discussed) on personal web sites of scientists involved into such investigations (see for instance the site of Prof. Olaf Deutshchmann: http://www.detchem.com) and companies and research groups developing commercial or semi-free software for kinetic simulations (e.g., www.kintecus.org, CHEMKINusergroup@reactiondesign.com).

A brief survey demonstrates that kinetic models taking into account both heterogeneous and homogeneous elementary steps potentially can serve as powerful tools for analysis of experimental data, development of catalytic systems, and selection of optimal reaction conditions. Despite a lack of information about kinetic parameters of numerous elementary reactions, more detailed kinetic schemes, even filled with preliminary (estimated) kinetic parameters, can offer very important mechanistic information. At the same time, shorter (trimmed) schemes can give misleading simulation results due to the ambiguity of routes to certain products in a complex reaction network. In other words, "fullness" of the scheme should always dominate over the use of "correct" values of kinetic parameters if both requirements cannot be executed simultaneously.

IV. Modeling and Experimentation

A. COMPARISON WITH EXPERIMENT

Experiment is the beginning and the end, the starting-point and final objective of any modeling. Despite the fact that the model is fundamentally not identical to the object of modeling, only experiment suggests the initial guess and provides primary data concerning the structure of the reaction scheme and the values of the kinetic parameters. In their turn, model verification and validation can be done only by comparison with experimental data.

Vedeneev et al. (1988a, b, c, d) regarded the model as acceptable if it describes the main peculiarities of the process and possesses some predictive power. Such target setting assumes the gradual upgrading of the model aimed at improving its quantitative agreement with various and regularly replenished experimental data.

In many cases modeling gives more definite and clearly interpretable results than experiment. It can also provide an understanding of the reasons for the discrepancy between different experiments. In modeling we have a possibility to set well-determined conditions (temperature, pressure, reactant flows, heat- and mass-transfer, including very complex regimes) and to vary them broadly. Also, a mapping of complex values, such as reaction (overall and elementary) rates, concentrations of reactive intermediates, pressure, flows, and even temperature, which is hardly accessible in experiment, can be obtained by adequate modeling. Such information is crucially important for optimization of various processes and their selected parameters. On the other hand, the design of complex experiments becomes much better grounded if it is based on preliminary modeling.

In the case of the processes of heterogeneous–homogeneous type, their intrinsic micro-chemical mechanisms can be revealed almost exclusively by modeling (see Sections I.C, II.D, III.E, and III.F).

Despite such deep interdependence, in practice the relationships between modeling and experimentation are not so cloudless for several reasons. One the one hand, any experiment carries various errors, uncertainties and effects of numerous non-accountable factors, and kinetic simulations in their turn are always non-ideal because of an inevitable incompleteness of the model and inaccuracy of kinetic parameters. This is why any deviations between compared experimental and simulated data would not be surprising. On the other hand, very often "good fits" to experimental results are reported for models, which by no means seem to be adequate for particular applications.

One of the most intriguing and paradoxical example of this kind was already mentioned in Section II.A, namely simulation of the SID. This gross parameter, which is very important for combustion kinetics, is very often used to prove the validity of kinetic models because of the ease and lack of ambiguity of its experimental determination. Since combustion itself is a high-temperature process, the SID values are also often simulated in the framework of kinetic schemes developed for the high-temperature regime of oxidation. As a rule, it is not a big problem to reach agreement between calculated SID values with experimental data, which would evoke wonder and disappointment rather than satisfaction. Why, the ignition takes place at low or moderate temperatures, whereas high-temperature models as a rule do not account for the reactions crucially important for these temperature ranges, but indeed unessential at temperatures of combustion, where they were developed. The existence of such examples suggests a high probability of reciprocal annihilation of errors in the course of simulations in the framework of complex kinetic schemes. This may cause very serious failures and lead to false conclusions.

Thus, the investigator is urged to treat more carefully the qualitative adequacy of the model for the particular experimentation.

The problem of adequacy is very serious also from the side of experimentation. Here we revert to the question about the goal, in this case the goal of comparison of experimental and simulated data, which determines

- The optimal parameters and criteria for comparison of calculated and experimental data;
- The main requirements of the experiment, which could be compared with results of modeling.

We have to confess that very often such questions are not taken into account and by no means does every experiment meet the minimal requirements for comparison with simulated data. Vast experience gained in the area of heat- and mass-transfer in combustion (Frank-Kamenetskii, 1969) indicates that the most appropriate for comparison with model calculations are relative parameters and criteria, which are less influenced by the particular experimental "hardware" and the scale of process.

As applied to the oxidative processing of light alkanes, and the formation of certain valuable products in particular, the switch to relative values would mean, for instance, the use of product distributions and yields instead of absolute concentrations and degrees of conversion instead of reaction time. This can be illustrated by the example given in Fig. 2. In this case—in methane-rich conditions—the level of conversion is controlled by oxygen. The use of oxygen conversion as an "independent" parameter is much more fruitful and convenient if characteristic times of processes in different conditions can differ very significantly. But what is even more important, such relative characteristics are much less sensitive to the particular values of kinetic parameters and, consequently, are more suitable for the analysis of the structure of the kinetic scheme.

Similar ideas will be further developed in the next section, along with some other criteria and requirements. In our opinion, a strict adherence to them would improve the efficiency of the interaction between experimentation and modeling. To conclude this section, let us formulate in brief the main tasks addressed concerning the comparison of modeling with experimental data as far as the optimization of the model targeted toward the studies of the reaction mechanism and process optimization over a wide range of parameters are concerned. In our opinion, such comparison must reveal the factors that have been underestimated and overestimated in the kinetic scheme. As to the values of kinetic parameters, they definitely can be "optimized", but this optimization should be based on exact physical and chemical (experimental and theoretical) arguments, but not on formal mathematical procedures.

B. Requirements of Experiment

It is clear now that the comparison of measured and simulated data suggests strict requirements not only for kinetic models, but also for the experiment. A common shortcoming of many experimental studies consists of incompleteness of the data. On the one hand, incompleteness in the description of reaction conditions does not allow for an adequate comparison with simulation, since the effect of some very important factors can be lost or at least distorted. On the other hand, irregular measurements of some parameters in the course of reaction, or even the absence of some important data, nullify the value of such studies for the comparison with modeling.

Any experiment suffers to some extent from the influence of various uncontrolled factors, such as mass- and heat-transfer regimes in different parts of the reaction system, irregularity of temperature fields and flow profiles, dead spaces, poor mixing, and independent flow of reactants, "wall chemistry" and its variation with time, etc. Another problem is a correlation between signals received from different sensors and real values of measured physical parameters. There is a voluminous literature on this subject; here we just mention the pitfalls related to the measurements of surface and gas temperatures in the course of reaction involving very active radical species.

In many cases in the reactor (and even in connecting tubing) reactions proceed at the conditions different from those recorded in the experimental protocol. Despite numerous attempts to avoid such effects of uncontrolled factors, the problem of an adequate description of reaction conditions is far from the exhaustive solution. Moreover, trying to solve some local problems we can modify the system to an extent that it may lose its identity. For example, Mantashyan and co-workers suggested (see Mantashyan, 1998 and literature cited therein) a so-called "wall-less" reactor as an approach to exclude the effect of surface processes. In this case, the reaction in the gas is initiated by a laser pulse at a distance from the reactor walls, and during some time the process can develop via a true homogeneous regime. However, we in fact dealing with the process affected by strong local heating in the zone where the laser beam interacts with the reacting gas and by sharp gradients of all the main physical parameters. In other words, trying to avoid some "disturbing" factor (reactor walls in particular) we introduce another one (laser beam) and substantially change the sequence of processes taking place in the reacting system.

Undoubtedly, such "differentiating" methodologies can be successfully used to highlight and isolate selected parts or features of the complex system and provide extremely valuable kinetic information. However, along with them the experimentation of "integrating" type is very important for verification and validation of kinetic models. Accordingly, the arrangement of such experiments should meet well-defined requirements. First of all, we have to define the parameters to be used for the most unambiguous comparison of experimental and simulated data.

For instance, a considerable number of studies in the area of partial oxidation are devoted to the optimization of target product(s) yield or selectivity at total oxygen conversion in hydrocarbon-rich conditions. However, one can insist that the results of such experiments are the most indefinite, since "total conversion" reflects only the sensitivity of the analytical method used (oxygen detection threshold). It does not characterize the stage of reaction from which a given yield is obtained. But the latter is crucial for the comparison with results of modeling, since some reactions of target and side products can substantially modify their distributions in "anaerobic" conditions. The solution of the problem can be found in the way of sampling from different sections of the reactor (e.g., along the axis in a flow system). Since in many cases such sampling is difficult, wide variations of gas flow (residence time) could be an alternative way to collect the data for the comparison with simulated results.

The use of conversion instead of time as an independent variable can be very distinctive and informative for model verification. Also, as can be seen from Fig. 2, the "yield (or selectivity) vs. conversion" relationship can serve as a characteristic function describing the behavior of the system under various conditions.

The optimal product distribution in many cases is an ultimate goal of applied and even basic research. From the standpoint of the subject under discussion, product distributions are among the most important criteria for the comparison of modeling with experiment. However, it is worth noticing that many experimental details can strongly affect the values used for comparison. For instance, the whole spectrum of components of reaction mixtures formed during alkane oxidation includes permanent gases (O_2, N_2, CH_4, CO, H_2), carbon dioxide, alkanes and unsaturated hydrocarbons C_2–C_4, condensable oxygenates, and water. If they are analyzed using gas chromatography, a wide range of analysis sensitivity towards different compounds is the least egregious problem. It is more important that they practically cannot be analyzed in one probe. Often gaseous compounds are analyzed on-line, whereas condensable products are first collected in cooled traps. During the collection (which may last relatively long) the condition in the reactor or reaction rate may change for several reasons, even if the process is studied in a stationary state. Moreover, condensation is never complete, and additional chemical processes (condensation, oligomerization) can proceed to some extent even in the trap, despite the low temperature and addition of inhibitors. All these factors hamper the correct determination of the mixture composition and assigning it to certain experimental conditions.

In our opinion, more definite correspondence between experiment and modeling can be established based on relative concentrations of selected (kinetically relevant) products and different conversions. Such products should be formed in comparable amounts and, if possible, analyzed in one probe (gaseous or liquid). Kinetic relevance means that there should be definite notions concerning the relationships between certain components in the proposed kinetic scheme: they should form from the same intermediate, or one from another, or have different relations at different stages of reaction. For example, in the case of methane oxidation, this could be the ratios of methanol and formaldehyde concentrations in a trapped liquid phase or $(CO+CO_2)/(C_2H_6+C_2H_4)$ in the gas. As to the CO_2-to-CO ratio, its variation with conversion can be very indicative of the contribution of reactor walls in gas-phase oxidation: carbon dioxide can form in homogeneous steps only via CO interaction with O-containing radicals (peroxy, hydroxy, alkoxy) or at relatively high temperatures with molecular oxygen. However, it can form directly from organic intermediates via a sequence of surface reactions (alkoxides → carboxylates → formiates → carbonates). Accordingly, if the fraction of CO_2 in the sum of CO_X does not tend to zero at low conversions, this is evidence for its heterogeneous formation and a reason for the corresponding modification of the kinetic scheme.

The completeness of the reaction mixture analysis is a very painful problem. In some cases the researcher even cannot guess which products can form and, correspondingly, what is the most adequate analytical methodology. More often the qualitative composition of the reaction mixture (excluding the most unstable intermediates) is known for the most part, but the complete quantitative analysis cannot be performed for technical reasons. In the latter case the chosen analysis always reflects some pre-assumptions about the kinetic (or mechanistic) relevance of different products. Unfortunately, a wrong initial guess in this case cannot be compensated for at later stages of investigation because the required information is irretrievably lost.

A typical example of this kind is a measurement of some "side" products during alkane oxidation. For instance, traditionally in experimental studies, in which gas chromatography is used as the main analytical method, water and hydrogen concentrations are not measured for different reasons. Quantitative measurement of water in the reaction mixture is difficult due to its high adsorbability. As to hydrogen, it is often believed that it does not form in oxidative processes. Moreover, the sensitivity of the most frequently used detectors—TCD (thermal conductivity detectors)—with respect to H_2 is very low (when helium is used as a carrier gas). As a result, whereas carbon balance is considered as one of the main criteria of the quality of the experimental procedure and setup, hydrogen balance is usually omitted. However, the concentration of molecular hydrogen and/or the H_2/H_2O (H_2/CO, H_2/olefins) ratio can serve as one of the main criteria indicating the routes of reaction and the nature of intermediates involved in the formation of various C-containing products (see, for instance, Sinev et al., 2003).

It can be asserted that a special experimental methodology should be developed for a proper verification and validation of modeling. Definite reaction conditions and complete analysis of the reaction mixture are the principal requirements for such experiments. Let us consider in more detail this requirement with reference to the reaction temperature. As it was mentioned in Section I.B, this is the most important parameter determining not only reaction rates and equilibrium states, but also the general type of oxidation process. Since oxidation is a substantially exothermic process, it proceeds to a considerable extent in a self-generated temperature regime, which can vary from isothermal to adiabatic, depending on the intensity of heat removal from the reaction zone. Correspondingly, very different temperature profiles (axial and radial) can exist in the reactor strongly affecting the reaction. In order to model the process correctly, one must have information about these profiles. Alternatively, they can be obtained as one of results of modeling, but in the latter case anyway the mapping of the temperature distribution is required in order to compare it with a calculated one. Such profiles can be obtained by placing multiple thermocouples into the reactor, or by measuring the temperature by one or a few thermocouples moving along the reactor in the axial and radial directions. However, one must take into account that the material of the thermocouple or of a cover can be active at the conditions of the oxidative process and affect (and modify) both the reaction itself and the results of temperature measurements.

Another possible solution of the problem is to perform the reaction at a very low initial oxygen concentration. Although in the case of alkane partial oxidation this is not favorable for obtaining high yields of target products (oxygenates, olefins, synthesis gas), an almost uniform temperature field can be achieved.

Coming back to the problem of surface reactions in homogeneous oxidation, let us mention briefly that their role can be revealed by varying the total pressure, reactor dimensions, wall material, and pre-treatment (see, for instance, Arutyunov et al., 2002; Burch et al., 1989).

Thus, one can conclude that reliable verification and validation of kinetic models require a special arrangement of experiments. In the ideal it includes a thorough description of reaction conditions, complete analysis of the reaction mixture, and use of adequate criteria for the comparison of experimental and calculated data. We must confess, however, that a consistent execution of these requirements is very time and resource consuming. This is why a pragmatic (trade-off) approach can be elaborated for the design of experiments. The following elements of this approach can be also employed for the selection of already published experimental data for verification of kinetic models:

- Low initial concentrations of oxygen for diminishing self-generated temperature gradients;
- Availability of the data for different conversions varied over a maximum possible range to construct characteristic curves "yield (or selectivity) vs. conversion";
- Availability of concentrations of products measured in one probe to diminish the distorting effect of sampling;
- Wide variation of experimental conditions (temperature, total pressure, initial concentrations, heterogeneous factors) to display possible qualitative effects revealing intimate peculiarities of kinetic scheme.

The above consideration does not purport to be full. We did not even mention here very powerful methods providing extremely valuable kinetic information, such as various spectroscopies (especially performed in *"in situ"* mode) and isotope tracing and labeling. Although these methods substantially improve the efficiency of model validation and elucidation of reaction pathways (see, for instance Mims *et al.*, 1994), they have not yet become very common in "kinetic experimentation for kinetic modeling". Instead, we focused on some problems typical for everyday practice and just traced possible solutions.

V. Some Examples and Comments

The examples given below illustrate and supplement the principles and general reasoning stated in the previous sections.

A. C_1–C_2 Joint Description

As we emphasized above, although the oxidation of methane is accompanied by the formation of a variety of compounds, C_2-hydrocarbons play among them a special role. First of all, recombination of methyl radicals, which are primary and in some cases the most abundant radicals in the system, leads to a quadratic termination of the chain process and to a large extent controls the overall reaction rate. As a result, ethane forms as one of the main and "early" products (in terms of the number of reaction steps leading to its formation). As its reactivity is somewhat higher than that of methane (due to a pronounced difference in C–H bond strength), it becomes involved in the process all along. This means that practically all elementary steps of the transformation of ethane (and consequently, ethylene and acetylene) with no doubt must be included into the kinetic scheme of methane oxidation. Some of these reactions play a very important role in the overall process (see, for instance, Section V.C). Thus, any model developed for methane oxidation that does not take into account the formation and subsequent reactions of C_2 hydrocarbons (e.g., Vedeneev et al., 1988a, b, c, d) does not meet the requirement of "fullness" and for most of applications cannot be considered as a good approximation.

On the other hand, if we start from ethane or any other C_2 hydrocarbon as an initial reactant, active intermediate species typical of the methane oxidation scheme appear in the reaction mixture at relatively early stages of the process. Among them are methyl radicals, which almost immediately produce methane by abstracting H-atom from the parent ethane molecule. In other words, the qualitative composition of the reaction mixture during C_2 hydrocarbon oxidation does not differ significantly from the one forming when methane as a starting reactant is oxidized.

In addition to the above reasoning, there is ample experimental evidence for the expediency of the joint analysis of C_1–C_2 hydrocarbon oxidation within a uniform kinetic scheme. One recent example of this kind was obtained recently during the oxidation of methane–ethane mixtures (Arutyunov et al., 2005). It was demonstrated that the α value characterizing the relative "burning out" of two hydrocarbons and expressed as

$$\alpha = \frac{([CH_4]_f/[C_2H_6]_f)}{([CH_4]_0/[C_2H_6]_0)} \quad (5.1)$$

where "0" and "f" subscripts designate initial and final concentrations, remains very close to 1 and just slightly varies over a wide range of initial methane-to-ethane ratios (Fig. 10), despite a significant difference in their reactivities. Such behavior can be adequately simulated in the framework of the joint kinetic scheme of C_1–C_2 hydrocarbon oxidation. On the contrary, in the course of oxidation of the methane mixtures with propane and butane, the latter compounds are burnt out first, which is indicated by the fast increase of methane concentration in the mixture (Fig. 11). It means that oxidation pathways of C_{3+} hydrocarbons are less conjugate with those of methane. As the first approximation, they can be neglected in a kinetic model of methane and ethane oxidation, unless the formation of higher hydrocarbons is among the main subjects of investigation (like in Marinov et al., 1996; Mims et al., 1994). Unlike that, oxidation of methane and ethane cannot be modeled separately.

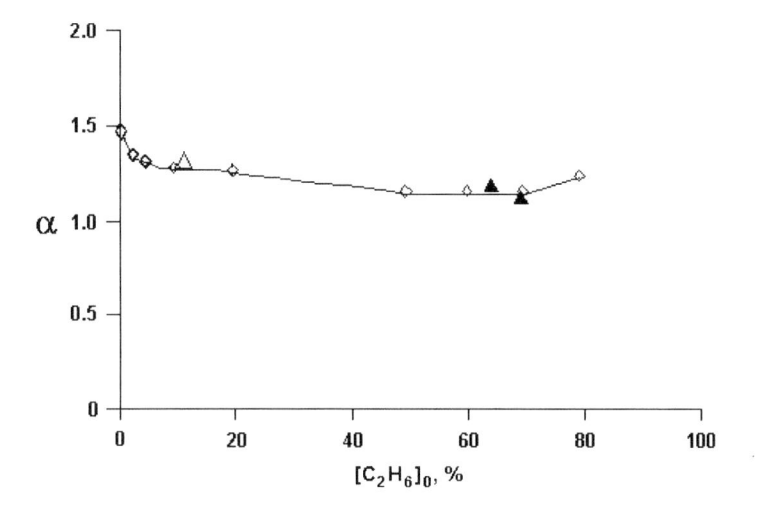

FIG. 10. Relative "burn out" of C_1–C_2 alkanes α as a function of initial ethane concentration. (-◇-)—Simulations ($T = 673$ K, $P = 70$ bar, $[O_2]_0 = 5\%$); (▲)—experiment ($T = 720$ K, $P = 25$ bar, $[O_2]_0 = 4$–6%); (△)—experiment ($T = 720$ K, $T = 35$ bar, $[O_2]_0 = 3\%$).

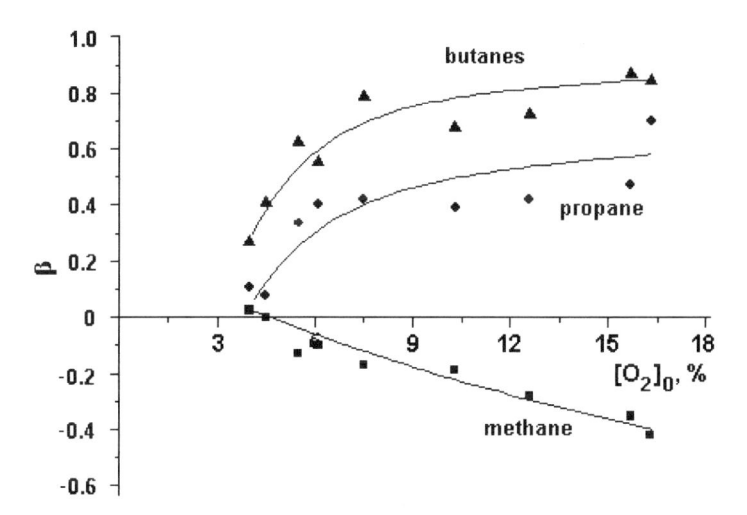

FIG. 11. Relative reactivities of methane and C_3–C_4 alkanes at different initial oxygen concentrations; $\beta = (C_{i,0} - C_{i,f})/C_{i,0}$. ($T = 800$ K, $P = 30$ bar). (■)—methane, (●)—propane, (▲)—butanes.

B. Expansion on Higher Hydrocarbons

If we accept the prohibition of varying or adjustment of rate constants as a ruling principle of model development, this opens the prospect for using the joint C_1–C_2 description as a basis for an expansion on some different areas. One possible direction is a description of catalytic processes, over oxide catalysts in particular, which was discussed in Section III.E. Another possibility is a step-by-step development of kinetic models of higher hydrocarbon oxidation. Of course, the expansion on C_3–C_4 alkane oxidation is the first aim. There are some factors making such expansion possible, including the deep analogy between corresponding reactions of C_1–C_2 and C_3–C_4 hydrocarbons and the similarity of intermediate species and main reaction routes of their oxidation.

However, there are also very important limitations. As we mentioned above, even in oxidation of methane and ethane many elementary reactions are not accessible for direct and detailed investigation. When we shift from ethane to propane, not only the number of carbon atoms in the molecule increases, but also the complexity of the reaction network. In particular, one may assume that even in propane oxidation the formation of complex oxygenate intermediates, including bi-radicals and complex peroxides, may take place. Although such compounds can play a principal role in kinetically relevant steps (such as chain-branching), up to now our knowledge about such compounds is negligible.

Another important difference between oxidation of C_1–C_2 and C_{3+} hydrocarbons is the appearance in the latter case of "degeneration" of the primary alkyl radicals. Already in the case of propane, the existence of two isomeric forms of propyl species (not always taken into account) can lead to substantial kinetic consequences because of the distinct difference in their thermochemistry and reactivity. Even certain reaction channels may vary depending on the isomeric form of propyl radicals. This factor may cause a substantial uncertainty especially in the case of modeling of catalytic oxidation due to a poor knowledge about thermochemistry and reactivity of surface active sites and chemisorbed species.

Nevertheless, there is a possibility to solve restricted tasks using approaches described above. In particular, the utility principle (see Section II.C) can be fruitfully utilized. For instance, such an approach was utilized by Vedeneev et al. (1997a, b) to describe the NTC behavior in propane oxidation. Generally speaking, if we are interested in the optimization of some particular product yield, we can select a block of reactions in which it is formed and further transformed and analyze it taking into account qualitative or even semi-quantitative notions about the reaction environment.

An important example of this kind is a contribution of cracking processes to partial oxidation of propane and higher hydrocarbons. In particular, in the case of catalytic propane ODH, the formation of lower hydrocarbons—first of all ethylene and methane—can substantially reduce propylene selectivity. The analysis of possible homogeneous and heterogeneous pathways of C–C bond breaking can provide valuable guidelines for further improvement of catalyst formulation and/or overall process design.

Figure 12 shows the lifetimes of *n*- and *iso*-propyl radicals in gas-phase reactions of H-atom and CH_3-radical elimination leading to the formation

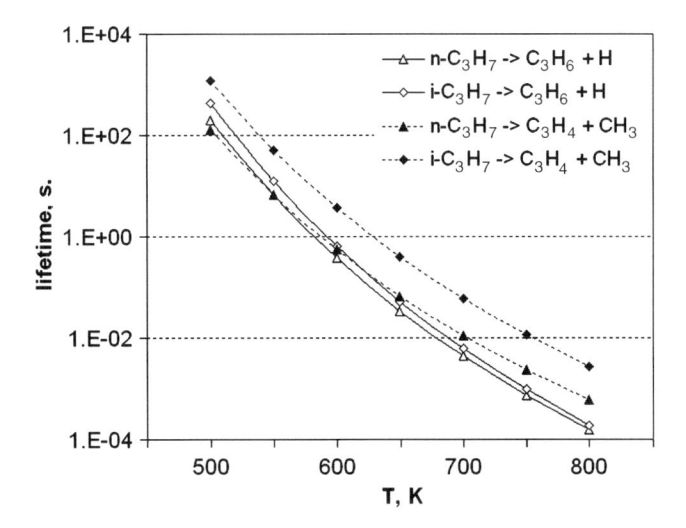

FIG. 12. Lifetimes of propyl radicals (*n*- and *iso*-) in the gas phase in reactions of H-atom and CH$_3$-radical elimination.

of propylene (C$_{3=}$) and ethylene (C$_{2=}$), respectively

$$C_3H_7 \rightarrow C_3H_6 + H \tag{23}$$

$$C_3H_7 \rightarrow C_2H_4 + CH_3 \tag{24}$$

If propyl radicals are generated into the gas phase by an active catalyst, the comparison of residence time with lifetimes given Fig. 12 can say whether or not such reactions can proceed while the gas mixture resides in the reactor. Also, if only these reactions are controlling the olefin distribution, the (C$_{3=}$)-to-(C$_{2=}$) ratio would be as shown in Fig. 13. The relative amount of propylene formed from both isomeric radicals is increasing with temperature. In fact, some additional amounts of propylene that can form in homogeneous reactions should be even higher due to several additional processes

$$C_3H_7 + X \rightarrow C_3H_6 + XH \tag{25}$$

where X is C$_3$H$_7$ and other radicals (and also maybe O$_2$—see Section V.C).

The same reaction can proceed via a participation of an oxidized surface active site

$$C_3H_7 + [O]_S \rightarrow C_3H_6 + [OH]_S \tag{26}$$

Propyl decomposition and the above two processes (reaction (25) and (26)) should increase the (C$_{3=}$)-to-(C$_{2=}$) ratio at rising temperature. However, experimental data indicate (see, e.g., Leveles, 2002) that, as a rule, the ethylene fraction in sum of C$_{3=}$ and C$_{2=}$ olefins increases.

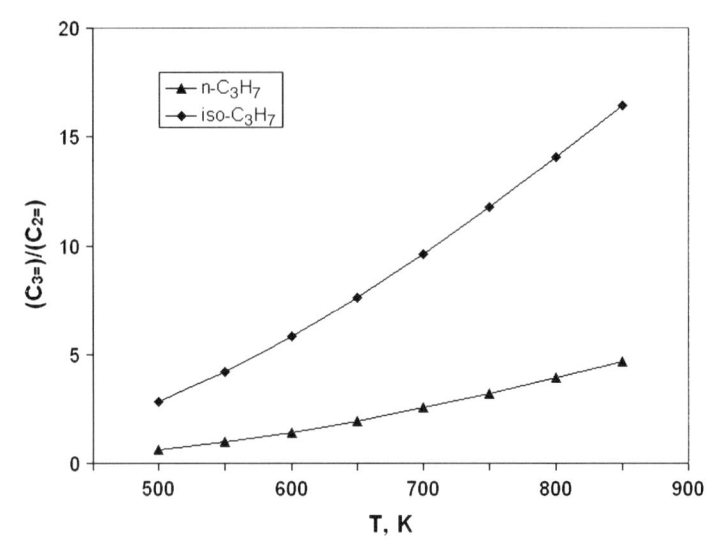

FIG. 13. Relative amounts of propylene and ethylene formed in homogeneous reactions of propyl radicals at different temperatures.

Methyl radicals formed simultaneously with ethylene can rapidly capture a hydrogen atom from any organic molecule in the gas (first of all, from propane) and transform it into methane. The latter is almost inert under the conditions of propane ODH. This means that methane-to-ethylene ratios close to 1 or above serve as evidence for predominantly homogeneous "anaerobic" degradation of the C–C–C skeleton. On the other hand, such degradation can also proceed via oxidative paths with the participation of molecular oxygen and/or oxidized (oxidative) intermediates. One of the most probable is an "aldehyde route"

$$C_3H_6 \, (+ RO_2 \text{ or } O_2) \to \ldots \to C_2H_5CHO \to C_2H_5 + CHO \to \text{total oxidation} \atop \downarrow \atop C_2H_4 \qquad (27)$$

Thus, a preliminary analysis of olefin production pathways can be performed based on the methane-to-ethylene ratio and on temperature dependence of the $(C_{3=})$-to-$(C_{2=})$ ratio. A more detailed elaboration can be reached from experiments with varied oxygen concentration and from the detailed analysis of the product distribution (including hydrogen formation). However, ethylene formation itself is strong evidence for the contribution of the radical route in product formation. The analysis of experimental data about product distribution during propane oxidation (Kondratenko et al., 2005) demonstrates that over rare-earth oxide catalysts radical route is prevailing in olefin formation. On the other hand, over supported V-containing catalysts, propylene forms predominantly via heterogeneous dehydrogenation of propane (or propyl radical intermediate), since over this catalyst only traces of ethylene are formed.

An examination of the nearest consequences of propyl radical transformation in the gas phase indicates that a substantial part of propane reaction proceeds within the above-discussed C_1–C_2 kinetic scheme. This fact opens up the prospect for the development of a propane oxidation scheme by the stepwise build-up of a C_1–C_2 joint description. In the course of this procedure special attention should be paid to the execution of the "fullness" principle. Although, as we mentioned above, in the case of C_{3+} hydrocarbon oxidation the contribution of several "mysterious" types of intermediates may substantially increase, the blocks accounting for the elementary reaction of "normal" species must be described as thoroughly as possible.

C. Reactions between Alkyl Radicals and Oxygen and Transformations of Alkylperoxy Radicals

The last statement of the previous section needs a more precise definition in some important cases, one of which will be discussed here. The central role of reactions between alkyl radicals and molecular oxygen and further transformations of alkylpreoxy species for combustion chemistry has been extensively discussed elsewhere (see, for instance, Miller and Klippenstein, 2001; Slagle *et al.*, 1986, and literature cited therein). Let us mention just some aspects of alkylperoxy chemistry and kinetics, which are of crucial importance for partial oxidation processes.

1. Alkylperoxy radicals are direct precursors of hydroperoxides—the key intermediates causing a degenerate chain-branching; thus, transformations of alkylperoxy species determine the overall kinetic features of the process.
2. Alkyl radicals are primary (and in many cases the most abundant) radical species during alkane oxidation, both homogeneous and catalytic heterogeneous–homogeneous (at least over oxides). Since the concentration of molecular oxygen in the system is much higher than that of any reactive radical up to very high conversions, namely reactions between alkyls and O_2 determine prevailing reaction pathways and control product distributions.
3. During methane oxidation, the equilibrium between formation and dissociation of methylperoxy (CH_3O_2) radicals determines the overall direction of the process. If CH_3O_2 formation dominates, oxygenates (methanol, formaldehyde) can form. On the contrary, in the condition of predominate dissociation (higher temperatures, lower oxygen partial pressures) the probability of CH_3 recombination and higher hydrocarbon formation (ODH selectivity) are increasing (see Section III.E).

In oxidative processing of higher alkanes, reactions under discussion also to a considerable degree determine the product distribution. Let us consider reactions of ethyl species as a representative example. First of all, as all reactions of C-centered radicals with oxygen, reaction between ethyl and O_2 can proceed as a reversible formation of ethylperoxy radical

$$C_2H_5 + O_2 \rightleftharpoons C_2H_5O_2 \tag{28}$$

The relative importance of different channels of further transformations of peroxy radicals determines the overall rate of reaction and selectivities to certain products. Reactions with any H-containing substance (first of all—with parent alkane) finally lead to degenerate chain-branching and exponential growth of the reaction rate

$$C_2H_5O_2 + RH \rightarrow C_2H_5OOH \rightarrow C_2H_5O + OH \tag{29}$$

Ethoxy species C_2H_5O are precursors for the formation of C_2-oxygenates. On the other hand, there is another (different from C–O bond dissociation) channel of decomposition

$$C_2H_5O_2 \rightarrow C_2H_4 + HO_2 \tag{30}$$

This reaction gives the product of another type—ethylene—and thus can play a key role in ethane ODH (including catalytic).

Additional problems arise due to the possibility of one more reaction, namely isomerization

$$C_2H_5O_2 \rightarrow CH_2CH_2OOH \tag{31}$$

Hydroperoxyethyl radicals formed can transform into ethylene oxide

$$CH_2CH_2OOH \rightarrow C_2H_4O + OH \tag{32}$$

or even react with one more oxygen molecule

$$CH_2CH_2OOH + O_2 \rightarrow OOCH_2CH_2OOH \tag{33}$$

Although this reaction is not among usually discussed in the literature, there are no serious arguments for its absence in nature. However, if we consider it and, consequently, further transformations of complex hydroperoxo–peroxocompounds, this may result in a serious change of the reaction rate (e.g., additional fast branching).

At last (not least!), ethyl radicals can react with oxygen giving ethylene and HO_2 radical directly ("direct H-abstraction")

$$C_2H_5 + O_2 \rightarrow C_2H_4 + HO_2 \tag{34}$$

Reaction (33) is moderately exothermic (about $-70\,kJ/mole$), so its activation energy should not exceed 15–20 kJ/mole. This means that it should be highly

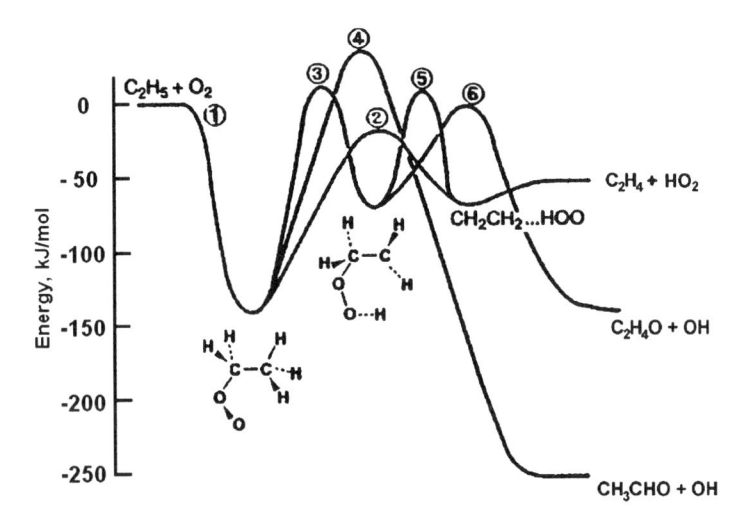

FIG. 14. Energetic diagram of possible channels of $C_2H_5 + O_2$ reaction and corresponding barriers.

competitive, even taking into account exothermicity and the non-activated character of alkylperoxy formation.

Despite extensive studies in this area, even for the reaction of ethyl radicals with oxygen a certain discrepancy remains between different authors. Separate kinetic experiments are better explained in the framework of different descriptions (see, for instance, Miller and Klippenstein, 2001; Naik *et al.*, 2004). The main problem consists of a very small difference in thermal effects and barriers of reactions (see Fig. 14). As shown by Naik *et al.* (2004), very small variations in parameters obtained by theoretical calculations can seriously improve the coincidence with experimental data. It is worth noticing that as a rule such a tiny difference in kinetic parameters is hardly distinguishable in kinetic experiments.

Simulations demonstrate, however, that variations in kinetic parameters of reactions under consideration lead to substantial consequences. Figure 15 shows how relatively small variations in the rate constant for reaction (30) influence the SID in methane–ethane mixtures. In such a reaction system (which models real compositions of natural gas) competition of different channels of ethyl-oxygen reaction overlaps (and very probably interferes) with methyl-oxygen chemistry. The latter is even somewhat qualitatively different: there are no variations in mono-molecular reactions of methylperoxy radicals at temperatures below 900 K (only dissociation to methyl and O_2) and all their bi-molecular reactions lead to branching as a nearest consequence. As to the ethyl-oxygen chemistry, it is much more "rich" and much less definite at the same time. So in this particular case, small variations in kinetic parameters lead to very substantial consequences.

One may expect the increase of complexity as soon as we move further and try to analyze reactions of C_3 radicals. We abstain from speculations concerning

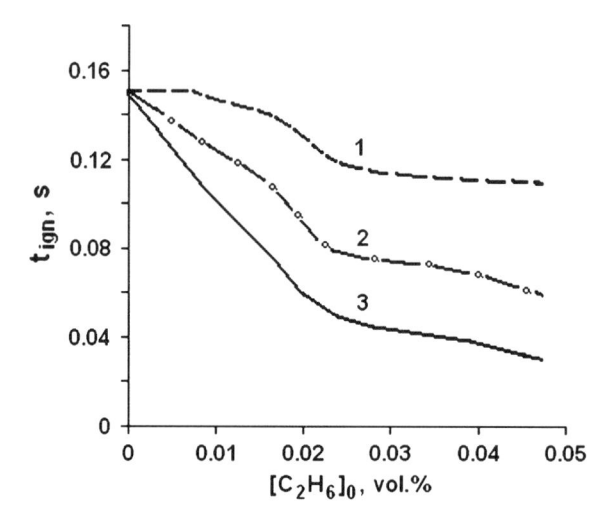

FIG. 15. Effect of initial ethane concentration on the self-ignition delay values (t_{ign}) in oxidation of methane–ethane mixtures; modeling at different values of reaction (28) rate constant ($T = 773$ K, $P = 70$ bar, $[O_2]_0 = 15.4\%$). (1) $k_{28} = 3 \times 10^5 \, s^{-1}$, (2) $k_{28} = 6 \times 10^5 \, s^{-1}$, (3) $k_{28} = 1.2 \times 10^6 \, s^{-1}$.

elementary reaction of C_3-peroxo radicals; this topic is still waiting for a detailed development.

D. CAPABILITIES OF PROCESS INFLUENCING, GOVERNING AND DESIGN

A very intriguing aspect of the reciprocal influence of homogeneous and heterogeneous processes is a possibility of intensification, or tuning gas-phase processes by precise (or directed) action of solid surfaces. At first glance, governing high-temperature oxidation processes in order to improve the formation of certain valuable products cannot be very prospective. Indeed, a catalyst containing very active surface sites can substantially accelerate the formation of initial free radicals (chain carriers), but the same effect can be also achieved using various physical methods. What is more important, even if the initiation rate could be substantially increased, the possibility of channeling the reaction of free radicals towards certain desired products is usually considered as doubtful.

One widely known example of such ineffective attempts is the direct oxidation of methane-to-methanol (DOMM) (see Arutyunov and Krylov, 1998 and literature cited therein). Whereas at atmospheric or somewhat higher pressures the introduction of catalysts (or homogeneous promoters) accelerates the process and increases product (mainly formaldehyde) yield to some extent, at increasing pressures their efficiency sharply drops. As a result, the methanol yield cannot be increased by any means above the level attainable in high-pressure homogeneous oxidation.

The kinetic simulations (Arutyunov, 2004; Arutyunov et al., 1999) based on the relatively simple homogeneous methane oxidation model (Vedeneev et al., 1988a, b, c, d) allowed resolving the origin of such behavior. It turned out that in the pressure range of several atmospheres the critical transition from slow radical chain mode to a fast quasi-stationary chain-branched mode occurs. In this high-pressure mode, the rate of radical generation in the endothermic homogeneous initiation reaction is a factor of $\sim 10^4$ slower than in chain-branching steps (Arutyunov, 2004). The rate of reaction is quasi-steady-state and does not turn into the chain explosion only because of radical decay via second-order termination steps. In order to be efficient under the conditions where this chain-branched mode is dominating, the rate of additional catalytic initiation must be comparable with that of fast branching reactions. Figure 16 shows the calculated methanol yield and reaction time vs. relative rate of additional methyl radical generation w_{eff} (normalized to "normal" homogeneous thermal initiation w_{therm}) imitating catalyst action. The calculations show that some increase in methanol yield can be expected, but at very intense additional injection of primary radicals. Besides that, this simulation does not account for the rise of heterogeneous decay of radicals and subsequent reactions of methanol, which could be also accelerated in the presence of catalyst capable of such strong initiation. The latter is the most probable explanation of the failure of any attempts to improve the DOMM process by heterogeneous catalysis.

Nevertheless, a more thorough analysis of kinetic schemes of oxidative transformations shows that there are additional possibilities, which have not been rationalized so far. Owing to the fundamentally non-linear nature of these

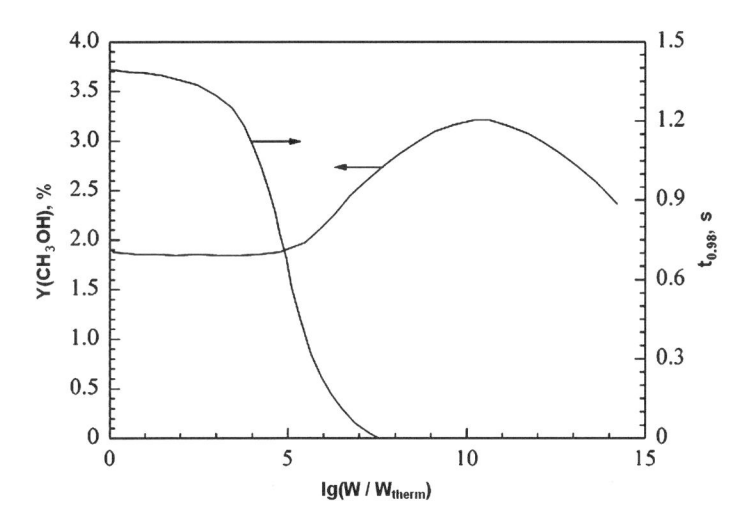

FIG. 16. Simulated methanol yield and time of 98% oxygen conversion ($t_{0.98}$) as functions of relative initiation rate (Arutyunov, 2004; Arutyunov et al., 1999). $T = 683$ K; $P = 100$ bar, $CH_4:O_2 = 19:1$.

reactions beginning very far from equilibrium and including multiple intermediates cross-linked by complex mutual transformations, there appears a principal possibility of affecting the overall process by relatively "tiny" influences. As a result, the system can attain substantially different steady states characterized by significantly different overall rates and product distributions.

The existence of multiple evolution paths is a fundamental feature of complex non-linear systems. Moreover, whereas far from the point of critical transition between two (or more) stationary states (bifurcation point) a limited impact can affect the behavior only quantitatively, near the bifurcation point even very small correction of one parameter is able to drastically change the direction of further evolution. It is worth noticing that light alkane oxidation belongs to that type of process in which various non-linear phenomena (such as cool flames, multiple and oscillating flashes, negative temperature reaction rate coefficients, hysteresises of different type) are taking place (Arutyunov et al., 1995b; Lewis and Elbe, 1987; Shtern, 1964; Sokolov et al., 1995; Vedeneev et al., 2000). Bearing this in mind, we can consider the modeling as a tool for searching the conditions for the reaction parameters and stages the most suitable for efficient forwarding the system evolution to a desired direction.

One example of such a type is described below. Figure 17 demonstrates simulated data of concentrations of different products at the early stages of methane oxidation. It is very clearly seen that characteristic times of coming into action for different species significantly differ. A selective "turning off" of particular intermediates may have dramatic consequences at later stages of reaction. In a more practical case—in a plug-flow reactor—such temporal development can be turned into a spatial evolution along the length of the reactor. The process can be influenced by placing catalytic material(s) into different cross-sections of the reactor and/or by varying the flow rate (in the latter case the location of catalyst can be unchanged). Schmidt (2001) presented indirect evidence for such a possibility: varying contact times in the millisecond range in methane oxidation over Pt leads to a substantial change of temperature and concentration profiles and to the shift in product distributions between preferential formation of synthesis gas ($CO+H_2$), oxygenates (formaldehyde), and C_2 hydrocarbons.

Hargreaves et al. (1990) also observed the shift from CO to C_2 hydrocarbons during methane oxidation over MgO catalyst at varying flow rate. However, in this case authors explained this effect in terms of the difference in kinetic orders of two reactions—OCM and oxidation to CO—with respect to CH_3 radical concentration, which in turn is changing at varied flow rate.

Similar effects can be achieved by changing the feed composition. Sinev et al. (1993) demonstrated that over the same phosphate catalysts it is possible to direct methane oxidation to different products by varying the initial oxygen concentration. Whereas at low oxygen content almost exclusive formation of ethane is observed, a gradual increase of oxygen concentration leads to a proportional increase of formaldehyde selectivity. It is important that the sum of

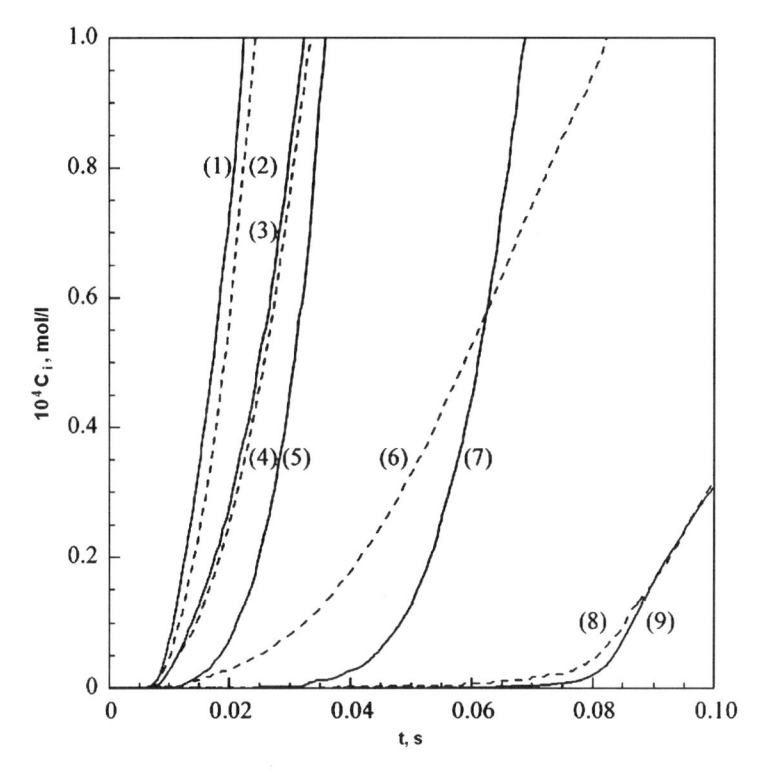

FIG. 17. Simulated concentration profiles of selected products on the initial stage of methane oxidation (Arutyunov, 2004) ($T = 733$ K, $P = 84$ bar, $CH_4:O_2 = 21:1$). (1) CH_2O, (2) H_2O, (3) H_2O_2, (4) CH_3OH, (5) CO, (6) H_2, (7) CO_2, (8) C_2H_6, (9) C_2H_5OH.

ethane and formaldehyde selectivities remains constant. This effect can be explained by shifting the equilibrium in reaction (5). However, there is an additional contribution to it from the catalyst side: at given oxygen concentration the ethane-to-formaldehyde ratio changes from one phosphate catalyst to another.

Although the above examples illustrate the possibility of influencing the formation of different products, all these experimental results were obtained independently of modeling and analysis of kinetic schemes. Indeed, the examples of modeling-driven directed process governing and design are scarce so far. This reflects the current state of the area and our understanding of the alkane oxidation mechanism and free radical chemistry in particular. However, one example of more advanced tuning of the reaction is given below.

While modeling the OCM process over oxide catalysts, it was demonstrated that a substantial fraction of methyl radicals (up to 95%) generated via reaction (12) are transformed back into methane in the reverse process

$$[OH]_S + CH_3 \rightarrow [O]_S + CH_4 \tag{35}$$

It was assumed that the yield of OCM products could be enhanced by increasing the efficiency of CH_3 recombination, which competes with reaction (35). Since the recombination of methyl radicals is a three-body process (see Section III.D), its efficiency can be increased by increasing total pressure, or by introducing an additional "inert" surface, which can play a role of third body. Indeed, it was demonstrated that the increase of inert gas (He or Ar) pressure at constant pressures of methane and oxygen leads to a substantial increase of OCM selectivity and yield (Sinev et al., 1996). Moreover, the addition of a 10-fold amount of various solid materials possessing a very low activity under the same conditions (quartz, fused MgO, Mg phosphate) to a relatively efficient OCM catalyst (Nd/MgO) led to a drastic increase (up to twofold) in the yield of OCM products (Sinev et al., 1997a, b).

Indeed, the area outlined in this section is in its infant state. Further development of kinetic models can play a significant role both in obtaining new information about the intrinsic mechanisms of processes under discussion and in its utilization for practical applications.

VI. Concluding Remarks

The main conclusion that could be derived from the above analysis is the ascertaining of the extraordinary complexity of the system we attempt to model. Even staying at the level of "micro-chemical" modeling we must accept as a fact the fundamental difference between the phenomenon and its model and, as a result, the impossibility to develop the description that reflects the reality in all its manifestations. A possible solution of the modeling problem, as applied to light alkane oxidative processing, can be found on the way of stepwise and successive execution of the basic principles stated below.

1. Precise formulation of the objective.
2. Elucidation of principal chemical and physical features of the system and relationships between its parts.
3. Compilation of kinetic scheme which meets the requirement of fullness. The latter assumes the accounting of all species and elementary reactions potentially relevant for the formation and transformations of certain product(s). We must emphasize here that the requirement of fullness includes the thermodynamic consistency as the essential constituent.
4. Determination of general kinetic features of each accounted elementary reaction (or groups of analogous reactions, e.g., "simple" homogeneous, pressure-dependent, heterogeneous, reactions in adsorbed layers, etc.). This includes, first of all, types of applicable kinetic description (e.g., mass action law, topochemical equations, probability of interaction on phase boundaries, etc.) and adequate form of representation of kinetic parameters.

5. Selection of the most reliable values of kinetic parameters available in literature, databases, etc. In the case of absence, parameters of certain reactions can be calculated or evaluated using various procedures. However, what should be avoided is fitting (or adjustment) of parameters to the experimental data in the framework of the same reaction scheme, which is going to be utilized for modeling also. The latter would mean the solution of the inverse kinetic problem, which in general (and in the most of particular cases of complex reactions consisting of hundreds of elementary steps) is a mathematically ill-conditioned (or even illegal) procedure.

Now let us raise the question—what the result of such modeling would mean? The answer is as follows: if the system under consideration is arranged as described by the model, its behavior would be as predicted by simulations. Taking into account all mentioned difficulties (first of all, imperfection of microchemical information and uncertainty of kinetic parameters), we have to confess that it is not realistic to expect good and moreover quantitative coincidence of simulated kinetics with some particular experimental data. Since any experimental result also cannot be considered as "absolute reality", the best way is to compare the results of modeling with experimental observations on a phenomenological level. In other words, at the current stage of understanding the comparison of observations and predictions is more fruitful in terms of phenomena and qualitative effects, rather than in numbers, approximation equations, or any other quantitative characteristics.

In spite of apparent "micro-chemical extremism", we greatly appreciate the efforts made by scientists and groups working in neighboring areas, such as measuring, calculation and evaluation of kinetic parameters, modeling of heat- and mass-transfer, development of simulation codes and software, etc. The problem lies in the plane of adequate use of achievements reached in one area for further development of the adjacent one and in solving the modeling problem as a whole.

ACKNOWLEDGMENTS

Authors gratefully acknowledged Dr. O.V. Sokolov, Mr. P.S. Stennikov and Ms. O.S. Moiseeva for their participation in kinetic simulations and Mr. V.V. Plenkin for his help in preparing the illustrations. A financial support from AURUS Marketing-Services GmbH (Vienna, Austria) is appreciated.

LIST OF SYMBOLS

A   pre-exponential factor in "three-parametric" form of Arrhenius equation (in mol, l, s.)

[A]	concentration of reactant (A), vol.%
a	constant (coefficient) in Polanyi–Semenov equation, kJ/mol
b	constant (coefficient) in Polanyi–Semenov equation (dimensionless)
C_i or $C(i)$	concentration of ith reactant, mol/l
D_i	coefficient of gas-phase molecular diffusion for ith particle, m^2/s
d	characteristic dimension (thickness) of one layer, m
E_a	activation energy, kJ/mol
h	Plank constant (6.6256×10^{-34} J s.)
K_j	equilibrium constant of jth reaction (in mol, l)
k_j	rate constant of jth reaction (in mol, l, s.)
k_0	rate constant of pressure-dependent reaction in low pressure limit (in mol, l, s.)
k_∞	rate constant of pressure-dependent reaction in high pressure limit (in mol, l, s.)
$k_{(+)}$	rate constant for forward reaction (in mol, l, s.)
$k_{(-)}$	reverse reactions must satisfy the connecting equation (in mol, l, s.)
k_{Di}	"diffusional" rate constant for ith particle, s^{-1}
L	gas gap thickness in spatially distributed system, m
m	reaction order with respect to ith reactant (dimensionless)
N_A	Avogadro number (6.0225×10^{23} mol^{-1})
P	total pressure, bar
R	gas constant (8.314 J/(mol K), or 0.08206 l atm/(mol K))
T	temperature, K
T_0	comparative temperature (as a rule, 273 or 298 K), K
t	time, s
$t_{0.1}$	time of reaching 10%-conversion, s
t_{ign}	self-ignition delay, s
W	rate of reaction, mol/(l s)
W_{tot}	total rate of methane activation, mol/(l s)
W_{hom}	rate of homogeneous activation of methane, mol/(l s)
$Y(P)$	yield of product P, %
Z	pre-exponential factor (for bi-molecular reactions—collision factor) in Arrhenius equation (in mol, l, s.)
ΔH	enthalpy change in the process (differences between enthalpy of formation of products and reactants), kJ/mol
$\Delta H^{\neq}$	enthalpy of formation of activated complex, kJ/mol
S_{LIM}	limiting selectivity to C$_2$-hydrocarbon formation
ΔS	change in the process (differences between standard entropy of products and reactants), J/(mol K)
$\Delta S^{\neq}$	entropy of formation of activated complex, J/(mol K)
α	relative "burning out" of two hydrocarbons in their mixture (dimensionless)
β	relative change of concentration of individual alkanes during the oxidation of their complex mixture (dimensionless)

χ	transmission coefficient (dimensionless)
μ	reduced molecular weight of colliding molecules, kg/mol
μ_g	molecular weight of gas particle, kg/mol
v	gas-to-surface collision frequency, s^{-1}
ϑ	concentration of surface sites per unit surface area, m^{-2}
σ	collision cross-section, m^2
σ_S	cross-section of the surface site, m^2
ξ	steric factor (dimensionless)

REFERENCES

Alekseev, B. V., "Mathematical Kinetics of Reacting Gases". Nauka, Moscow (1988).
Allison, T. C. *Fuel Chemistry Preprints* **50**, 96 (2005).
Aparicio, L. M., Rossini, S. A., Sanfilippo, D. G., Rekoske, J. E., Treviño, A. A., and Dumesic, J. A. *Ind. Eng. Chem. Res.* **30**, 2114 (1991).
Arutyunov, V. S. *Russ. Chem. Bull.* **51**, 2170 (2002).
Arutyunov, V. S. *J. Nat. Gas Chem.* **13**, 10 (2004).
Arutyunov, V. S., Basevich, V. Ya., and Vedeneev, V. I. *Ind. Eng. Chem. Res.* **34**, 4238 (1995a).
Arutyunov, V. S., Basevich, V. Ya., Vedeneev, V. I., and Krylov, O. V. *Kinet. Catal.* **40**, 382 (1999).
Arutyunov, V. S., Basevich, V. Ya., Vedeneev, V. I., Parfenov, Yu. V., and Sokolov, O. V. *Russ. Chem. Bull.* **45**, 45 (1996).
Arutyunov, V. S., Basevich, V. Ya., Vedeneev, V. I., and Sokolov, O. V. *Kinet. Catal.* **36**, 458 (1995b).
Arutyunov, V. S., and Krylov, O. V., "Oxidative Conversion of Methane". Nauka, Moscow (1998).
Arutyunov, V. S., Rudakov, V. M., Savchenko, V. I., and Sheverdenkin, E. V. *Theor. Found. Chem. Eng.* **39**, 516 (2005).
Arutyunov, V. S., Rudakov, V. M., Savchenko, V. I., Sheverdenkin, E. V., Sheverdenkina, O. G., and Zheltyakov, A. Yu. *Theor. Found. Chem. Eng.* **36**, 472 (2002).
Atkinson, R., Baulch, D. L., Cox, R. A., Hampson, R. F.Jr., Kerr, J. A., and Troe, J. *J. Phys. Chem. Ref. Data* **18**, 881 (1989).
Azatyan, V. V. *Uspekhi Khimii (Russ. Adv. Echem.)* **54**, 33 (1985).
Azatyan, V. V. *Zh. Fiz. Khimii (Russ. J. Phys. Chem.)* **79**, 397 (2005).
Baranek, P., Pinarello, G., Pisani, C., and Dovesi, R. *Phys. Chem. Chem. Phys.* **2**, 3893 (2000).
Batiot, C., and Hodnett, B. K. *Appl. Catal. A* **137**, 179 (1996).
Baulch, D. L., Cobos, C. J., Cox, R. A., Esser, C., Frank, P., Just, Th., Kerr, J. A., Pilling, M. J., Troe, J., Walker, R. W., and Warnatz, J. *J. Phys. Chem. Ref. Data* **21**, 411 (1992).
Baulch, D. L., Cobos, C. J., Cox, R. A., Frank, P., Hayman, G., Just, Th., Kerr, J. A., Murrells, T., Pilling, M. J., Troe, J., Walker, R. W., and Warnatz, J. *J. Phys. Chem. Ref. Data* **23**, 847 (1994).
Baulch, D. L., Cox, R. A., Crutzen, P. J., Hampson, R. F.Jr., Kerr, J. A., Troe, J., and Watson, R. T. *J. Phys. Chem. Ref. Data* **11**, 327 (1982).
Baulch, D. L., Cox, R. A., Hampson, R. F.Jr., Kerr, J. A., Troe, J., and Watson, R. T. J. *J. Phys. Chem. Ref. Data* **13**, 1259 (1984).
Bengaard, H. S., Nørskov, J. K., Sehested, J., Clausen, B. S., Nielsen, L. P., Molenbroek, A. M., and Rostrup-Nielsen, J. R. *J. Catal.* **209**, 365 (2002).
Beretta, A., and Forzatti, P., 6th Natural Gas Conversion Symposium, Alaska, USA, June 17–22, *in* "Proceedings (on CD ROM)" (E. Iglesia, J. J. Spivey, and T. H. Fleisch Eds.), Article 31 (2001).

Beretta, A., Forzatti, P., and Ranzi., E. *J. Catal.* **184**, 469 (1999).
Bobyshev, A. A., and Radzig, V. A. *Kinetika i Kataliz (Russ. Kinet. Catal.)* **29**, 638 (1988).
Bobyshev, A. A., and Radzig, V. A. *Kinetika i Kataliz (Russ. Kinet. Catal.)* **31**, 925 (1990).
Bogoyavlenskaya, M. L., and Kovalskii, A. A. *Zh. Fiz. Khimii (Russ. J. Phys. Chem.)* **20**, 1325 (1946).
Borisov, A. A., Skachkov, G. I., and Troshin, K. Ya. *Chem. Phys. Reports* **18**, 1665 (2000).
Bristolfi, M., Fornasari, G., Molinari, M., Palmeri, S., Dente, M., and Ranzi, E. *Chem. Eng. Sci.* **47**, 2647 (1992).
Buda, F., Bounaceur, R., Warth, V., Glaude, P. A., Bournet, R., and Battin-Leclerc, F. *Combust. Flame* **142**, 170 (2005).
Bui, P. A., Vlachos, D. G., and Westmoreland, P. R. *Surf. Sci.* **385**, L1029 (1997).
Burch, R., Squir, G. D., and Tsang, S. C. *J. Chem. Soc., Faraday Trans. I.* **85**, 3561 (1989).
Buyevskaya, O. V., Rothaemel, M., Zanthoff, H., and Baerns., M. *J. Catal.* **146**, 346 (1994).
Bychkov, V. Yu., Krylov, O. V., and Korchak, V. N. *Kinet. Catal.* **43**, 94 (2002a).
Bychkov, V. Yu., Sinev, M. Yu., and Korchak, V. N. 3rd Workshop on C1-C3 Hydrocarbon Oxidation, July 14–17, 1997, Krasnoyarsk, Russia. Book of Abstracts, paper A7 (1997).
Bychkov, V. Yu., Sinev, M. Yu., Korchak, V. N., Aptekar, E. L., and Krylov, O. V. *Kinet. Catal.* **30**, 1142 (1989).
Bychkov, V. Yu., Tulenin, Yu. P., and Korchak, V. N. *Kinet. Catal.* **44**, 353 (2003).
Bychkov, V. Yu., Tulenin, Yu. P., Krylov, O. V., and Korchak, V. N. *Kinet. Catal.* **43**, 775 (2002b).
Cattolica, R. J., and Schefer, R. W., 19th Symposium on Combustion. Proceedings. The Combustion Institute, Pittsburgh, PA, 311 (1982).
Cohen, N., and Westberg, K. R. *J. Phys. Chem. Ref. Data* **20**, 1211 (1991).
Couwenberg, P. M. PhD Thesis, Eindhoven University of Technology, Eindhoven (1994).
Curran, H. J. *Prepr. Pap.-Am. Chem. Soc. Div. Fuel Chem.* **49**, 263 (2004).
Davis, M. B., Pawson, M. D., Veser, G., and Schmidt, L. D. *Combust. Flame* **123**, 159 (2000).
Deng, Y. Q., and Nevell, T. G. *Faraday Discuss* **105**, 33 (1996).
de Smet, C. R. H., de Caroon, M. H. J. M., Berger, R. J., Marin, G. B., and Schouten, J. C. *AIChE J.* **46**, 1837 (2000).
Deutschmann, O., Behrendt, F., and Warnatz, J. *Catal. Today* **21**, 461 (1994).
Deutschmann, O., and Schmidt, L. D. *AIChE J.* **44**, 2465 (1998).
Deutschmann, O., Schwiedernoch, R., Maier, L. I., and Chatterjee, D., 6th Natural Gas Conversion Symposium, Alaska, USA, June 17–22, *in* "Proceedings (on CD ROM)" (E. Iglesia, J. J. Spivey, and T. H. Fleisch Eds.), Article 41 (2001).
Driscoll, D. J., Campbell, K. D., and Lunsford, J. H. *Adv. Catal.* **35**, 139 (1987).
Dyer, M. J., Pfefferle, L. D., and Crosley, D. R. *Appl. Optics* **29**, 111 (1990).
Emanuel, N. M., and Knorre, D. G., "Chemical Kinetics Textbook". 4th Ed. Vys'shaya Shkola, Moscow (1984).
Evlanov, S. F., and Lavrov, N. V., *in* "Scientific Bases of Hydrocarbon Catalytic Conversion," p. 210. Naukova Dumka, Kiev (in Russian) (1977).
Eyring, H., Lin, S. H., and Lin, S. M., "Basic Chemical Kinetics". Wiley, New York, Chichester, Brisbane, Toronto (1980).
Fairbrother, D. H., Peng, X.-D., Stair, P. C., Fan, J., and Trenary, M., Symposium on Methane and Alkane Conversion Chemistry, 207th National Meeting, ACS, San Diego, CA, p. 280 (1994).
Fang, T., and Yeh, C. T. *J. Catal.* **69**, 227 (1981).
Firsov, E. I., Sinev, M. Yu., and Shafranovsky, P. A. *J. Electron Spectrosc. Relat. Phenomena* **54–55**, 489 (1990).
Frank-Kamenetskii, D. A., "Diffusion and Heat Transfer in Chemical Kinetics". Plenum Press, New York (1969).
Frenklach, M., Wang, H., Yu, C.-L., Goldenberg, M., Bowman, C. T., Hanson, R. K., Davidson, D. F., Chang, E. J., Smith, G. P., Golden, D. M., Gardiner W. C., and Lissianski, V. (1995) http://www.me.berkeley.edu/gri_mech

Frolov, S. M., Basevich, V. Ya., Neuhaus, M. G., and Tatschl, R., *in* "Advanced Computation & Analysis of Combustion" (G. D. Roy, S. M. Frolov, and P. Givi Eds.), ENAS Publishers, Moscow (1997).
Fukuhara, Ch., and Igarashi, A. *Chem. Eng. Sci.* **60**, 6824 (2005).
Gardiner, W. C., Jr. (Ed.), "Combustion Chemistry". Springer-Verlag, New York, Berlin, Heidelberg, Tokyo (1984).
Garibyan, T. A., Mantashyan, A. A., and Nalbandyan, A. B. *Doklady Akademii Nauk SSSR (Russ.: Reports of the Academy of Sciences of the USSR)* **186**, 1114 (1969).
Garibyan, T. A., Mantashyan, A. A., and Nalbandyan, A. B. *Armyanskii Khimicheskii Zhurnal (Russ.: Armenian Chemical Journal)* **24**, 4 (1971).
Garibyan, T. A., and Margolis, L. Ya. *Catal. Rev.- Sci. Eng.* **31**, 35 (1989–90).
Gershenzon, Yu. M., and Purmal, A. P. *Uspekhi Khimii (Russ. Adv. Chem.)* **59**, 1729 (1990).
Gershenzon, Yu. M., Zvenigorodskii, S. G., and Rozenshtein, V. B. *Uspekhi Khimii (Russ. Adv. Chem.)*, **59**, 1601 (1990).
Gilbert, R. G., Luther, K., and Troe, J. *Ber. Bunsen. Phys. Chem.* **87**, 169 (1983).
Glasstone, S., Laidler, K. J., and Eyring, H., "The Theory of Rate Processes". New York and London (1941).
Gorodetskii, V. V., Elokhin, V. I., Bakker, J. W., and Nieuwenhuys, B. E. *Catal. Today* **105**, 183 (2005).
Gorokhovatskii, Ya. B., Kornienko, T. P., and Shalya, V. V., "Heterogeneous-Homogeneous Reactions". Thechnika, Kiev (1972).
Green, W. H. *Ind. Eng. Chem. Res.* **40**, 5362 (2001).
Hall, R. B., Castro, M., Kim, C.-M., Chen, J., and Mims, C. A., Symposium on Methane and Alkane Conversion Chemistry, 207th National Meeting, ACS, San Diego, CA, p. 282 (1994).
Hammer, B., and Nørskov, J. K. *Adv. Catal.* **45**, 71 (2000).
Hansen, E. W., and Neurock, M. *Chem. Eng. Sci.* **54**, 3411 (1999).
Hargreaves, J. S. J., Hutchings, G. J., and Joyner, R. W. *Nature (London)* **348**, 428 (1990).
Hatano, M., Hinson, P. G., Vines, K. S., and Lunsford, J. H. *J. Catal.* **124**, 557 (1990).
Heitnes, K., Lindberg, S., Rokstad, O. A., and Holmen, A. *Catal. Today* **21**, 471 (1994).
Hickman, D. A., and Schmidt, L. D. *J. Catal.* **138**, 267 (1992).
Hickman, D. A., and Schmidt, L. D. *Science* **259**, 343 (1993).
Hoebink, J. H. B. J., Couwenberg, P. M., and Marin, G. B. *Chem. Eng. Sci.* **49**, 5453 (1994).
http://www.asponews.org
http://www.detchem.com
http://www.eia.doe.gov
http://www.kintecus.org
http://www.nist.gov/srd/nist17.htm
Hu, Y. H., and Ruckenstein, E. *Ind. Eng. Chem. Res.* **37**, 2333 (1998).
Huff, M. C., and Schmidt, L. D. *J. Phys. Chem.* **97**, 11815 (1993).
Isagulyants, G. V., Belomestnykh, I. P., Vorbeck, G., and Perregaard, J. *Rossiiskii Khimicheskii Zhurnal (Russ. Chem. J.)*, **44**, 2/69 (2000).
Ito, T., and Lunsford, J. H. *Nature* **314**, 721 (1985).
Kee, R. J., Rupley, F. M., Meeks, E., and Miller, J. A. CHEMKIN-III: A FORTRAN Chemical Kinetics Package for the Analysis of Gasphase Chemical and Plasma Kinetics. Sandia National Laboratories, Unlimited Release. Printed May (1996).
Keller, G. E., and Bhasin, M. M. *J. Catal.* **73**, 9 (1982).
Kiperman, S. L., "The Fundamental Chemical Kinetics in Heterogeneous Catalysis". Khimiya, Moscow (1979).
Kislyuk, M. U., Tret'yakov, I. I., Savkin, V. V., and Sinev, M. Yu. *Kinet. Katal.* **41**, 71 (2000).
Knyazev, V. D., and Tsang, W. *J. Phys. Chem.* **103**, 3944 (1999).
Kolios, G., Frauhammer, J., and Eigenberger, G. *Chem. Eng. Sci.* **56**, 351 (2001).

Kondratenko, E., Mulla, S. A. R., Buevskaya, O. V., and Baerns, M. 5th World Congress on Oxidation Catalysis, Program, Abstracts, Sapporo, Japan, p. 342 (2005).
Kondratiev, V. N., "Rate Constants of Chemical Reactions. Reference Book". Nauka, Moscow (1970).
Kondratyev, V. N., and Nikitin, E. E., "Kinetics and Mechanism of Gas-Phase Reactions". Nauka, Moscow (1974).
König, D., Weber, W. H., Poindexter, B. D., McBride, J. R., Graham, G. W., and Otto, K. *Catal. Lett.* **29**, 329 (1994).
Krylov, O. V., Kislyuk, M. U., Shub, B. R., Gesalov, A. A., Maksimova, N. D., and Rufov, N. Yu. *Kinetika i Kataliz (Russ. Kinet. Catal.)* **13**, 598 (1972).
Kvenvolden, K. A. *Chem. Geol.* **71**, 411 (1988).
Larin, I. K., and Ugarov, A. A. *Khimicheskaya Fizika (Russ. Chem. Phys.)* **21**, 58 (2002).
Latkin, E. I., Elokhin, V. I., and Gorodetskii, V. V. *Chem. Eng. J.* **91**, 123 (2003).
Leveles, L. Oxidative conversion of lower alkanes to olefins, PhD Thesis, The University of Twente, Enschede, The Netherlands (2002).
Lewis, B., and Elbe, G., "Combustion, Flames and Explosions of Gases". Acad. Press, Orlando (1987).
Lichanot, A., Larrieu, C., Zicovich-Wilson, C., Roetti, C., Orlando, R., and Dovesi, R. *J. Phys. Chem. Solids* **60**, 855 (1999).
Liebmann, L. S., and Schmidt, L. D. *Appl. Catal. A: Gen.* **179**, 93 (1999).
Lin, C.-H., Campbell, K. D., Wang, J.-X., and Lunsford, J.-H. *J. Phys. Chem.* **90**, 534 (1986).
Lindemann, F. *Trans. Faraday Soc.* **17**, 598 (1922).
Lunsford, J. H. *Langmuir* **5**, 12 (1989).
Mallens, E. P. J., Hoebink, J. H. B. J., and Marin, G. B. *Stud. Surf. Sci. Catal.* **81**, 205 (1994).
Mallens, E. P. J., Hoebink, J. H. B. J., and Marin, G. B. *Catal. Lett.* **33**, 291 (1995).
Mantashyan, A. A. *Combust. Flame* **112**, 261 (1998).
Marinov, N. M., Pitz, W. J., Westbrook, C. K., Castaldi, M. J., and Senkan, S. M. *Combust. Sci. Tech.* **116–117**, 211 (1996).
McCarty, J. G., McEwen, A. B., and Quinlan, M. A. *Stud. Surf. Sci. Catal.* **55**, 405 (1990).
Melvin, A., "Natural Gas: Basic Science and Technology". Adam Hilger, Bristol and Philadelphia (1988).
Mhadeshwar, A. B., and Vlachos, D. G. *J. Phys. Chem. B* **109**, 16819 (2005).
Miller, J. A., and Klippenstein, S. J. *Int. J. Chem. Kinet.* **33**, 654 (2001).
Mims, C. A., Mauti, R., Dean, A. M., and Rose, K. D. *J. Phys. Chem.* **98**, 13357 (1994).
Naik, C. V., Carstensen, H.-H., and Dean, A. M. *Prepr. Pap.-Am. Chem. Soc., Div. Fuel Chem.* **49**, 437 (2004).
Neurock, M. *J. Catal.* **216**, 73 (2003).
Neurock, M., Wasileski, S. A., and Mei, D. *Chem. Eng. Sci.* **59**, 4703 (2004).
Nibbelke, R. H., Scheerová, J., de Croon, M. H. J. M., and Marin, G. B. *J. Catal.* **156**, 106 (1995).
Ozkan, U. S., Kumthekar, M. W., and Karakas, G. *J. Catal.* **171**, 67 (1977).
Palmer, M. S., Neurock, M., and Olken, M. M. *J. Am. Chem. Soc.* **124**, 8452 (2002a).
Palmer, M. S., Neurock, M., and Olken, M. M. *J. Phys. Chem. B* **106**, 6543 (2002b).
Peskov, N. V., Slinko, M. M., and Jaeger, N. I. *Chem. Eng. Sci.* **58**, 4797 (2003).
Pfefferle, L. D., Griffin, T. A., Dyer, M. J., and Crosley, D. R. *Combust. Flame* **76**, 325 (1989).
Pfefferle, L. D., Griffin, T. A., and Winter, M. *Appl. Optics* **27**(15), 3197 (1988).
Pfefferle, L. D., and Pfefferle, W. C. *Catal. Rev.-Sci. Eng.* **29**, 219 (1987).
Pfefferle, W. C. Belgian Pat. 814752, 8 Nov. (1974); U.S. Pat. 3928961, 30 Dec. (1975).
Politenkova, G. G., Sinev, M. Yu., Tulenin, Yu. P., Vislovskii, V. P., and Cortés Corberán, V. 5th European Congress on Catalysis, 2–7 Sept., Limerick, Ireland. Abstracts, paper 5O17 (2001).
Polyakov, V. M. *Uspekhi Khimii (Russ. Adv. Chem.)* **17**, 351 (1948).
Radzig, V. A. *Colloids Surf., A: Physicochemical and Engineering Aspects* **74**, 91 (1993).
Reyes, S. C., Iglesia, E., and Kelkar, C. P. *Chem. Eng. Sci.* **48**, 2643 (1993).

Reyes, S. H., Sinfelt, J. H., Androulakis, I. P., and Huff, M. C., 6th Natural Gas Conversion Symposium, Alaska, USA, June 17–22, 2001, *in* "Proceedings (on CD ROM)" (E. Iglesia, J. J. Spivey, and T. H. Fleisch Eds.), Article 81 (2001).

Satterfield, C. N., "Mass Transfer in Heterogeneous Catalysis". M.I.T., Cambridge, MA (1970).

Satterfield, C. N., "Heterogeneous Catalysis in Practice". McGraw-Hill Book Company, New York (1980).

Schmidt, L. D., 6th Natural Gas Conversion Symposium, Alaska, USA, June 17–22, 2001, *in* "Proceedings (on CD ROM)" (E. Iglesia, J. J. Spivey, and T. H. Fleisch Eds.), Article 1 (2001).

Sellers, H., and Shustorovich, E. *Surf. Sci.* **504**, 167 (2002).

Semenov, N. N., "Some Problems in Chemical Kinetics and Reactivity". Pergamon Press, London (1959).

Shafranovsky, P. A. PhD Thesis, Semenov Institute of Chemical Physics, Moscow (1988).

Shalgunov, S. I., Zeigarnik, A. V., Bruk, L. G., and Temkin, O. N. *Russ. Chem. Bull.* **48**, 1876 (1999).

Shi, C., Hatano, M., and Lunsford, J. H. *Catal. Today* **13**, 191 (1992).

Shtern, V. Ya., "Oxidation of Hydrocarbons". Pergamon Press, Oxford, London, New York (1964).

Shustorovich, E. *Adv. Catal.* **37**, 101 (1990).

Shustorovich, E., and Sellers, H. *Surf. Sci. Rep.* **31**, 1 (1998).

Shustorovich, E. M., and Zeigarnik, A. V. *Russ. J. Phys. Chem.* **80**, 4 (2006).

Silberova, B., Burch, R., Goguet, A., Hardacre, C., and Holmen, A. *J. Catal.* **219**, 206 (2003).

Sinev, M. Yu. *Catal. Today* **13**, 561 (1992).

Sinev, M. Yu., Symposium on Methane and Alkane Conversion Chemistry, 207th National Meeting, ACS, March 13–18, San Diego, CA, USA, preprints (1994).

Sinev, M. Yu. *Catal. Today* **24**, 389 (1995).

Sinev, M. Yu. *J. Catal.* **216**, 468 (2003).

Sinev, M. Yu., Free radicals as intermediates in catalytic oxidation of light alkanes: New opportunities, *Res. Chem. Inter.* **32**, 205 (2006).

Sinev, M. Yu., and Bychkov, V. Yu. *Kinetika i Kataliz (Russ. Kinet. Catal.)* **34**, 309 (1993).

Sinev, M. Yu., and Bychkov, V. Yu. *Kinetika i Kataliz (Russ. Kinet. Catal.)* **40**, 906 (1999).

Sinev, M. Yu., Bychkov, V. Yu., Korchak, V. N., and Krylov, O. V. *Catal. Today* **6**, 543 (1990).

Sinev, M. Yu., Fattakhova, Z. T., Tulenin, Yu. P., Stennikov, P. S., and Vislovskii, V. P. *Catal. Today* **81**, 107 (2003).

Sinev, M. Yu., Korchak, V. N., and Krylov, O. V. *Kinetika i Kataliz (Russ. Kinet. Catal.)* **28**, 1376 (1987).

Sinev, M. Yu., Korchak, V. N., Krylov, O. V., Grigoryan, R. R., and Garibyan, T. A. *Kinetika i Kataliz (Russ. Kinet. Catal.)* **29**, 1105 (1988).

Sinev, M. Yu., Margolis, L. Ya., Bychkov, V. Yu., and Korchak, V. N. *Stud. Surf. Sci. Catal.* **110**, 327 (1997a).

Sinev, M. Yu., Margolis, L. Ya., and Korchak, V. N. *Russ. Adv. Chem.* **64**, 373 (1995).

Sinev, M. Yu., Setiadi, S., and Otsuka, K. *Mendeleev Commun.* 10 (1993).

Sinev, M. Yu., Tulenin, Yu. P., Kalashnikova, O. V., Bychkov, V. Yu., and Korchak, V. N. *Catal. Today* **32**, 157 (1996).

Sinev, M. Yu., Tulenin, Yu. P., and Korchak, V. N. 5th European Workshop on Methane Activation, 8–10th June, University of Limerick, Book of Abstracts, 51 (1997b).

Slagle, I. R., Ratajczak, E., and Gutman, D. *J. Phys. Chem.* **90**, 402 (1986).

Sokolov, O. V., Arutyunov, V. S., Basevich, V. Ya., and Vedeneev, V. I. *Kinet. Catal.* **36**, 290 (1995).

Smith, G. P., Golden, D. M., Frenklach, M., Moriarty, N. W., Eiteneer, B., Goldenberg, M., Bowman, T., Hanson, R. K., Song, S., Gardiner, Jr. W. C., Lissianski, V. V., and Qin, Z. http://www.me.berkeley.edu/gri_mech/

Smudde, G. H., Min, Yu., and Stair, P. C. *J. Am. Chem. Soc.* **115**, 1988 (1993).

Stewart, P. H., Larson, C. W., and Golden, D. *Combust. Flame* **75**, 25 (1989).

Suleimanov, A. I., Aliev, S. M., and Sokolovskii, V. D. *React. Kinet. Catal. Lett.* **31**, 291 (1986).
Temkin, O. N. *Russ. Chem. J.* **44**, 4/58 (2000).
Temkin, O. N., Zeigarnik, A. V., and Bonchev, V. D., "Chemical Reaction Networks. A Graph—Theoretical Approach". CRC Press, Boca Raton, USA (1996).
Tong, Y., and Lunsford, J. H. *J. Am. Chem. Soc.* **113**, 4741 (1991).
Tsang, W., and Hampson, R. F. *J. Phys. Chem. Ref. Data* **15**, 1087 (1986).
Tulenin, Yu. P., Sinev, M. Yu., and Korchak, V. N. 11th Int. Congress on Catalysis. June 30–July 5, Baltimore, ML, USA. Proceedings, Po-275 (1996).
Tulenin, Yu. P., Sinev, M. Yu., Savkin, V. V., and Korchak, V. N. *Catal. Today* **91–92**, 155 (2004).
Tulenin, Yu. P., Sinev, M. Yu., Savkin, V. V., Korchak, V. N., and Yan, Y. B. *Kinetika i Kataliz (Russ. Kinet. Catal.)* **40**, 405 (1999).
Tulenin, Yu. P., Sokolov, O. V., Sinev, M. Yu., Savkin, V. V., Arutyunov, V. S., and Korchak, V. N. EUROPACAT-V. 2–7 Sept., Abstracts. Book 4. Paper 5-P-12 (2001).
Van Santen, R. A., and Neurock, M. *Catal. Rev.- Sci. Eng.* **37**, 557 (1995).
Vedeneev, V. I., Arutyunov, V. S., and Basevich, V. Ya. *Chem. Phys. Reports* **16**, 459 (1997a).
Vedeneev, V. I., Arutyunov, V. S., Basevich, V. Ya., Parfenov, Yu. V., and Bernatosyan, S. G. *Chem. Phys. Reports* **19**, 94 (2000).
Vedeneev, V. I., Goldenberg, M. Ya., Gorban', N. I., and Teitel'boim, M. A. *Kinet. Catal.* **29**, 1 (1988a).
Vedeneev, V. I., Goldenberg, M. Ya., Gorban', N. I., and Teitel'boim, M. A. *Kinet. Catal.* **29**, 8 (1988b).
Vedeneev, V. I., Goldenberg, M. Ya., Gorban', N. I., and Teitel'boim, M. A. *Kinet. Catal.* **29**, 1121 (1988c).
Vedeneev, V. I., Goldenberg, M. Ya., Gorban', N. I., and Teitel'boim, M. A. *Kinet. Catal.* **29**, 1126 (1988d).
Vedeneev, V. I., Krylov, O. V., Arutyunov, V. S., Basevich, V. Ya., Goldenberg, M. Ya., and Teitel'boim, M. A. *Appl. Catal. A* **127**, 51 (1995).
Vedeneev, V. I., Romanovich, L. B., Basevich, V. Ya., Arutyunov, V. S., Sokolov, O. V., and Parfenov, Yu. V. *Russ. Chem. Bull.* **46**, 2006 (1997b).
Vislovskiy, V. P., Suleimanov, T. E., Sinev, M. Yu., Tulenin, Yu. P., Margolis, L. Ya., and Cortés Corberán, V. *Catal. Today* **61**, 287 (2000).
Wagner, A. F., and Wardlaw, D. M. *J. Phys. Chem.* **92**, 2462 (1988).
Warnatz, J., Rate coefficients in the C/H/O system, *in* "Combustion Chemistry" (W. C. Gardiner, Jr. Ed.), Springer-Verlag, New York, Berlin, Heidelberg, Tokyo (1984).
Williams, F. A., Reduced chemical kinetics for turbulent hydrocarbon-air diffusion flames, *in* "Advanced Computation & Analysis of Combustion" (G. D. Roy, S. M. Frolov, and P. Givi Eds.), ENAS Publishers, Moscow (1997).
Xu, M., Shi, C., Yang, X., Rosynek, M. P., and Lunsford, J. H. *J. Phys. Chem.* **96**, 6395 (1992).
Zabarnick, S. *Fuel Chem. Preprints* **50**, 88 (2005).
Zanthoff, H., and Baerns, M. *Ind. Eng. Chem. Res.* **29**, 2 (1990).
Zeigarnik, A. V., and Shustorovich, E. M. *Russ. J. Phys.* Chem. in press.
Zeigarnik, A. V., and Valdes-Perez, R. E. *J. Comput. Chem.* **19**, 741 (1998).
Zerkle, D. K., Allendorf, M. D., Wolf, M., and Deutschmann, O. *J. Catal.* **196**, 18 (2000).
Zhang, X., Lee, C. S. M., Hayward, D. O., and Mingos, D. M. P. *Appl. Catal. A: Gen.* **248**, 129 (2003).
Zhang, X., Lee, C. S. M., Hayward, D. O., and Mingos, D. M. P. *Catal. Today* **105**, 283 (2005).
Zhdanov, V. P., Pavlicek, J., and Knor, Z. *Catal. Rev.- Sci. Eng.* **30**, 501 (1988).

KINETIC METHODS IN PETROLEUM PROCESS ENGINEERING

Pierre Galtier[*]

Institut Français du Pétrole, IFP-Lyon, BP.3, F-69390 Vernaison, France

I. Introduction	260
II. Kinetic Modelling by Single-Events	269
A. Introduction	269
B. Bifunctional Catalysis Mechanisms	270
III. Generation of Reaction Networks	271
A. Computer Representation of Species and Chemical Reactions	273
B. Results Obtained for Computer Generation of Networks from $nC8$ to $nC15$	273
IV. Kinetics by Single-Events	273
A. Single-Events Microkinetic Concept	273
B. Separation of Chemical and Structural Contributions	275
C. Free Enthalpies of Reactants and Activated Complexes	276
D. Thermochemical Restrictions and Constraints	278
V. Late Lumping Kinetic Model	279
A. Three-Phase Model	280
B. Catalytic Act	280
C. Composition of the Reaction Intermediates	281
D. Lumping by Families	281
E. Lumped Kinetics	282
F. Summing Up	285
VI. Extrapolation to Heavy Cuts	286
A. Estimation of Kinetic Parameters	286
B. Extension to Heavy Paraffins	289
C. Extrapolation Capacities in Number of Carbon Atoms: Heavy Paraffinic Waxes	294
VII. Perspectives	299
References	302
Further Reading	304

[*]Corresponding author. E-mail: pierre.galtier@ifp.fr

Abstract

Kinetic modelling of petroleum processes is particularly difficult given the complexity of the feedstocks. Different lumping strategies are proposed according to the petroleum cuts type. As a paradox, a microkinetic method does allow a "late" lumping method for the case of distillates and bifunctional heterogeneous catalysis. The construction of a so-called "single-events" model consists in: (a) an automated generation of the complete network of reaction elementary steps; (b) a reduction in the number of kinetic constants; (c) a rigorous *a posteriori* relumping. Furthermore, an alternative late lumping method, not requiring network generation, has been developed. The extrapolation ability of the model is demonstrated by simulations of Fischer–Tropsch waxes hydrocracking. Finally, the perspectives of kinetic modelling by single-events are given.

I. Introduction

Thermodynamics (far from and close to equilibrium) is the mother of all sciences (Prigogine, 1996). Applied to gas–liquid equilibria, it was expressed very generally as follows: the equations of state f(P, V, T, N, parameters) = 0. The latter could be, e.g., cubic or based on statistical physics. These equations of state are sufficiently generic to serve as a calculation basis for practically all separation unit operations (distillation, adsorption, extraction). It led to the development of commercial flowsheeting[1] software capable of simulating process diagrams, and the sequencing of their steps, for virtually all hydrocarbon-containing mixtures. In these programs, however, the reaction section modules available to "Process Engineering" professionals remain extremely poor and simplified.

The design of chemical reactors (Trambouze *et al.*, 1984) is based primarily on the laws of chemical kinetics. These laws are only general as regards their elementary steps, by definition of first order with respect to the reaction intermediates. When these elementary steps are poorly understood, which is the case in particular for heterogeneous reactions, no general expression is available which would make calculations of the degree of conversion of reactions even slightly generic. This situation explains why simulation methods have not been developed for chemical reactors as systematically as for the other unit operations; the reactors remain based on the use of apparent kinetics, with no intrinsic physical meaning, and contingent on each particular application.

Nevertheless, Chemical Kinetics is the key for development of processes in general (Froment and Bischoff, 1990), and more especially for the development

---

[1]FLOWSHEETER = Process diagram calculation software package. Examples: ASPEN™, PROII™, HYSIS™.

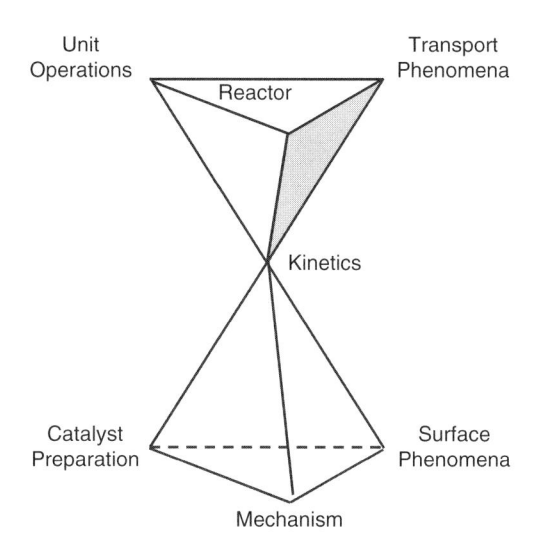

FIG. 1. Catalytic reaction engineering.

of *catalytic processes*. This essential science bridges the gap between two major disciplines: Catalysis and Process Engineering. It is a fundamental activity founded on the description that can be made of the catalytic act (physicochemistry of solids, surface sites, elementary reaction mechanisms). Combined if necessary with the transfer laws, chemical kinetics is the applied tool enabling this contact catalysis (design, dimensioning and extrapolation of catalytic reactors) to be implemented, Fig. 1 (Marin *et al.*, 2000).

It allows the transition from the nanoscale of molecular processes to the mesoscale of unit operations (Charpentier, 2000).

The methodology of the kinetic approach is well understood when a molecular analysis of the feedstock is available. This is usually the case for the manufacturing processes of the major petrochemical intermediates. Although generally costly and time consuming, a complete kinetic study is possible if the method is systematically applied to each constituent identified at molecular level in the feedstock. The domain of refining is much more complex and the method is difficult to apply.

Refining operations (of which at least 80% are catalytic processes) are designed to transform and purify various petroleum cuts, derived from primary distillation, into the various fuels and refined products. As a reminder, a list of the main refining units is shown as a simplified refining scheme in Figs. 2 and 3.

Petroleum cuts, derived from crude oils, mainly consist of a continuum of paraffins,[2] naphthenes[3] and aromatic compounds, which are traditionally

---
[2] Paraffin = alkane.
[3] Naphthene = (alkyl-) cyclo-alkane.

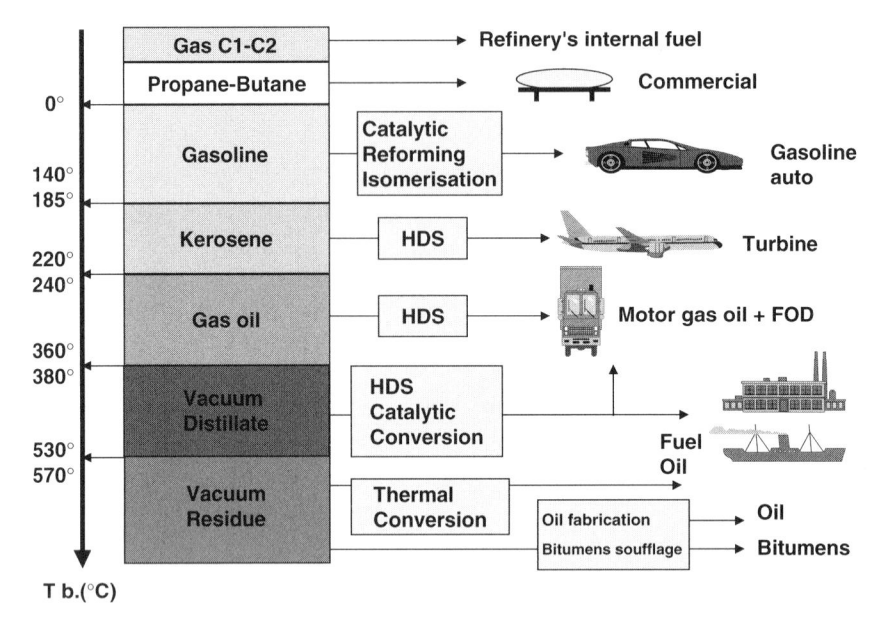

FIG. 2. Refining.

classified, depending on their distillation intervals, into various "gasoline", "diesel", "distillate" and "residue" cuts, from the lightest C5 compounds to the heaviest C50+ compounds, Fig. 4.

Petroleum cuts are extremely complex mixtures, however, which may contain several thousand chemically different molecules. Table I (Read, 1976) indicates the number of possible isomers according to the number of carbon atoms, thereby giving an idea of the complexity of the mixtures involved.

Analysis of feedstocks and effluents undeniably represents the main "bottleneck" for kinetic modelling of complex mixtures.

In addition to traditional monitoring analyses and standardised measurement of the various petroleum properties, modern analytical models now also include determination of the chemical composition of feedstock and recipes.

(1) In Fig. 5, the advanced (but difficult) analytical techniques developed at the IFP[4] (Fafet and Magné-Drisch, 1995) involve preparative liquid chromatography to separate the saturated and aromatic extracts, on adsorbents such as silica gels [the S.A.(R.(A.)) method[5]].
(2) Each extract is then analysed separately by coupling gas chromatography (GC) to mass spectrometry (MS).

---
[4]IFP = Institut Français du Petrole, BP.3, F-69390 Vernaison, France.
[5]S.A.(R.(A.)) = Saturates, Aromatics, (Resins, (Asphaltenes)).

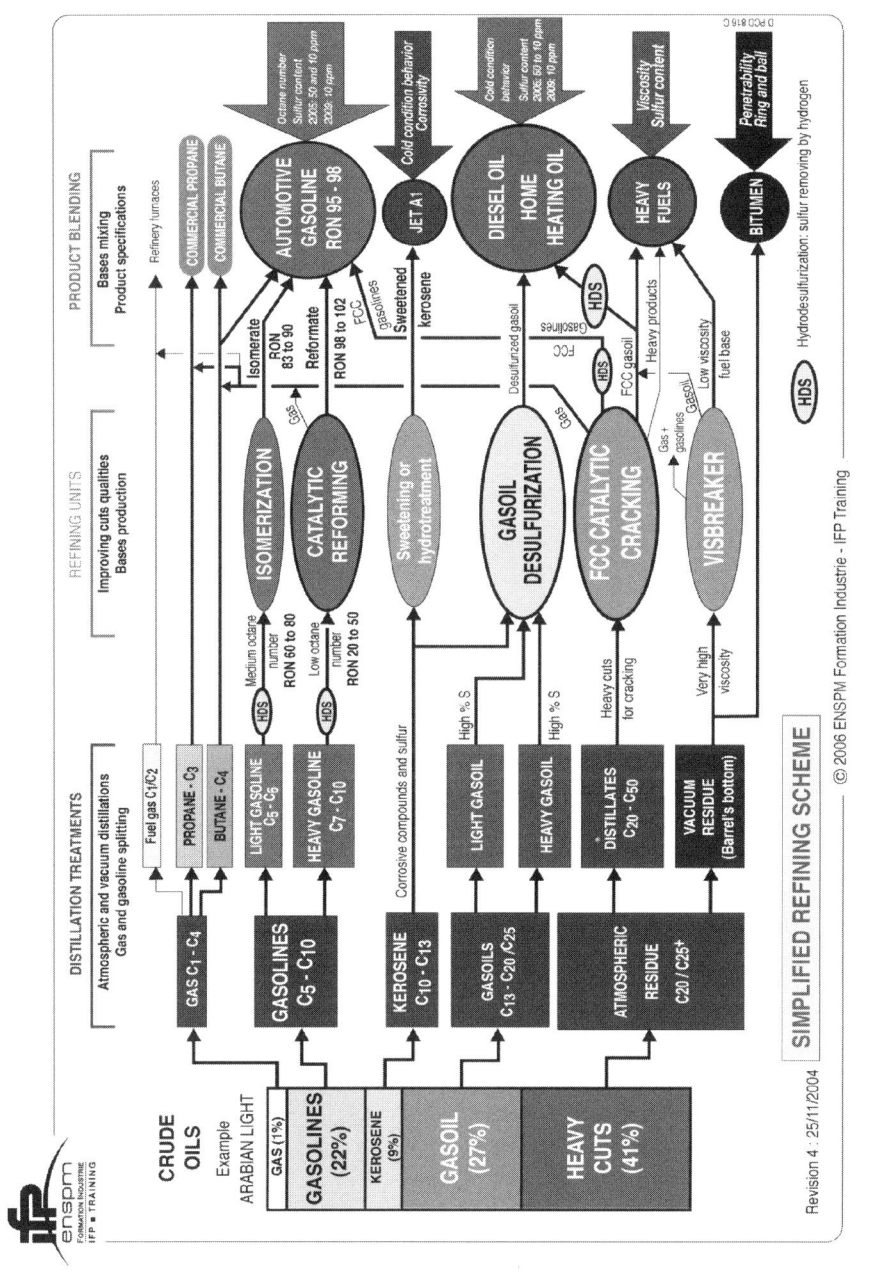

FIG. 3. Simplified refining scheme.

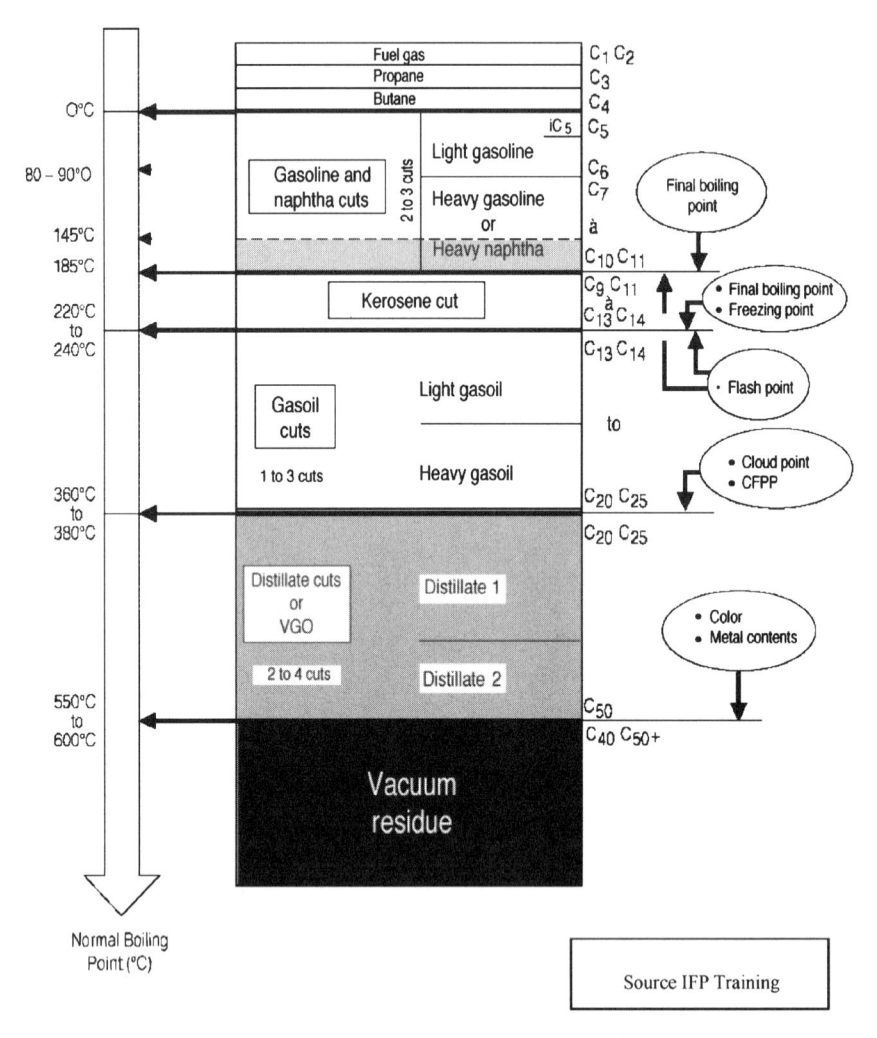

FIG. 4. Crude oil fractionation—volatility and cut points constraints.

This method gives the distribution by family (paraffins, isoparaffins, naphthenes with one or more rings, etc., aromatics) and by number of carbon atoms (up to a maximum of C33 for saturated compounds and up to C22 for aromatic compounds).

For example, Fig. 6 provides the weighted analysis, as detailed as possible (Hillewaert, 1986), obtained for a vacuum gas oil (VGO) after partial hydrogenation, with a density at 300 K of 0.832 kg/m^3 and a distillation interval ranging from 300 to 480+.[6]

---

[6] 480+ = Petroleum cut whose distillation interval exceeds a temperature of 480°C.

TABLE I
Structural Isomers of Different Hydrocarbon Types

Carbon number	Paraffins	Olefins	Alkylbenzenes
5	3	5	–
10	75	377	22
15	4,374	36,564	2,217
20	366,319	4,224,993	263,381
25	36,797,588	536,113,477	33,592,349

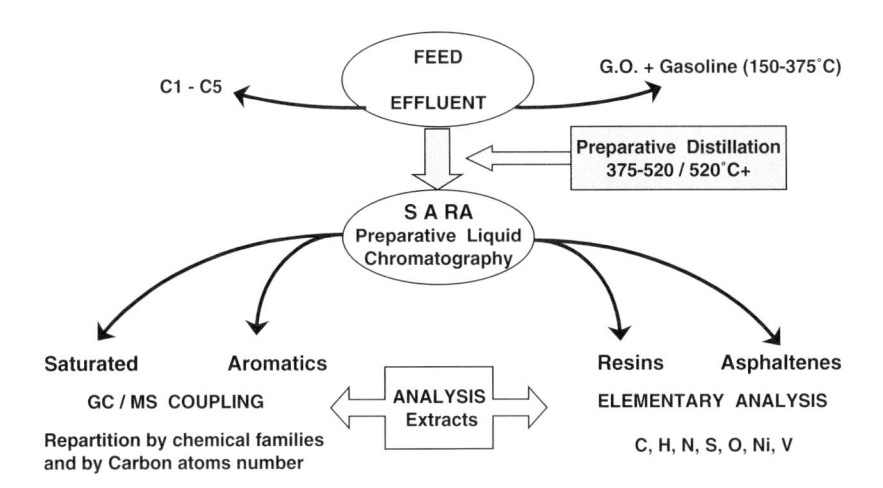

FIG. 5. Analytical set.

Despite their sophistication, it is clear that these advanced analysis methods can only be applied to the various light, middle and heavy "distillates", according to their origin ("atmospheric", "vacuum"). For petroleum cuts involving "residues", this type of pseudo molecular description cannot be obtained, so a way must be found to "reconstruct" it.

Since there is no detailed analysis at molecular level, kinetic modelling of complex mixtures raises a problem of methodology which clearly demands a scientific approach (Astarita and Sandler, 1991; Sapre and Krambeck, 1991). Petroleum cuts contain so many molecules and isomers, with such large variations in structure and chemical composition that their physico-chemical properties vary almost continuously with the number of carbon atoms.

Several attempts have been made to construct a "kinetics of continuous reactions" (Aris and Gavalas, 1966; Astarita and Ocone, 1988). These mathematical developments mainly use either distribution functions to represent variations in molecule reactivity according to their chain lengths or boiling points for example (Aris, 1989), or the method of moments (Kodera et al., 2000;

nC	NPA(wt%)	MPA (wt%)	DPA (wt%)	TPA(wt%)	MNA(wt%)	DNA(wt%)	TNA (wt%)	TEN (wt%)
14	0	0	0	0	1.24	0.047	0	0
15	0.002	0	0	0	2.33	0.54	0	0
16	0.016	0.002	0.0002	0	2.714	1.51	0	0
17	0.077	0.002	0.0002	0	2.352	1.441	0.323	0
18	0.167	0.035	0.004	0	1.738	1.046	0.711	0
19	0.365	0.138	0.014	0	1.057	0.647	0.298	0.018
20	0.729	0.302	0.030	0	0.86	0.781	0.364	0.196
21	1.01	0.518	0.052	0	1.116	1.163	0.611	0.447
22	1.43	0.827	0.083	0	1.482	1.4	0.66	0.723
23	1.56	1.15	0.115	0	1.92	1.77	0.873	0.793
24	1.86	1.45	0.145	0	2.38	2.22	1.24	0.873
25	2.07	1.56	0.156	0	2.79	2.46	1.37	0.693
26	1.65	1.31	0.132	0	2.59	2.27	1.09	0.533
27	1.27	1.23	0.124	0	2.38	2.11	0.897	0.439
28	0.9	1.01	0.101	0	2.07	1.93	0.615	0.395
29	0.538	0.657	0.066	0	1.55	1.52	0.403	0.235
30	0.343	0.387	0.039	0	1.15	1.14	0.163	0.118
31	0.183	0.243	0.024	0	0.712	0.608	0.09	0.082
32	0.115	0.148	0.015	0	0.405	0.248	0.092	0.06
33	0.04	0.064	0.006	0	0.236	0.015	0.125	0.047
Sum	14.325	11.062	1.106	0	33.072	24.866	9.925	5.652

FIG. 6. Composition of a hydrogenated VGO (NPA, MPA, DPA, TPA: normal-, monobranched-, dibranched and tribranched paraffins; MNA, DNA, TNA, TEN: mono-, di-, tri-, tetra-naphtenes).

Wang *et al.*, 1995). Strictly speaking, it is therefore a continuous approximation of a discrete system. Although interesting and skilful, these formal approaches only apply to certain cases of thermal cracking reactions in homogeneous phase. The analytical solutions found remain limited, however, to reaction networks, which are purely parallel or purely consecutive, as well as to reaction orders assumed to be simple. In order to generalise them to the mechanisms and complexity of surface reactions, the distribution functions to be used should not be empirical (Gamma Laws, etc.) but should systematise the available chemical knowledge of these reaction networks. The "single-events" detailed kinetic modelling method, which we will describe below, would form a preliminary means of constructing distribution functions respecting the physico-chemical understanding acquired of the relative reactivities for these reaction sets.

The approach chosen, privileged by the IFP "Kinetics and Modelling" group, aims to be much more respectful of the chemistry and reaction mechanisms involved.

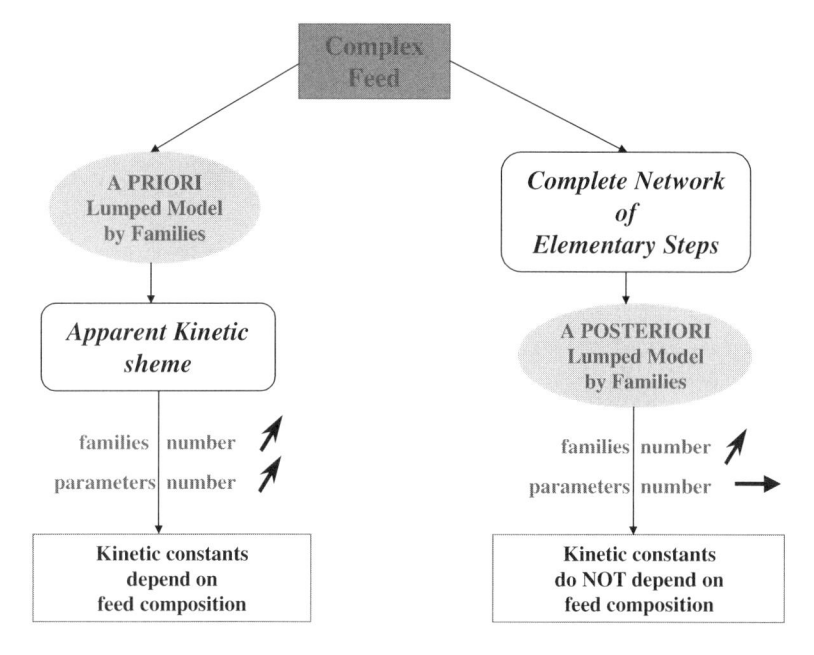

FIG. 7. Kinetics modelling.

Since modelling complex processes involves numerous reactions and molecules, a certain degree of simplification is required; the objective is always to reduce the number of compounds by *lumping*. However, like (Nigam *et al.*, 1991), we identify two different *lumping* approaches, Fig. 7.

(1) Early lumping:
This approach is applied naturally by the chemist on the basis of his know-how and/or intuition. Several models in the literature describe lumping operations carried out according to distillation cuts (Verma *et al.*, 1996), structural considerations (Quann and Jaffe, 1992) or functional considerations (paraffins, olefins, nitrogenated or sulphurated molecules). Lumping, often based on the available analyses, considerably reduces the number of constituents and can be used to produce a kinetic model comprising few parameters.

The problem with this type of model is that it frequently depends on the composition of the feedstock; in this case, an expensive experimental data bank is required to obtain a set of kinetic parameters which, to say the least, are apparent and dependent on the feedstock composition.

In addition, since it is always tempting to refine the model by subdividing the lumped families, the number of kinetic constants between all these lumpings increases at an almost exponential rate with the number of pseudo-constituents considered, since they are apparent constants.

To model hydrotreatment (HDT) processes, however, we have extended this approach to all reactions hydrodesulphurisation (HDS), hydrodenitrogenation (HDN) and hydrogenation of aromatics (HDA), since it is the only one possible in the case of sulphide catalysis (Bonnardot, 1998; Magné-Drisch, 1995).

(2) Late lumping:

In more favourable cases, for example acid and/or bifunctional catalysis, we will see that a different approach can be considered.

There is in fact sufficient knowledge available on the elementary steps involved in the reaction mechanisms to allow automatic generation of the entire network and the reaction intermediates, which run into several tens of thousands (Baltanas and Froment, 1985). In retrospect, and given a few fairly reliable assumptions, *late lumping* of the species and the reaction intermediates is possible (Vynckier and Froment, 1991).

In addition, the apparent kinetic constants between these families so lumped are *strictly* expressed according to a small number of elementary and fundamental kinetic constants. Their number remains finite, irrespective of the number of carbon atoms in the molecule; furthermore, due to their intrinsic nature, it may be possible to determine them from simple model molecules.

This method is therefore doubly *generic*:

(a) Firstly, it can be used to *extrapolate* the kinetic constants according to the length of the hydrocarbon chains.
(b) Secondly, it can be generalised to all processes using acid catalysis, in other words 80% of the refining operations.

We decided to validate it under *conditions representative* of various industrial processes: isomerisation, catalytic reforming, hydrocracking and fluid catalytic cracking (FCC).

(3) Reconstruction of feedstocks; case of heavy cuts:

The two previous techniques are obviously limited when it comes to refining heavy feedstocks (520+); in practice, they can only be applied to distillates.

With heavy cuts containing residues, there is insufficient analytical data to provide the detailed composition of feedstock and effluents. The aim (Neurock *et al.*, 1989, 1990, 1994) in this case is therefore to artificially reconstruct an equivalent molecular population, whose global properties (density, viscosity, etc.) are the same as those measured on the actual feedstock.

The studies conducted at the IFP by the "Kinetics and Modelling" research group are structured around the three previously defined axes, summarised in Fig. 8.

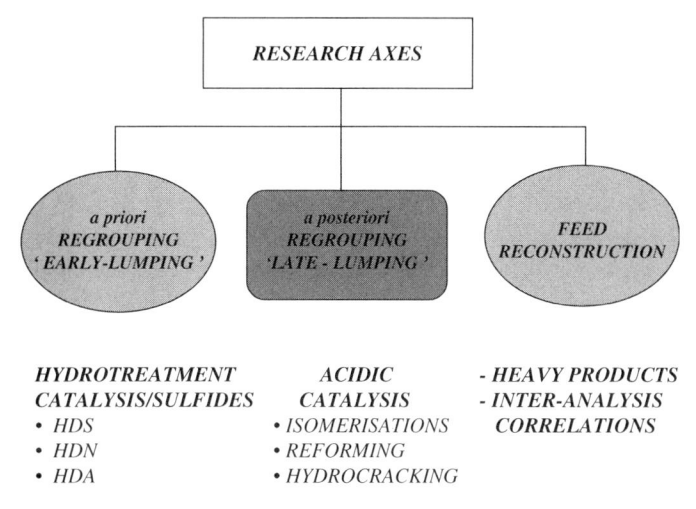

FIG. 8. Research axes.

II. Kinetic Modelling by Single-Events

A. INTRODUCTION

The creation of models that are as representative as possible is a major step in the industrial development of refining processes. The models created can be used firstly to optimise the operating conditions: "design" of the unit, choice of reactor, determination of optimum running conditions. Secondly, they can be used to predict the performance of the unit.

To optimise the modelling studies—which may be lengthy and costly—detailed predictive models must be developed which can be adapted to evolutions in the process, i.e. be extrapolated to different feedstocks, different operating conditions, etc.

Modelling based on the single-events theory[7] and developed in radical chemistry, or in acid catalysis, meets these requirements since it allows detailed prediction of the yields output from the units. This methodology, developed in the "Laboratorium voor Petrochemische Techniek" at Ghent University (Froment, 1991), consists in constructing a reaction network which, although exhaustive, is described by a limited and uncoupled number of kinetic constants independent of the number of atoms in the molecules. The behaviour of complex feedstocks can therefore be predicted on the basis of studies conducted on model molecules. This method can be applied to several refining processes, as demonstrated by a series of studies currently in progress at the IFP

---

[7] = "Single-Events MicroKinetics" (SEMK) method.

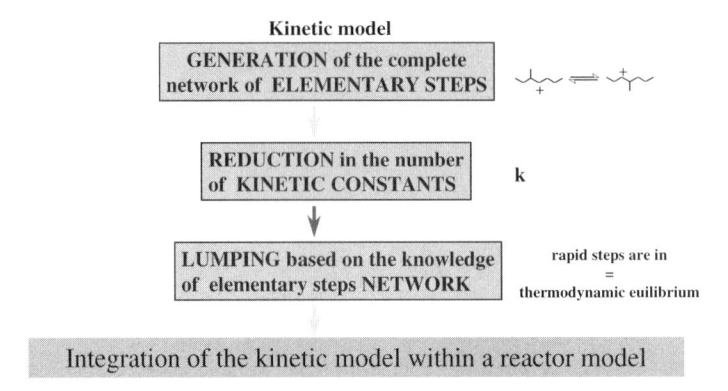

FIG. 9. "Single-Events" modelling.

(isomerisation, reforming, hydrocracking, etc.) (Cochegrue, 2001; Schweitzer, 1998; Valery, 2002).

The "Single-Events" methodology can be used to produce a highly detailed representation of the reaction mechanisms occurring in the acid phase of the catalyst. It can be used to describe all the reactions involving carbocations adsorbed on acid sites and derived from paraffins and naphthenes in the feedstock. The rates of formation and disappearance of all the hydrocarbons are expressed according to the kinetics of formation and disappearance of these carbocations and can be lumped by chemical family, if necessary. Figure 9 schematically represents the different stages of the methodology.

B. BIFUNCTIONAL CATALYSIS MECHANISMS

Most refining catalysts consist of two active phases: a metallic phase and an acid phase, Fig. 10.

- *The metallic phase* protects the catalyst and allows the formation of unsaturated reaction intermediates for the acid phase.
- *The acid phase* forms carbocations whose carbon skeletons undergo modifications (ring opening/closing, cracking, isomerisation, etc.), Fig. 11.

All reactions involving primary ions are neglected, considering their low thermochemical stability compared with secondary and tertiary carbenium ions.

For example in Fig. 12, the *n*-hexane hydrocracking reaction network—the last which can be drawn entirely by hand—contains 6 paraffins, 14 olefins and 10 carbocations, and involves the following elementary reactions:

- 14/14 hydrogenations/dehydrogenations
- 18/18 protonations/deprotonations

KINETIC METHODS IN PETROLEUM PROCESS ENGINEERING   271

• **Reaction Mechanisms**

$$\text{Saturated Hydrocarbons} \circlearrowright \text{Hydrogenolysis}$$

$$\text{Hydrogenation} \updownarrow \text{Dehydrogenation}$$

-------- **Insaturated Hydrocarbons** -------- **Metallic Phase**
                                                 Acidic Phase

$$\text{Deprotonation} \updownarrow \text{Protonation}$$

Carbenium Ions

Reactions on the acidic surface

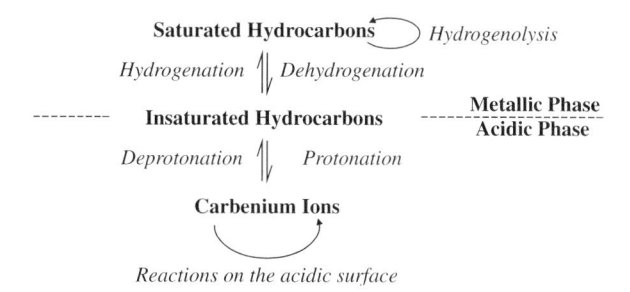

FIG. 10. Bifunctional catalysis—reaction mechanisms.

• **Reaction Mechanisms**

   ◆ Saturated carbenium ions
      ☞ Rearrangement or shift reaction (HS, MS, ES, intra ring alkyl shift (IRAS))
      ☞ Isomerisation by the intermediaries PCP
      ☞ Beta Cracking (acyclic, exocyclic, endocyclic)
      ☞ Deprotonation

   ◆ Unsaturated carbenium ions
      ☞ Allylic Resonance
      ☞ Rearrangement or shift reaction (HS)
      ☞ Cyclization
      ☞ Deprotonation

FIG. 11. Acidic phase—reaction mechanisms.

- 8 hydride shifts (HS)
- 4 methyl shifts (MS)
- 20 isomerisations by a protonated cyclopropane (PCP)
- 2 isomerisations by a protonated cyclobutane (PCB)
- 1 β-scission cracking reaction

III. Generation of Reaction Networks

Generation is carried out according to chemical criteria. A molecule undergoes a sequence of reactions which leads to new products, respectively undergoing

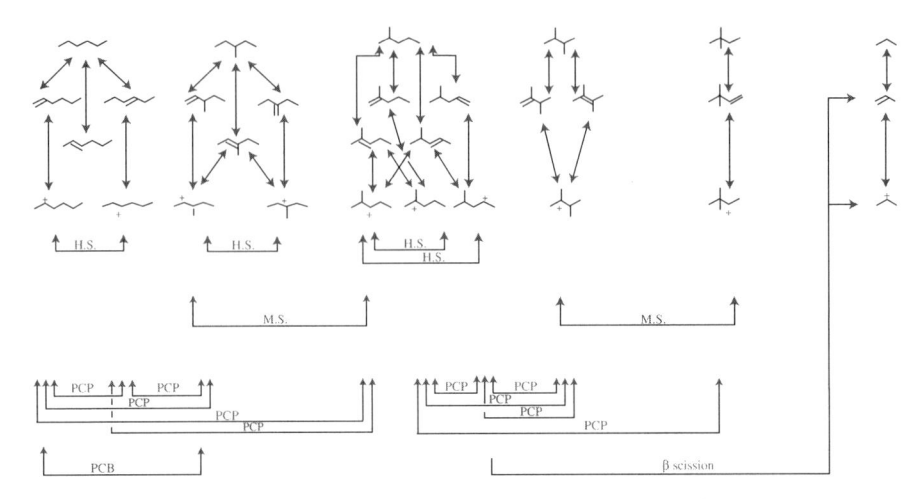

FIG. 12. Elementary steps network for n-hexane cracking.

- **Computing by generation algorithm**
  - ☞ Repeating a limited number of elementary steps (tree-form scheme)
  - ☞ Avoiding duplication of same reaction pathways

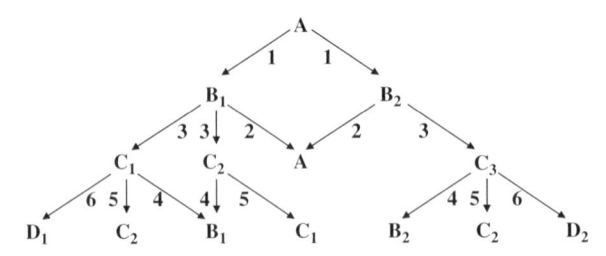

FIG. 13. Elementary steps network generation.

another sequence depending on their type and structure. We then obtain a tree structure. As an example in Fig. 13, from a body A, we obtain the following generation:

This method can be used to take into account all types of reaction which are chemically possible. Although there are thousands of elementary steps, the reaction network actually consists of repetitions and combinations of a limited number of reaction types.

A computer tool is required to construct a network of complex reactions. The algorithm used for this purpose precludes possible duplication of reaction species and pathways during their generation.

A. Computer Representation of Species and Chemical Reactions

The hydrocarbon species are represented by Connectivity Matrices (Clymans and Froment, 1984). The carbon–hydrogen bonds are not shown explicitly. In contrast, the carbon skeleton (containing no hetero atoms) can be mathematically represented by Boolean matrices. If carbon 1 is bonded to carbon 2, matrix elements (1;2) and (2;1) are assigned a value of 1. This is illustrated in the example shown in Fig. 14.

B. Results Obtained for Computer Generation of Networks from $nC8$ to $nC15$

The Fig. 15 indicates the change in the *number of species* generated (paraffins, olefins, ions, in Fig. 15a) and the *number of reactions* (hydrogenations, protonations, HS, MS, ethyl shiftes (ES), PCP branching, PCB branching, beta scissions, in Fig. 15b) according to the *number of carbons* of a single hydrocracked normal paraffin.

The above examples concern only aliphatic molecules. Cyclic molecules have similar curves. With cyclic molecules, however, there are more products and more reactions.

IV. Kinetics by Single-Events

The concept of single-events lies upstream from the notion of elementary step.

A. Single-Events Microkinetic Concept

The reaction network consists of elementary steps. Each step is the result of one or more "single-events" (Baltanas *et al.*, 1989; Vynckier and Froment, 1991).

● **Numerical Representation of reactions**
☞ Operation on the boolean matrix

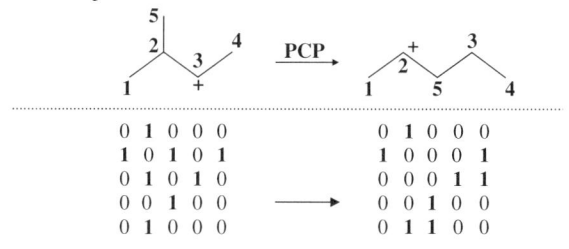

$$\begin{matrix} 0 & 1 & 0 & 0 & 0 \\ 1 & 0 & 1 & 0 & 1 \\ 0 & 1 & 0 & 1 & 0 \\ 0 & 0 & 1 & 0 & 0 \\ 0 & 1 & 0 & 0 & 0 \end{matrix} \longrightarrow \begin{matrix} 0 & 1 & 0 & 0 & 0 \\ 1 & 0 & 0 & 0 & 1 \\ 0 & 0 & 0 & 1 & 1 \\ 0 & 0 & 1 & 0 & 0 \\ 0 & 1 & 1 & 0 & 0 \end{matrix}$$

FIG. 14. Network generation by computer.

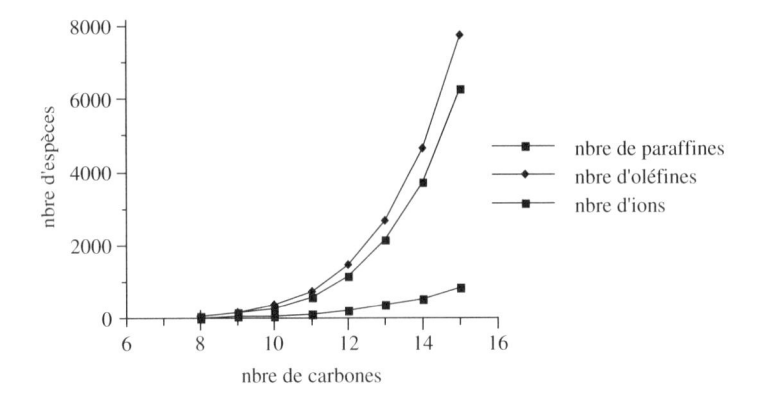

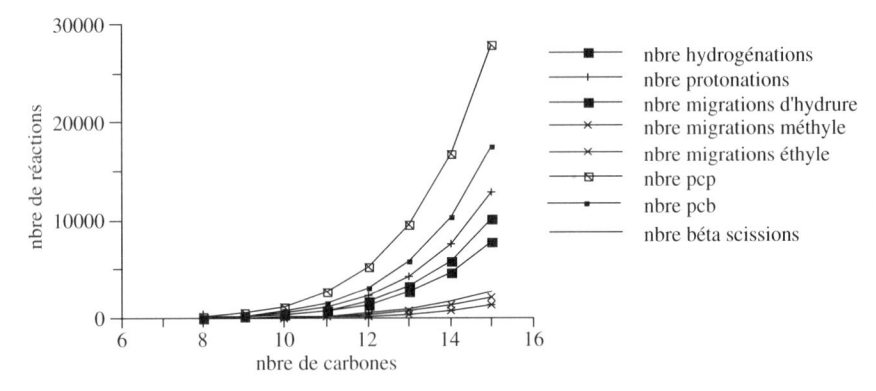

FIG. 15. Results of networks generation.

The following figure shows an isomerisation equilibrium by MS between two secondary ions. Two methyls can shift in the direction A → B,[8] but only one in the direction B → A. The example of Fig. 16[9] illustrates the importance of symmetries in the reaction rates.

We are now at the root of the principles on which the single-events method is based. The rate constant of an elementary step depends on two classes of factor (Van Raemdonck and Froment, 1989):

- The specific reactivity of the active centre (more precisely, of the activated complex). This reactivity is defined by the reaction type (in our example, an MS) and by the type of reactant and product ions (in our example, secondary ions).
- The multiplicity of structural elements involved in the elementary step. In our example, this is the number of methyl groups likely to shift (two for A and one for B).

---
[8] A = the methyl_2 hexyl_3 ion and B = the methyl_3 hexyl_2 ion.
[9] "Branchement(s)" means "Branching(s)" and "Migration(s)" means "Shift(s)".

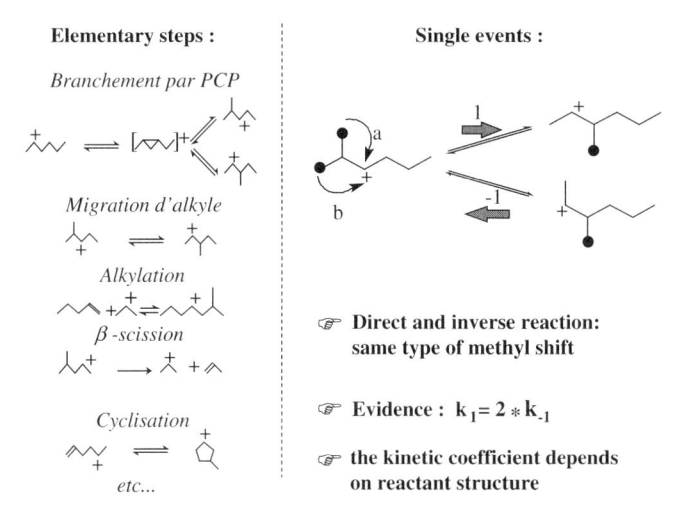

KINETIC METHODS IN PETROLEUM PROCESS ENGINEERING   275

FIG. 16. Elementary steps/Single-events.

The kinetic constant of each elementary step will therefore be the product of a single-events number n_e and an intrinsic constant k_{reac} which depends on the reaction type and on the types (m,u) of the reactant and product ions:

$$K = n_e * k_{reac}(m, u)$$

The following paragraph shows how to calculate the single-events number n_e from the symmetries of the reactant and of the activated complex.[10]

B. SEPARATION OF CHEMICAL AND STRUCTURAL CONTRIBUTIONS

In Fig. 17, we will start from the activated complex theory and the Eyring law and decompose the free enthalpy into its *intrinsic* and *symmetry components*.

The standard entropy $S°$ is the sum of several contributions, translation, internal rotation, external rotation and vibration. Each rotational contribution can be broken down into an intrinsic term $S°_{int}$ and a logarithmic term taking into account the symmetry number σ of the compound (if the molecule is optically active, the standard entropy must also be corrected by adding an asymmetry term related to the chirality: 2^n).[11] We can therefore express above

---
[10]Remark: The number of single-events could also be calculated using the notion of statistical factors (Bishop and Laidler, 1965, 1969). The statistical factor of a reaction is equal to the number of products that a reaction can form if a distinction is made between the atoms of each reactant. In our example, two methyl groups can shift in one direction and only one in the other direction: $k_1 = 2 * k_{-1}$.

[11]n = number of chiral carbon atoms.

• **Elementary constant (Eyring law)**

$$k = \frac{k_B T}{h} e^{\frac{\Delta S^\#}{R}} e^{-\frac{\Delta H^\#}{RT}} \qquad \Delta S_\#^0 = S_{int\#}^0 - S_{intA}^0 + R \ln \frac{\sigma_{glob}^A}{\sigma_{glob}^\#}$$

$$k = \left(\frac{\sigma_{glob}^A}{\sigma_{glob}^\#}\right)\left(\frac{k_B T}{h} e^{\frac{S_{int\#}^\circ - S_{intA}^\circ}{R}} e^{-\frac{H^\#}{RT}}\right)$$

Hypothesis: activated complex

Number of « Single Events »   Intrinsic constant

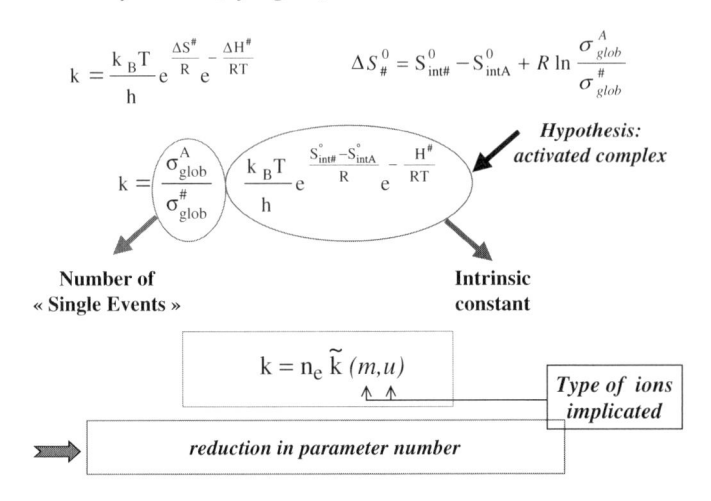

$$k = n_e \tilde{k}(m, u)$$

Type of ions implicated

reduction in parameter number

FIG. 17. Single-events.

the activation entropy $\Delta S_\#^\circ$ which is equal to the entropy difference between the activated complex and the reactant.

The elementary kinetic constant k is therefore the product of a term calculated from a difference of intrinsic free enthalpies [12] and a ratio of symmetry numbers, which we will call: single-events number: $n_e = \sigma_{glob}^A / \sigma_{glob}^\#$.[13]

Since the Eyring relation can be used to link a free activation enthalpy to a kinetic constant, the single-events method can be described in two different ways:

- Either as a model composed of single-events number and intrinsic kinetic constants (see above).
- Or as an "energy" model[14] for the reactants and activated complexes composed of intrinsic free enthalpy components and symmetry number (see below).

C. Free Enthalpies of Reactants and Activated Complexes

The single-events theory, in its "energy form" is therefore an alternative to the molecular modelling approach. Instead of using *ab initio* calculations, the

---

[12] The intrinsic free enthalpy is the free enthalpy minus the symmetry entropic contributions.
[13] $\sigma_{glob} = \sigma_{int} * \sigma_{ext}/2^n$
[14] The free enthalpy is homogeneous with an energy.

single-events method proposes a model of free enthalpy levels for ions and activated complexes, which involves few parameters. The idea remains the same: calculate the free activation enthalpy to calculate the kinetic constant of each elementary step.

1. Paraffins and Olefins

The free enthalpy of a paraffin or an olefin is determined by the Benson group contribution method (Benson, 1976). This method is traditionally used to calculate the formation enthalpies in gaseous phase at a given temperature of a molecule by describing it as structural atomic groups. Amongst other things, this provides a way of calculating the equilibrium constant of the hydrogenation/dehydrogenation reactions between the paraffins and olefins of the model.

2. Ions

To calculate the free enthalpy of an ion, we need to know its symmetry number and its intrinsic free enthalpy. The symmetry number can be calculated methodically and algorithmically. The intrinsic free enthalpy is obtained from the assumptions of the single-events theory. For an ion, the intrinsic free enthalpy calculations involve two parameters:

- The ion type (secondary or tertiary).
- The number of carbon atoms.

a. Intrinsic free enthalpy and ion type. Several formation enthalpies were determined further to studies conducted by (Brouwer, 1980) in superacid media and confirmed by *ab initio* calculations (Lenoir and Siehl, 1990).

The ions shown on the Fig. 18 all have seven carbon atoms. We observe that the formation enthalpy depends neither on the position of the charge nor on the neighbouring groups, but only on the primary,[15] secondary or tertiary character of the ion. The single-events theory accepts that observations made for formation enthalpy are also true for the intrinsic free enthalpy of formation.

Assumption 1. For a given number of carbon atoms, the intrinsic free enthalpy of formation of an ion depends only on its type (secondary or tertiary).

It means the stability of an ion depends only on the degree of substitution of the carbon atom carrying the charge.

---

[15] 181.8 kcal/mol for a primary ion. This higher value corresponds to greater instability and explains why the formation of primary ions is generally neglected compared with that of secondary- and *a fortiori* tertiary-ions.

278   PIERRE GALTIER

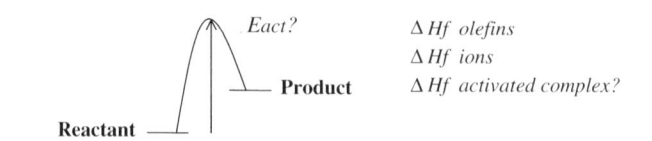

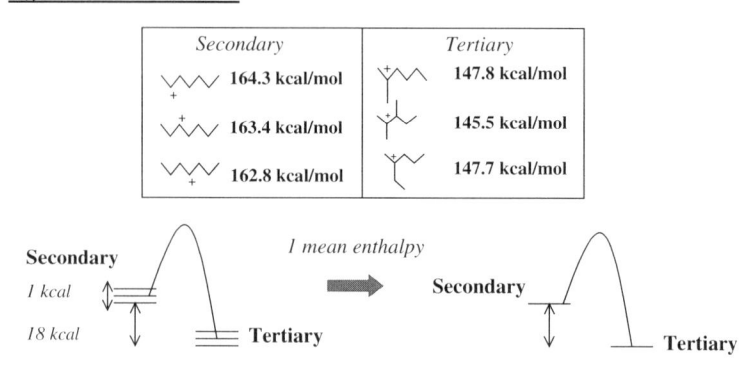

FIG. 18. Hypothesis 1 on carbocations enthalpy.

b. Intrinsic free enthalpy and number of carbon atoms. The single-events theory is also based on the assumption that only the nature of the reaction centre and its immediate environment will have an impact on the rate:

Assumption 2. The intrinsic free enthalpy of activation of an event depends only on the nature of the event and the type of ion(s) involved (Van Raemdonck and Froment, 1989; Vynckier and Froment, 1991).

The number and position of the carbon atoms surrounding the active centre have only a minor influence, since the stability of a double bond, a carbocation or an activated complex is primarily modified by the inductive effect, Fig. 19.

D. THERMOCHEMICAL RESTRICTIONS AND CONSTRAINTS

In addition to the first assumptions of the single-events theory, we must also include the various thermodynamic relations linking in particular the direct and inverse elementary constants through the thermochemical equilibrium constants. This is detailed in all the studies, and more completely by (Verstraete, 1997, Chapter 11).

These assumptions and thermodynamic constraints considerably reduce the number of kinetic parameters to be estimated; this is one of the main advantages of this theory compared with the representation and modelling of complex feedstocks.

Independence on carbons number

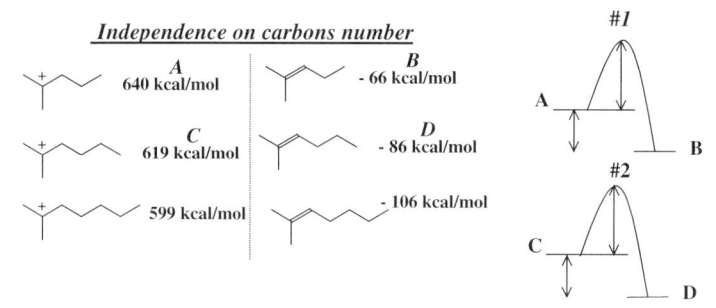

A 640 kcal/mol   B −66 kcal/mol
C 619 kcal/mol   D −86 kcal/mol
   599 kcal/mol   −106 kcal/mol

#1: A → B
#2: C → D

Hypothesis : same difference between reactants and products
⇒ same difference between activated complexes

$$\tilde{k}(\text{réactif}, \text{produit})$$

⟹ **reduction in parameters number**

FIG. 19. Hypothesis 2 on activated complexes enthalpy.

- **Required S. E. constants** (acidic phase)
  - ◆ Protonation: $k_{Pr}(s)$ $k_{Pr}(t)$
  - ◆ Deprotonation: $k_{De}(s;O_i)$ $k_{De}(t;O_i)$
  - ◆ Isomerisation : $k_{iso}(s;s)$ $k_{iso}(s;t)$
                        $k_{iso}(t;s)$ $k_{iso}(t;t)$
  - ◆ Cyclisation: $k_{Cyc}(s;s)$ $k_{Cyc}(s;t)$
                        $k_{Cyc}(t;s)$ $k_{Cyc}(t;t)$
  - ◆ Cracking: $k_{Cr}(s;s,O_i)$ $k_{Cr}(s;t,O_i)$
                        $k_{Cr}(t;s,O_i)$ $k_{Cr}(t;t,O_i)$

- **Total : 57559 kinetics constants** (C_{11} network)

FIG. 20. Reduction in number of parameters.

In the example given on Fig. 20, only 16 intrinsic parameters are required to define all the kinetic constants regarding the 57,559 constitutive pathways of a catalytic reforming reaction network, including molecules with up to 11 carbon atoms.

V. Late Lumping Kinetic Model

Another—important—advantage of the single-events method is that it allows *rigorous* lumping by chemical families, using a few simple additional assumptions.

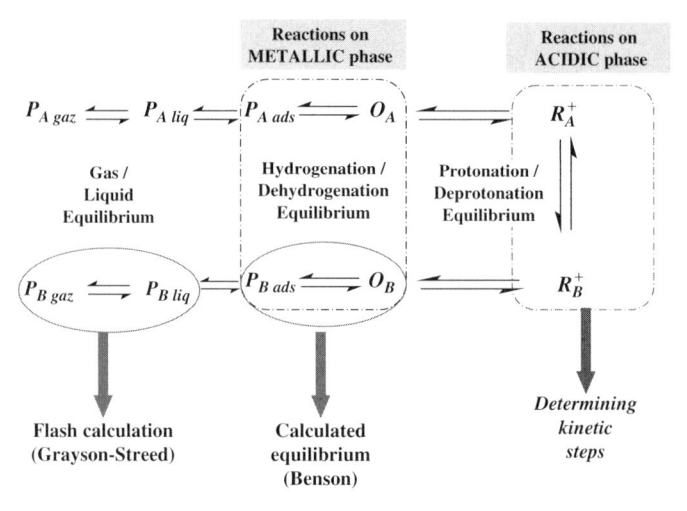

FIG. 21. Reaction model.

Although the single-events theory defines a kinetic model of elementary reactions occurring on the acid phase, this model must be included in a more general framework in order to model access to this acid phase. We will therefore consider the reaction path of a paraffin in Fig. 21:

A. THREE-PHASE MODEL

The three-phase model takes into account the possible appearance of a liquid phase in the reactor.

Assumption 3. The associated gas–liquid equilibrium is generally achieved.

Under these conditions—the solid is considered as being completely wetted (by capillary action)—the activity of a compound will be considered as equal to its liquid concentration. The gas–liquid partition of the compounds is expressed using Henry coefficients He[16] whose values are obtained by a standard Flash calculation (Grayson–Streed method if hydrogen is present).

B. CATALYTIC ACT

All reactions on the acid phase are preceded by a series of input/output steps: possible physisorption, hydrogenation/dehydrogenation and protonation/

---
[16] $He_{P_i} = P_{P_i}/[P_i]$; $He_{O_{ij}} = P_{O_{ij}}/[O_{ij}]$; $He_{H_2} = P_{H_2}/[H_2]$; the non-idealities, concerning hydrogen in particular, were not taken into account here, considering the lower activity of the sulphide catalysts.

deprotonation. These steps are considered as being fast enough to be in equilibrium (Vynckier and Froment, 1991).

Assumption 4. The possible steps of physisorption, hydrogenation/dehydrogenation, shift between the metallic and acid sites, and protonation/deprotonation, are in equilibrium.

C. COMPOSITION OF THE REACTION INTERMEDIATES

The free enthalpies of formation and therefore the equilibrium constants between paraffins, olefins and ions are known: we can therefore calculate the olefin and ion concentrations. Given the operating conditions (high temperature and high hydrogen pressures), the quantities of olefins and ions are always negligible compared with those of paraffins.[17]

Assumption 5. In addition to the steady state—the quasi-steady state approximation (QSSA) applies locally to ions and olefins.

Consequently, the rates of appearance of these reaction intermediates are zero and those of paraffins take the following form, Fig. 22:
$r_{reac}(\{m\} \to \{p\})$ = double summation of the reaction rates "reac" where m and p are reactant and product carbocations (summation indices m and p)
$\{m\}$ = summation set counted on the reaction network generated.

D. LUMPING BY FAMILIES

Various experimental observations (Svoboda *et al.*, 1995; Weitkamp, 1982) demonstrated that the reaction rates of methyl (Me) or ethyl (Et) group shifts are much faster than those of PCP and PCB isomerisation reactions and cracking reactions. Consequently, when a compound is formed, all its isomers with the same number of branches are instantaneously formed by shift reactions. All compounds with the same number of carbon atoms and the same number of branches are therefore in thermodynamic equilibrium, Fig. 23.

For example, a given paraffin will not react alone, as would be expected due to its specific structure; the reactivity observed will be that of the family of paraffins with the same numbers of carbon atoms and branches. The notion of reactive paraffin is therefore replaced by the notion of reactive paraffin family. Figure 24 summarizes the above.

---

[17]Remark: The single-events theory itself does not make any presumptions regarding the nature of the elementary steps which could be kinetically determining for the rate(s) on the acid phase. Only the introduction of Assumption 4 leads to a certain number of restrictions, by excluding the protonation/deprotonation steps.

- **Formation rate of paraffins**

$$R(Pi) = r_{cr}(\{s\} \to \{j\}) - r_{alk}(\{j\} \to \{t\})$$
$$+ r_{isom}(\{o\} \to \{m\}) - r_{isom}(\{m\} \to \{o\})$$
$$+ r_{cr}(\{p\} \to \{m\}) - r_{cr}(\{m\} \to \{p\})$$
$$+ r_{alk}(\{q\} \to \{m\}) - r_{alk}(\{m\} \to \{q\})$$
$$+ r_{end}(\{r\} \to \{n\}) - r_{cyc}(\{n\} \to \{r\})$$

- **Intermediaries concentrations**

* Q.S.S.A.

+ site balance

$R(R_m^+) = 0$

$R(Oj) = 0$

* Equilibria

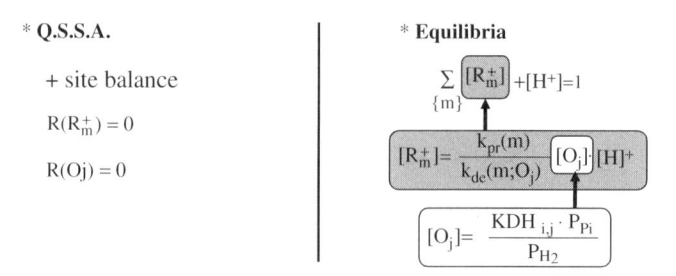

$$\sum_{\{m\}} [R_m^+] + [H^+] = 1$$

$$[R_m^+] = \frac{k_{pr}(m)}{k_{de}(m;O_j)} [O_j] \cdot [H^+]$$

$$[O_j] = \frac{KDH_{i,j} \cdot P_{Pi}}{P_{H_2}}$$

FIG. 22. Kinetic equations.

Assumption 6. Compounds[18] with the same number of carbon atoms and the same number of branches are lumped in the same-reactivity-family.

The Fig. 25 shows a simple example of lumpings used in the case of isomerisation of molecules with seven carbon atoms:

E. LUMPED KINETICS

Based on the previous assumptions, we can *rigorously and explicitly* determine an analytical expression of the apparent kinetic constants, between all the lumped families ($F_x \to F_y$) according to intrinsic elementary kinetic constants as specified in the "Single-Events" theory (Cochegrue, 2001; Schweitzer, 1998; Valery, 2002). The same applies for the denominator DEN (see next Fig. 26), in which the sum of the terms expresses the competitive chemisorption of all secondary and tertiary carbocations on the acid sites. The apparent kinetics so obtained formally return Langmuir–Hinshelwood expressions, traditional

---

[18] This means that a paraffin, an olefin and an ion with the same numbers of carbon atoms and branches belong to the same family. We have seen however (see Assumption 4) that thermodynamically the paraffins are the most stable compounds, to such an extent that all compounds in a family are mixed with all the paraffins in this family.

MB = MonoBranched; DB = DiBranched; TB = TriBranched and over

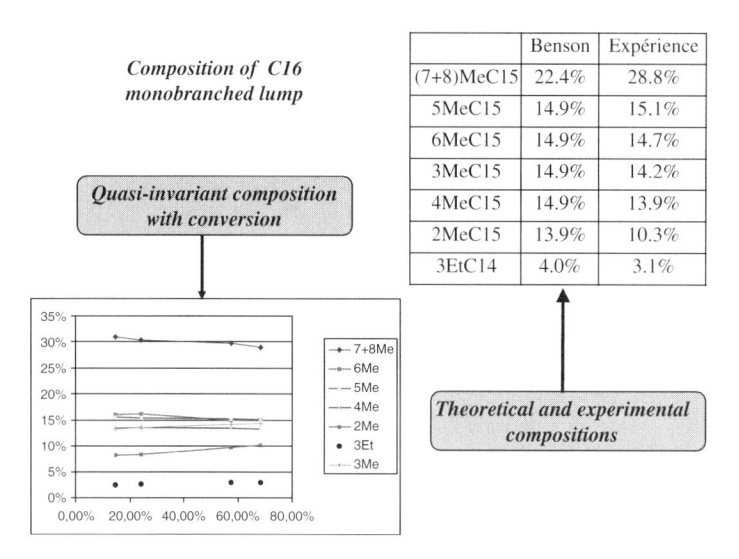

KINETIC METHODS IN PETROLEUM PROCESS ENGINEERING

Composition of C16 monobranched lump

	Benson	Expérience
(7+8)MeC15	22.4%	28.8%
5MeC15	14.9%	15.1%
6MeC15	14.9%	14.7%
3MeC15	14.9%	14.2%
4MeC15	14.9%	13.9%
2MeC15	13.9%	10.3%
3EtC14	4.0%	3.1%

Quasi-invariant composition with conversion

Theoretical and experimental compositions

FIG. 23. Consistency from experiment and hypothesis.

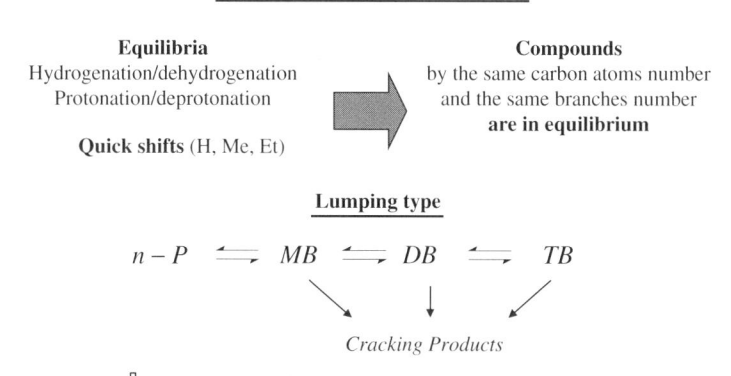

Hypothesis that lead to Lumping

Equilibria
Hydrogenation/dehydrogenation
Protonation/deprotonation

Quick shifts (H, Me, Et)

Compounds
by the same carbon atoms number
and the same branches number
are in equilibrium

Lumping type

$$n-P \rightleftharpoons MB \rightleftharpoons DB \rightleftharpoons TB$$

Cracking Products

↳ *Lumping by families in thermodynamic equilibrium*

FIG. 24. Lumping.

in catalytic heterogeneous kinetics, but in this case generalised to the case of chemical families lumped by number of carbon atoms, by number of branches, and therefore having reached their equilibrium composition.

Written in this way, each global kinetic constant can be broken down into a sum of generally four terms. Each term is itself the product of intrinsic elementary constants—in common factor—and a (double) sum of contributions usually known as "Lumping Coefficient"—in white in Fig. 26—see also Section VI.B.1. There are as many lumping coefficients as there are different

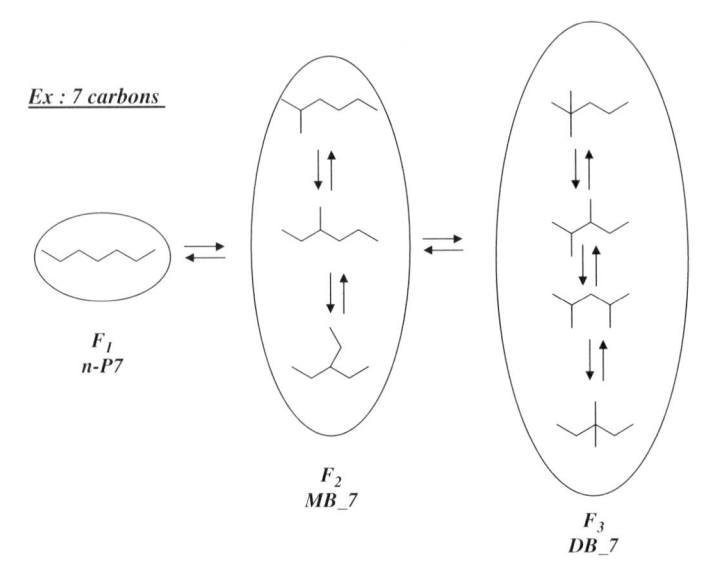

FIG. 25. Lumping by families in equilibrium.

reactions and reaction types (s,s), (s,t), (t,s), (t,t),[19] involved in the transformation of chemical families F_1 into F_2.

These "lumping coefficients" contain no kinetic parameters and can be calculated *independently* using:

the knowledge, through its *a priori* generation, of the complete reaction network, in order to enumerate the set $\{S_1\}$ of elementary steps involved in the transformation $F_1 \rightarrow F_2$.

the calculation of the single-events numbers n_e associated with all these steps; note that this initial calculation (see Section IV.B) involves determining the symmetry numbers σ of the ions, and activated complexes, derived from all the constituents of the lumpings F_1 and F_2.

the prior calculation of thermochemical equilibria (Benson group contribution method) to determine the molar fractions y_i of the lumped families—see Fig. 25, as well as the hydrogenation/dehydrogenation equilibrium constants between paraffins and olefins forming the lumpings F_1 and F_2—for given operating conditions (P,T).

Today, these calculations are easily carried out on modern workstations with no CPU limits through the use of a suitable computer architecture.

---

[19] (s,s) = secondary–secondary
(s,t) = secondary–tertiary
(t,s) = tertiary–secondary
(t,t) = tertiary–tertiary

Lumped Kinetics

$$R_{(F_1 \to F_2)} = \frac{k_{(F_1 \to F_2)} \cdot P_{F_1} - k_{(F_2 \to F_1)} \cdot P_{F_2}}{DEN}$$

Calculations from the complete network

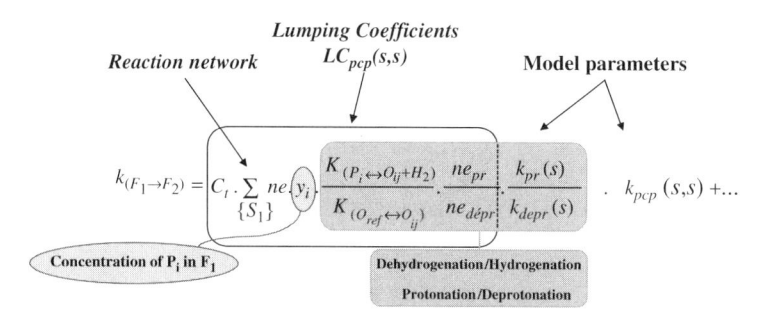

$$k_{(F_1 \to F_2)} = C_t \cdot \sum_{\{S_1\}} ne\left(y_i\right) \frac{K_{(P_i \leftrightarrow O_{ij}+H_2)}}{K_{(O_{ref} \leftrightarrow O_{ij})}} \cdot \frac{ne_{pr}}{ne_{d\acute{e}pr}} \cdot \frac{k_{pr}(s)}{k_{depr}(s)} \cdot k_{pcp}(s,s) + ...$$

- Reaction network
- Lumping Coefficients $LC_{pcp}(s,s)$
- Model parameters
- Concentration of P_i in F_1
- Dehydrogenation/Hydrogenation
- Protonation/Deprotonation

Fig. 26. Regrouped kinetics.

Paraffins	nP	11
	moP	8
	diP	7
	triP	5
Naphtènes	N	2
	SN	6
	DN	5
	TN	4
	TeN	3
Dinaphtènes	DiN	3
	SDiN	3
	DDiN	2
	TDiN	1
Aromatiques	A	1
	SA	5
	DA	4
	TA	3
	TeA	2
Naphténo-aromatiques	NA	2
	SNA	2
	DNA	1
Diaromatiques	DA	1
	SDA	1
		82

- **Constituents:**
  - 22 for the C7 network
  - 82 for the C11 network

- **Reactions:**
  - 59 for the C7 network
  - 483 for the C11 network

- **Parameters:**
  - 20 kinetic constants, 3 adsorption constants for the C7 network
  - 23 kinetic constants, 6 adsorption constants for the C11 network

Fig. 27. Lumped reforming network.

F. Summing Up

To conclude, we will give the example of a lumped network, built as part of the thesis of (Cochegrue, 2001) and including cyclic and acyclic molecules up to C11, to represent a complex reaction network of catalytic reforming, Fig. 27.

This example demonstrates the benefit of using the single-events theory in the *reduction of kinetic networks*.

We can also appreciate the utility of this method in *reducing the number of kinetic parameters*, not only in terms of their number but also in terms of their meaning;

Remember that we are concerned here with parameters that are intrinsic by construction and independent of the nature and composition of the feedstock; as a result, they will be easier to determine from simple model molecules; this last point will be demonstrated in all the following sections.

VI. Extrapolation to Heavy Cuts

By construction, the single-events theory introduces kinetic parameters which are not only intrinsic but, above all, independent of the feedstock. This first property offers a twofold advantage:

Firstly, these fundamental parameters can be determined from measurements taken on model molecules, often simpler to implement experimentally and analyse.

Secondly, using the estimations of the kinetic parameters obtained, we can predict—directly—the behaviour of heavy feedstocks, which are much more complex in terms of the number of carbon atoms and the number of branches.

In this chapter, we will illustrate this double extrapolation capacity of the single-events theory, using the example of hydrocracking[20]/hydroisomerisation of paraffinic cuts, Fig. 28.

A. ESTIMATION OF KINETIC PARAMETERS

1. Hydrocracking of n-Hexadecane (nC16 = Model Molecule)

n-Hexadecane was chosen as model molecule since it is relatively easy to implement and obtain as a pure body. Its reaction network—non-exhaustive on Fig. 29—is representative since it includes all the elementary steps involved in the hydrocracking/hydroisomerisation of heavy paraffinic cuts. After reduction, just six kinetic parameters (two for isomerisation, four for cracking) are required to represent this type of network.

The model derived from the single-events theory is generally more detailed than that which can be really obtained by analysing recipes: with *n*C16 for example, the single-events theory—generalised up to six branches—will use seven

---

[20]Remark: The objective here is simply to represent the chemical transformations occurring in the 2nd reactor.

Catalytic Cracking Process under Hydrogen

Aim: Conversion of heavy cuts to fuels (gas oil)

Operating Conditions
- High Pressure (100-150 bar)
- Temperature from 350 to 400°C
- Excess of hydrogen (H$_2$/oil from 1000 to 2000 Nl$_{H2}$/l$_{feed}$)

Advantages	Drawbacks
High quality products	Investments
Low deactivation	Operating Costs

FIG. 28. Hydrocracking processes.

5 isothermal mass balances
from 5 to 12 days per m.balance

For 1 balance, 32 concentrations

⟹ **C16 families**:
  nC16
  monobranched
  multibranched

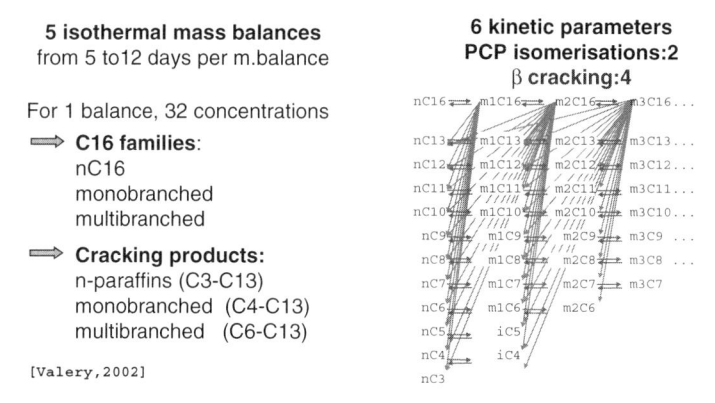

⟹ **Cracking products:**
  n-paraffins (C3-C13)
  monobranched (C4-C13)
  multibranched (C6-C13)

[Valery, 2002]

6 kinetic parameters
PCP isomerisations: 2
β cracking: 4

FIG. 29. Hydrocracking of nC16—experiment and model.

families per number of carbon atoms, whereas analytically, only three families can be measured (n-alkane, monobranched isomers and multibranched isomers). Cracking by β-scission forms alkanes with 3 to 13 carbon atoms. The analyses can be used to determine a total of 32 independent observables in each recipe of the five isothermal balances carried out on a pilot unit in continuous operation.[21]

a. Model vs. experiment comparison

(1) Yields and C16 fraction
 There is good agreement between the conversion, isomerisation and cracking yield calculations (full lines) and the experimental data (dots), Fig. 30.

---
[21]Remark: Allow 5 to 12 days per balance.

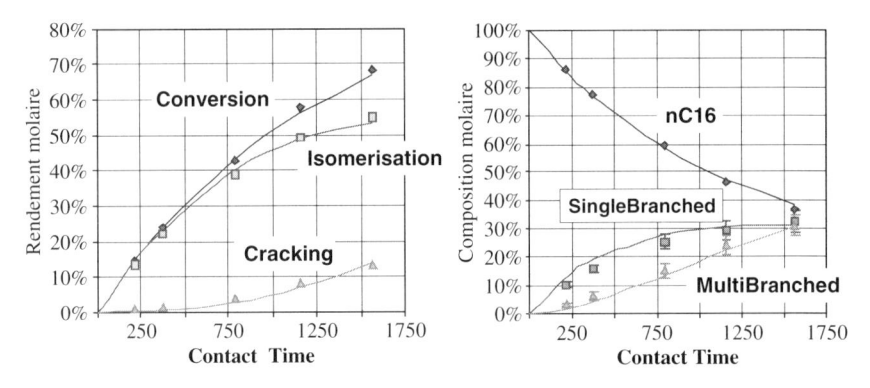

FIG. 30. Parameters estimation—yields and C16 fractions.

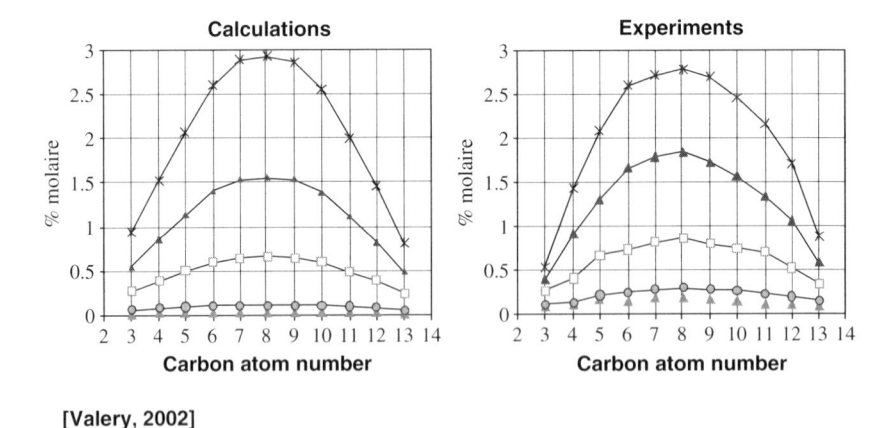

[Valery, 2002]

FIG. 31. Parameters estimation—cracking products distribution.

Describing the composition of the C16 fraction in terms of its linear, monobranched and multibranched paraffins provides an understanding into the evolution of the isomerisation. We also observe the appearance of monobranched compounds as primary products and multibranched compounds as secondary products.

(2) Cracking products

The calculated composition of cracking products depends not only on the model parameters, but also on the reactants, i.e. the variation in composition of the C16 fraction throughout the reactor.

The Fig. 31 shows the experiment—calculation comparisons for the sum—by number of carbon atoms from C3 to C13—of all the paraffins produced during cracking, for five increasing contact times.

B. Extension to Heavy Paraffins

The single-events theory uses computer algorithms to generate the exhaustive network of reactions and species involved (see Section III). The number of molecules and reactions in the generation process increases with the number of carbon atoms of the molecules considered.

When implementing this methodology, however, current storage capacities and the CPU time required to generate the networks quickly become limiting factors for large molecules (number of carbon atoms greater than or equal to 20).

For the hydrocracking process, we must therefore find a way of directly calculating the lumping coefficients between chemical families (see Section V), without having to generate the reaction network (no storage of molecules and reactions), Fig. 32.

The methodology applied is based on *factorisation of lumping coefficients* into several elements which can be calculated independently:

- A first Section (VI.B.1) describes the factorisation of the equation giving the expression of lumping coefficients,
- The second Section (VI.B.2) will describe how each element of this reformulation is calculated.

1. Simplified Equation of Lumping Coefficients

The equation of lumping coefficients is the product of a sum of inverse symmetry numbers and an entire series of thermodynamic terms (see Section V.E). This complex sum is generally calculated by adding its component terms after generating a reaction network. If we examine the problem which led to this complex sum from a different angle, we can determine another equation which is as rigorous but formally simpler.

Exponential growth of reaction networks size
⇩
Classical method (generation then lumping)
too slow - limited to C20 and 3 branches
⇩
Original and fast calculation method
for lumping coefficients
⇩
WITHOUT GENERATION of ELEMENTARY STEPS NETWORK

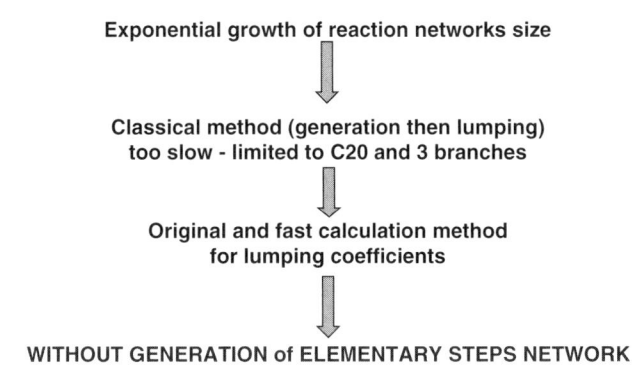

FIG. 32. Extrapolation to heavy reactants "the Real Problem".

$$LC_{isom}^{F_1 \to F_2}(m,u) = C_t.C_{sat}.b_i^{liq} \cdot \sum_{\substack{\{isom(m,u)\} \\ \{F_1 \to F_2\}}} \left[y_i \cdot \frac{K_{(P_i \leftrightarrow O_{ij}+H_2)}}{K^*_{(O_{ref} \leftrightarrow O_{ij})}} \cdot \frac{ne_{pr}}{ne_{dépr}} \cdot ne_{pcp} \right]$$

$$\Downarrow$$

$$LC_{isom}^{F_1 \to F_2}(m,u) = C_t.C_{sat}.b_1^{liq} \cdot e^{\left(\frac{\Delta G^*(O_{ref})+\Delta G(H_2)-\Delta G(F_1)}{RT}\right)} \cdot \sum_{\substack{\{isom(m,u)\} \\ \{F1 \to F2\}}} \frac{1}{\sigma_{isom}^{\#}}$$

- $\Delta G^*(O_{ref})$ — easy calculation by Benson's method
- Free Enthalpy of reactive paraffins
- Inverse numbers of activated complexes symmetries

[Valery, 2002]

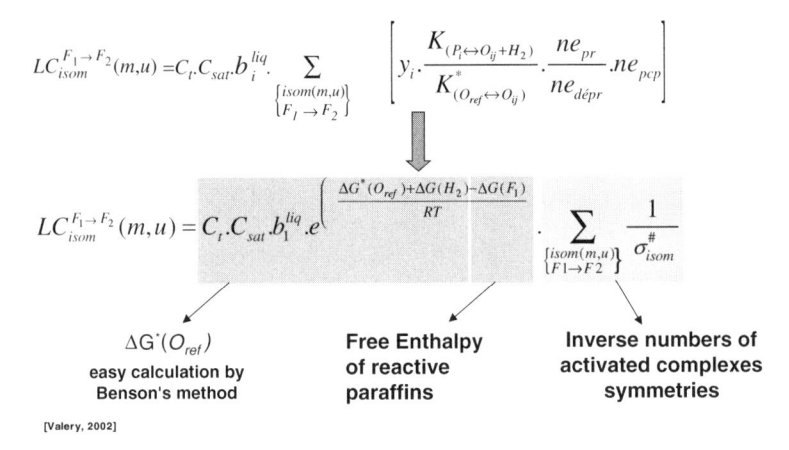

FIG. 33. Extrapolation of calculations—factorised equation.

On the basis of the simplified (in the mathematical meaning) equation of Fig. 33, we can identify a calculation logic; all lumping coefficients can be broken down into a product of two terms describing the respective free enthalpies of the reactive paraffins and activated complexes involved.[22] These two terms are calculated irrespective of the number of carbon atoms and the number of branches. The calculation is performed using recursive series and is therefore extremely fast (approximately two minutes for a C30 network limited to eight branches); this is discussed in Paragraph VI.B.2 below.

At this stage, it is important to understand that the summations no longer concern the elementary reactions but the activated reaction intermediates; *consequently, there is no need to generate the elementary step network* (Martens, 2000; Schweitzer, 1998; Thybaut, 2002; Valery, 2002; Vynckier, 1994).

2. Calculation of the Activated Complex Component

We now need to calculate the sum of the reciprocals of the symmetry numbers of the activated complexes for all reactions included within a given lumping coefficient. This calculation requires two calculation principles:

First principle: rather than generating and counting the reactions, simply generate the activated complexes involved and calculate the symmetry numbers.

Second principle: find how many reactions are likely to correspond to each complex generated.

---

[22] After mathematical simplification, we are only left—*in fine*—with the reciprocals of the symmetry numbers of the activated complexes (see Fig. 33).

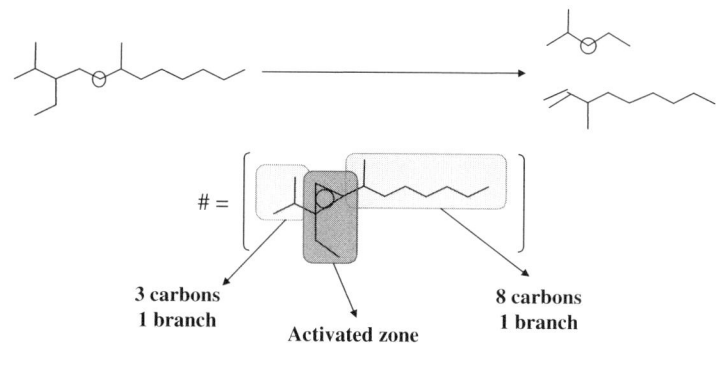

FIG. 34. Structure of an activated complex.

a. Description of an activated complex. In this section, we will take the simplest example: cracking by β-scission.

The first point to note is that cracking a compound with *nc* carbon atoms and *nb* branches by β-scission involves an activated complex with *nc* carbon atoms and *nb* branches, as shown on the following Fig. 34.

All activated complexes—written #—will be broken down into three parts:

a "*left*" side chain, written A;
an activated zone, written ZA;
a "*right*" side chain, written B.

The associated global symmetry number is equal to: $\sigma_\# = \sigma_A \sigma_{ZA} \sigma_B \sigma_{ext(A,ZA,B)}$[23]

b. Sets of reactions and complexes.

$\left\{ \begin{array}{c} react(m,u) \\ F_o \to F_p \end{array} \right\}$ Set of reactions react(*m*,*u*) leading from reactive family F_o to family F_p.

$R = \left\langle \begin{array}{c} ^\# react(m,u) \\ F_o \to F_p \end{array} \right\rangle$ Set of activated complexes of reactions react(*m*,*u*) leading from reactive family F_o to family F_p.

$CA(R)$ Set of all triplets $A-ZA-B$ forming an activated complex of the set R.

---
[23] With cracking, the active zone is, by definition, asymmetric. Consequently, the possibilities of external symmetries are eliminated. In addition, in this case each activated complex corresponds to one and only one reaction type—in our example, secondary–secondary. The same is not true for PCP isomerisation, whose activated complexes include *nc* carbon atoms and *nb*–1 branches, and which requires some additional developments (external symmetries, etc. Valery, 2002).

$n_{corr(A,ZA,B)}$ Number of times where an activated complex $A-ZA-B$ appears in the set R.

The sum to be calculated concerns the set of reactions and we will replace it by a sum over the set of activated complexes in Eq. (1). We therefore need to introduce $n_{corr(A,ZA,B)}$ to count the activated complexes as many times as required. We then obtain:

$$\sum_{ikl \in \left\{\begin{array}{c} react(m,u) \\ F_o \to F_p \end{array}\right\}} \frac{1}{\sigma_\#} = \sum_{\# \in R} \frac{1}{\sigma_\#} = \sum_{(A,ZA,B) \in CA(R)} \frac{n_{corr(A,ZA,B)}}{\sigma_A \cdot \sigma_{ZA} \cdot \sigma_B \cdot \sigma_{ext(A,ZA,B)}} \quad (1)$$

The problem of calculating the symmetry properties of the activated complexes is therefore uncoupled and reduced to calculating these properties for the equivalent side chains.

1. Activated zones. An activated complex has a limited number of carbon atoms involved in the activated zones of the molecules; the other carbon atoms form the side chains.

Taking the special case above: secondary–secondary cracking by β-scission of ions from a family of tribranched paraffins with 16 carbon atoms, into monobranched olefins with five carbon atoms and into monobranched ions with 11 carbon atoms. Two types of complex are involved in this type of elementary reaction (see Fig. 35):

1st type ZA_1: the activated zone carries an ethyl group; the side chains then share $16-5 = 11$ carbons and $3-1 = 2$ branches

2nd type ZA_2: the activated zone carries a methyl group; the side chains then share $16-4 = 12$ carbons and $3-1 = 2$ branches

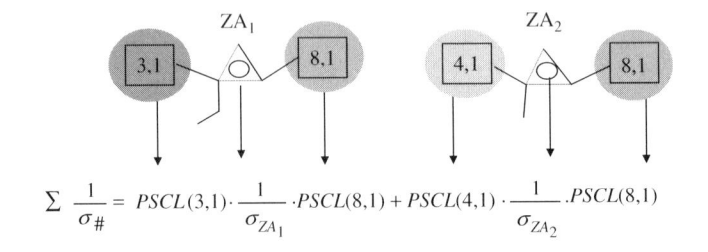

$$\sum \frac{1}{\sigma_\#} = PSCL(3,1) \cdot \frac{1}{\sigma_{ZA_1}} \cdot PSCL(8,1) + PSCL(4,1) \cdot \frac{1}{\sigma_{ZA_2}} \cdot PSCL(8,1)$$

Calculation principles:
1. Registration of homogous complexes
2. Using of properties of the equivalent side chains

[Valery, 2002]

FIG. 35. Cracking (s,s) C16,3 → C5,1 + C11,1—homologous activated complexes.

2. *Side chains.* The speed of the hydride, methyl and ethyl shift reactions is the basis of the lumping by number of carbon atoms and by number of branches. Consequently, as soon as a compound with a given number of branches is formed, the other isomers with the same number of branches are formed immediately. This is also true for the side chains of an activated complex, enabling us to define the notion of equivalent side chains.

Side chains with the same number of carbon atoms and the same number of branches are said to be *equivalent*.

Using recursive series, the symmetry properties of the side chains can be calculated step by step (Valery, 2002). These series $U(np,nc,nb)$ are used to obtain the sum of the reciprocals of the symmetry numbers of the side chains with nc carbon atoms and nb branches $PSCL(nc,nb)$.[24]

3. *Homologous activated complexes*

Activated complexes with the same 'activated zone' and two 'equivalent side chains' two-by-two are said to be *homologous*.

By definition, homologous activated complexes have the same number of carbon atoms and the same number of branches. They are therefore involved in the same lumping coefficient. Consequently, each lumping coefficient has a sum of sums of the reciprocals of the symmetry numbers of each class of homologous activated complexes as a factor—Fig. 35 as cracking example.

3. *Summary of the Method*[25]

The objective is to calculate, in the reformulated lumping coefficient (of Section VI.B.1), the sum of the reciprocals of the symmetry numbers of the activated complexes: $\sum_{ikl \in \{{react(m,u) \atop F \to G}\}} 1/\sigma_{\#_{ikl}}$ according to the breakdown of Eq. (1) (see Section VI.B.2).

After listing the activated zones and "homologous complexes" involved in a given reaction type, the method is based on a threefold observation, see Fig. 36:

The previous problem of calculating a sum on the activated complexes is therefore "*replaced*" by the problem of calculating sums on the equivalent side chains. The PSCL series can then be defined and calculated by:

$$PSCL(nc, nb) = \sum_{A \in \{{nc\ carbones \atop nb\ branchements}\}} \frac{1}{\sigma_A} = \sum_{np=1}^{nc} U(np, nc, nb)$$

A summary of the applied methodology is given in Fig. 37.

---

[24] **PSCL**(*nc,nb*) = side chain symmetry properties (**P**ropriétés de **S**ymétries des **C**haînes **L**atérales). $PSCL(nc, nb) = \sum_{A \in \{{nc\ carbones \atop nb\ branchements}\}} 1/\sigma_A = \sum_{np=1}^{nc} U(np, nc, nb)$ where np = main chain length.

[25] An alternative method was developed at Ghent University (Martens, 2000; Martens and Marin, 2001).

Example: **Beta Scission (s,s)**

= A ⟨H⁺⟩ B

Observation 1: A and B are sharing nc-4 carbons and nb branches

For ex.
nc = 16
nb = 2

Observation 2: $\sigma_\# = \sigma_A * \sigma_{ZA} * \sigma_B$

Observation 3: $\sum \dfrac{1}{\sigma_\#} = \dfrac{1}{\sigma_{ZA}} \cdot \left[\sum_{A \in \{\substack{7\ carbones \\ 1\ branchement}\}} \dfrac{1}{\sigma_A} \right] \cdot \left[\sum_{B \in \{\substack{5\ carbones \\ 1\ branchement}\}} \dfrac{1}{\sigma_B} \right]$

PSCL sums calculated by recursive series

[Valery, 2002]

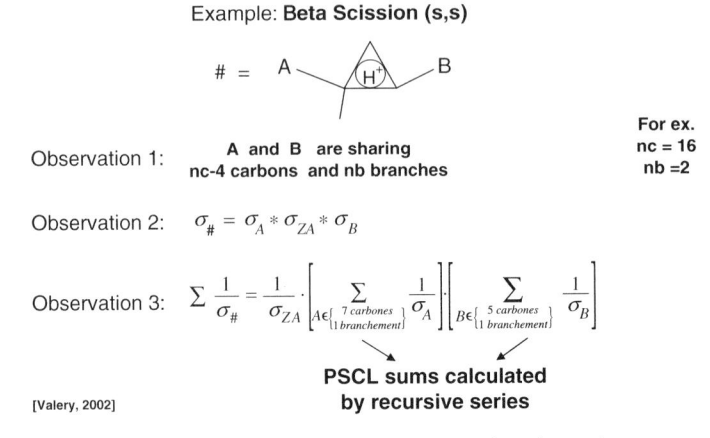

FIG. 36. Extrapolation of calculations—activated complexes.

C. EXTRAPOLATION CAPACITIES IN NUMBER OF CARBON ATOMS: HEAVY PARAFFINIC WAXES

Paraffinic waxes are a mixture of linear and branched paraffins.

Apart from the very large number of elementary steps it contains, the reaction network of heavy paraffins is the same as that of *n*-hexadecane. The apparent kinetic network is unchanged: the linear paraffins are isomerised, and then cracked.

A set of kinetic parameters was produced using the experimental data obtained on *n*-hexadecane. The reaction rates of all the acyclic paraffins can therefore be calculated. This Section VI.C will compare the calculation results with the experimental data obtained on heavy feedstocks. The experiments conducted on heavy paraffinic waxes can be used to test the extrapolation capacities with respect to the number of carbon atoms (from *n*C16 to *n*C20–*n*C33).

1. Hydrocracking of n-Paraffin Mixtures Composed of 20–30 Carbon Atoms

This feedstock, supplied by the company Schuman Sasol, is a cut of a Fischer–Tropsch effluent. NMR and infrared spectroscopy analyses revealed no traces of olefins or alcohols. The impurities (13% of the total) were therefore interpreted as being monobranched isomers.

Some of this feedstock was fractionated on a vacuum distillation unit at IFP into four other narrower cuts, centred on an increasing number of carbon atoms from C24 to C27/C28. These cuts cannot be used to accurately determine the reaction kinetics of a given linear paraffin, but may reveal possible mixture effects or steric hindrance.

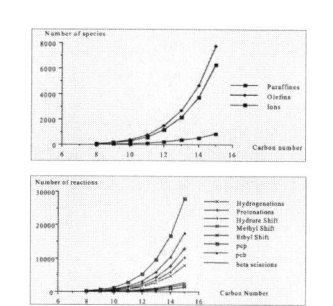

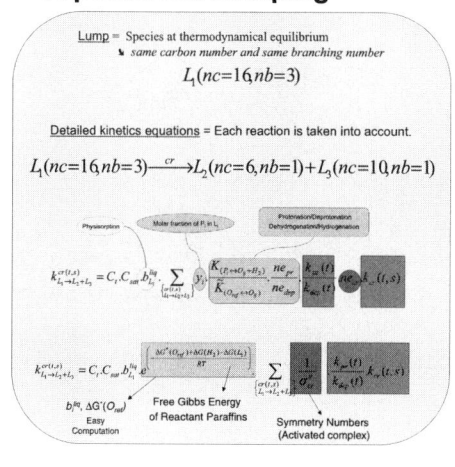

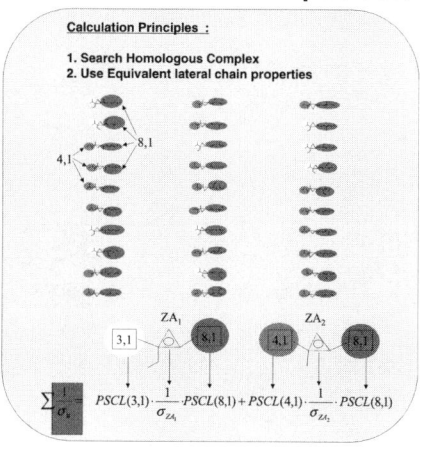

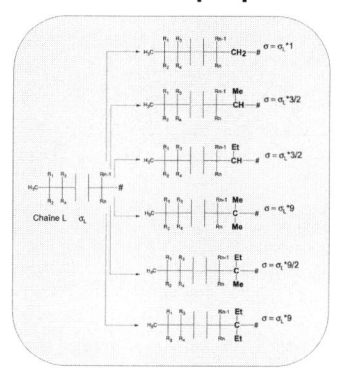

FIG. 37. Single-events modelling—extension to large networks (Please see Color Plate Section in the back of this book).

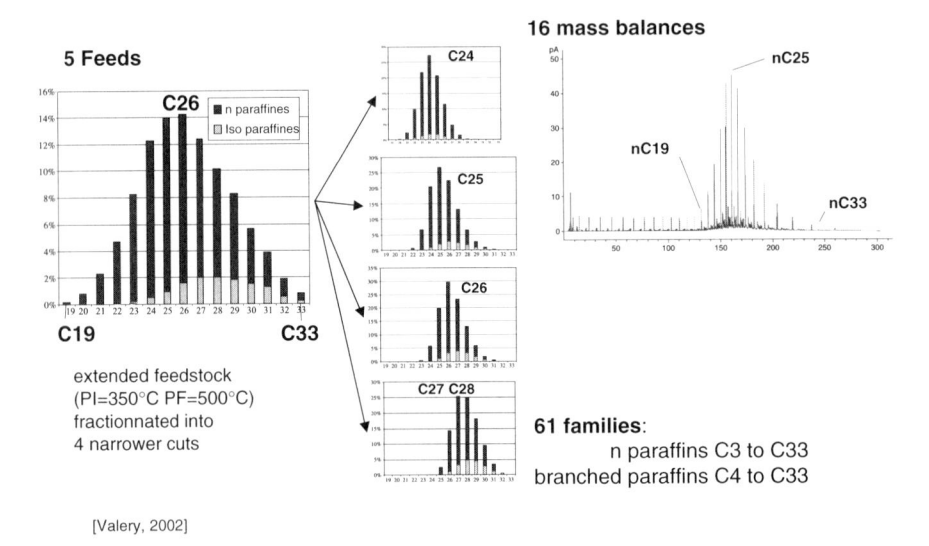

FIG. 38. Hydrocracking of paraffinic waxes—feeds and analysis.

For the paraffinic wax recipes, 61 observables can be deduced from the chromatograms, Fig. 38.

The molar composition in linear paraffins with 3 to 33 carbon atoms, i.e. 31 observables.[26]

The molar composition in isomers with 4 to 33 carbon atoms, i.e. 30 observables.[27]

a. Experimental results. Considering all the paraffinic feedstocks, 16 isothermal balances have been produced for increasing contact times:

The Fig. 39—evolution of molar concentrations (in %) of recipes resulting from the hydrocracking of C20–C30 waxes—illustrates the reactivity of paraffinic waxes:

At low contact time, the concentration in heavy isomers increases (larger white area) and the cracking products are distributed uniformly between the C3 and C19 hydrocarbons.

At high contact time, the heavy paraffins disappear (smaller blue area) and the distribution of cracking products is deformed by overcracking.

---

[26] In these recipes, it is no longer possible to measure a conversion since it is impossible to distinguish between the linear paraffins produced by cracking and those which have not been converted.

[27] It is no longer possible to distinguish between monobranched–multibranched compounds by analysing the heavy paraffin mixtures.

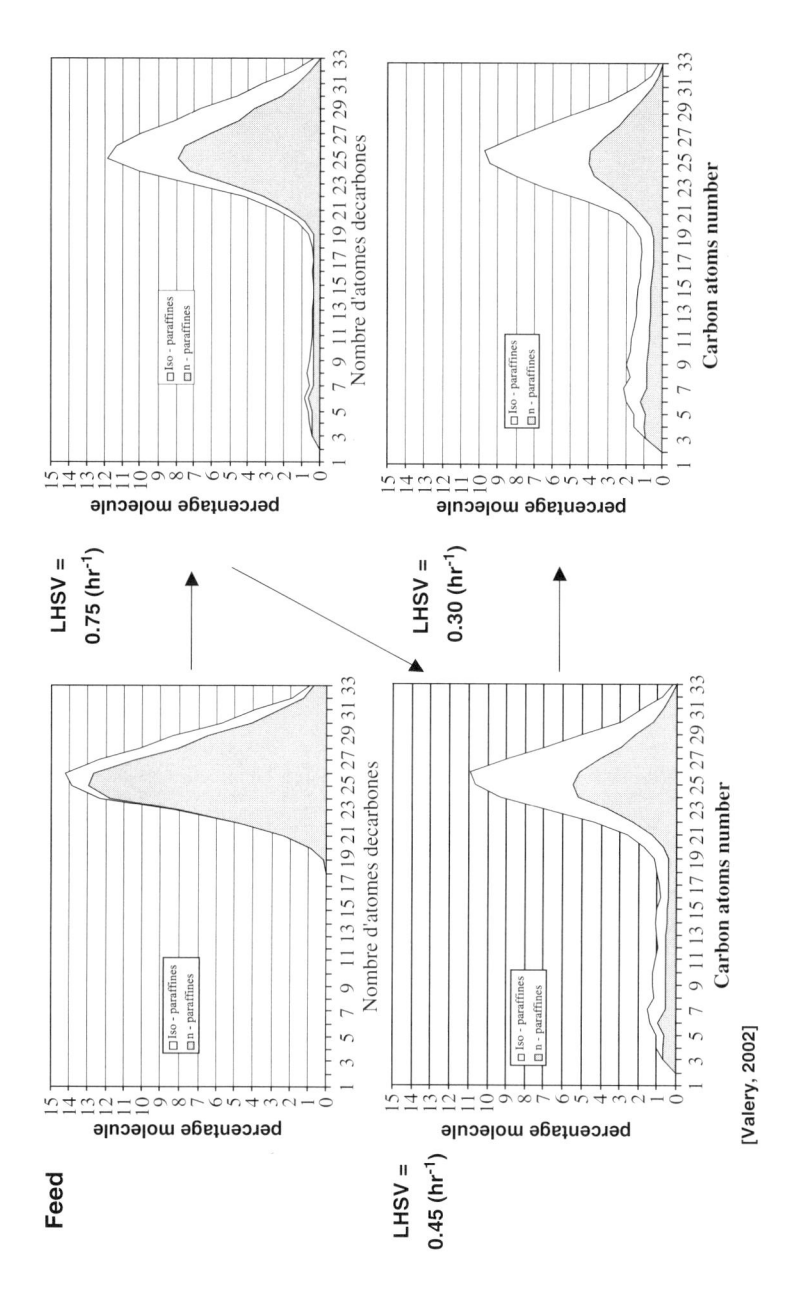

FIG. 39. Hydrocracking of paraffinic waxes—experimental results.

b. *Extrapolation vs. experiment comparisons*

(1) Comparisons of Global Quantities
   On Fig. 40, the global quantities used to compare experiments (dots) and extrapolated calculations (full lines) are:

   Weight of linear paraffins with 20 or more carbon atoms (written %nC20+)

   Weight of isomerised paraffins with 20 or more carbon atoms (written %iC20+)—approaching an isomerisation yield

   Weight of paraffins with less than 20 carbon atoms—approaching a cracking yield.

The fact that there is good agreement between experiments and extrapolated calculations obtained on the evolution of the percentages by weight of heavy linear paraffins, irrespective of the initial mixture, indicates:

   that for up to 30 carbon atoms there is no steric limitation in the catalyst.
   that the physisorption and gas–liquid equilibria used clearly represent the activities of the paraffins in various mixtures.

We will now examine the details of the point circled on Fig. 40.

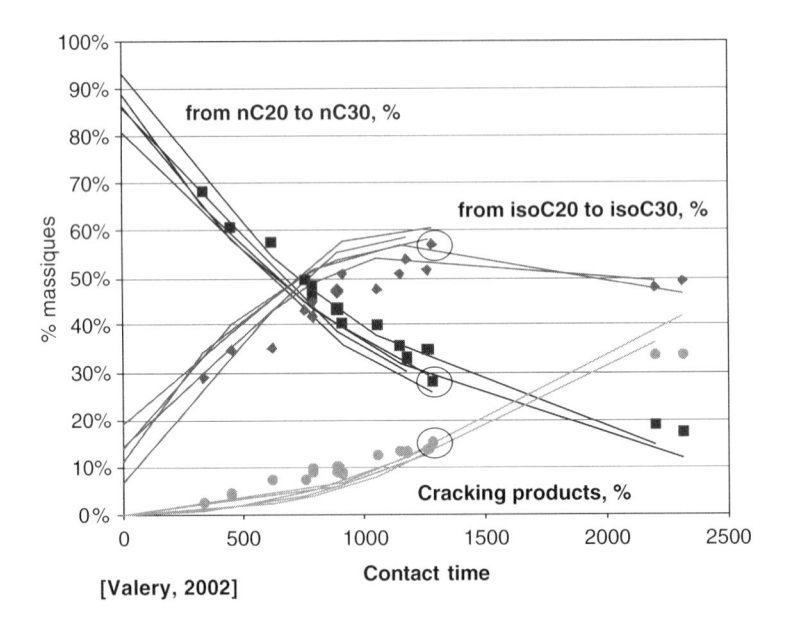

FIG. 40. Experimental—extrapolation: global comparison.

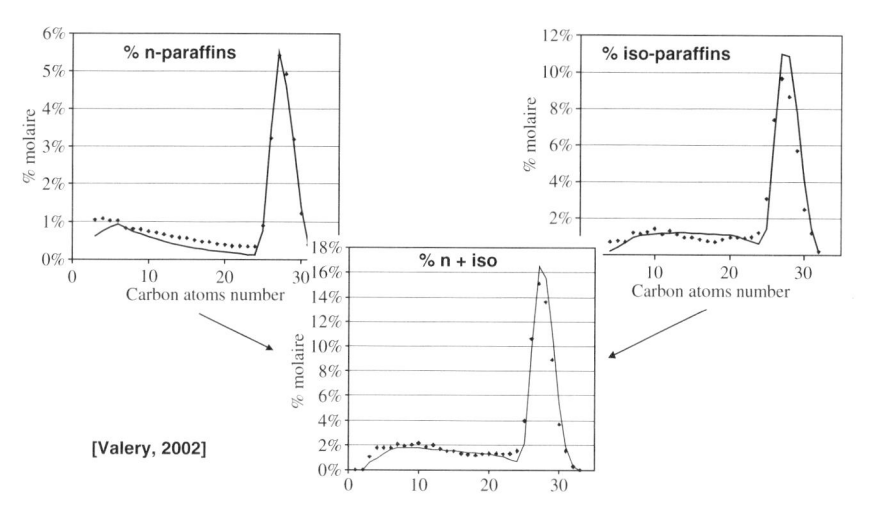

FIG. 41. Experimental—extrapolation: detailed comparison.

(2) Detailed Comparisons

On Fig. 41, we can observe the marked absence of a maximum at C13 (rupture in the middle of the reactive paraffin chain), unlike the cracking of C16s which displays a maximum at C8. These characteristic phenomena are therefore perfectly predicted and taken into account by the single-events theory calculations.

VII. Perspectives

The principles of the single-events methodology, then of its extrapolation to heavy cuts, have already been summarised in Paragraphs V.F and VI.B.3. The method provides a detailed and rigorous kinetic model, regarding the elementary steps, from which we must draw all the consequences.

Considering the potential interest of this method and its generic character, the IFP decided to apply it and validate it for nearly all the major refining processes involving acid catalysis, alone and/or bifunctional. Examples are given in Fig. 42. Specific adaptations are required each time to take into account the characteristics of the transformations but also of the complex hydrocarbon feedstocks involved for each of these processes.

In line with the IFP's vocation, we kept in mind that these validations should only be carried out under *conditions* truly representative of *industrial* operations, both as regards the operating conditions (temperature, partial pressures, contact times, etc.) and the actual reaction media (number of phases, real feedstocks,

- **EXPANSION of S.E. METHODOLOGY APPLICATIONS**
  - ISOMERISATION(s)
  - REFORMING
  - HYDROISOMERISATION
  - HYDROCRAQUAGE

- **INDUSTRIAL CONDITIONS**

- **CYCLIC COMPOUNDS**

Fig. 42. Perspectives.

- **DEACTIVATION / COKING**
  - FCC / microbalance T.E.O.M.
- **FORM SELECTIVITY**
  - restriction(s) during generation
- **CATALYST**
  - specific ACIDITY contribution
  - MOLECULAR MODELING
- **METALLIC CATALYSIS**
  - Fischer-Tropsch synthesis
- **HETERO-ATOMS**

Fig. 43. Extensions.

impurities, deactivation, etc.). In this respect, we benefit from the know-how of the IFP "processes" teams.

The next important development concerns *cyclic compounds*; for these compounds, the choice of model molecules and their availability will be determining. Amongst other things, the existing rules and calculation codes will have to be adapted and possibly new ones created.

The longer term objective is to extend the scope of the single-events method, in particular to improve the way that the *nature and structure of the catalyst* are taken into account.

These items on Fig. 43 represent *research directions* to continue developing and extending the scope of the single-events theory.

(1) The first extension to be considered concerns *catalyst deactivation and coke formation*. This is studied in two cases:
   - Firstly, for the FCC reaction, in which the coke formation kinetics is the same order of magnitude as the kinetics of the main reactions,
   - Secondly, for the regenerative catalytic reforming reaction (with Pt–Sn/ Al$_2$O$_3$ catalyst), and, in this case, prediction of coke formation is essential for the design and dimensioning of the reactors at the centre of the process.

   For this approach, the use of a novel experimental tool, the TEOM (Tapered Element Oscillating Microbalance) inertial microbalance, for continuous recording of weight increases in a catalytic (micro)-bed, opens new perspectives regarding the problem of determining deactivation functions. In collaboration with Ghent University, our aim is therefore to produce a detailed and not simply cursory description of the kinetics of coke formation, according to a methodology derived from the single-events approach.

(2) One limitation of the single-events theory consists in considering the surface intermediates, carbocations and/or activated compounds, as if they were free species or radicals. In the important case of small-pore zeolitic catalysts, this limited representation of the catalytic act is no longer valid since the *shape selectivity* problems, introduced by the geometry and confinement of zeolitic cages, are ignored.

   This type of restricted, steric selectivity must therefore be introduced into the "single-events" methodology. Fortunately, a certain number of exclusion rules can be introduced at the reaction network generation step, to take into account this type of restriction.

(3) Apart from taking into account these possible steric limitations, the chronic weakness of kinetic modelling in heterogeneous catalysis lies in the absence of a direct relation between the catalyst type (composition, structure, morphology, etc.) and its reactivity. In practice, the *nature and structure of the catalyst* are only involved through the values of kinetic constants. These values vary from one catalyst to another and, in principle, must be re-estimated whenever the catalyst is changed.
   - An initial step was carried out by taking into account the *catalyst acidity*. A recent study (Thybaut, 2002; Thybaut *et al.*, 2001) demonstrated how microkinetics considerations (Dumesic *et al.*, 1993) can be used to restrict the specific contribution of the acid strength of the catalytic sites to the stability of adsorbed carbocations. The standard protonation enthalpy term varies with the number of carbon atoms and the strength of the acid sites.
   - In the longer term, we must be able to predict the values of the kinetic constants *ab initio* according to the type and morphology of the catalyst. It is now possible to determine these structure/reactivity relations thanks to the rapid development of *molecular modelling* techniques, which will become an essential tool in the near future.

(4) Although currently limited to acid and/or bifunctional catalysis, the approach most frequently applied in the refining and petrochemistry processes

is the "single-events" approach. The main reason behind this suitability is based on the chemistry of carbocations, whose reaction mechanisms are relatively well known and described. Our goal is to extend the scope of this "single-events" method to the no less important case of *metal catalysis*. In this case there is much less agreement regarding the chemical mechanisms; one of the obstacles to be crossed concerns the adsorption mode of the molecules which, for a metal, may be multi-site. This partly explains why we chose a monometallic catalyst and the Fisher–Tropsch synthesis reaction to tackle this study (thesis in progress).

(5) Lastly, the "single-events" theory, which was historically designed for the activation of carbon–carbon bonds, does not currently cover the reactivity of C–S and C–N bonds. Some computer models have been produced to represent the possible presence of *hetero atoms* in hydrocarbon structures. Avenues are therefore open for a very wide field of application, that of HDT reactions and sulphide catalysis. They must nevertheless be based on an in-depth, improved description of the heterolytic mechanisms, also studied (Blanchin *et al.*, 2001) under IFP supervision.

These outlines demonstrate—if need be—that much research work is still essential before kinetic modelling using the "single-events" methodology can become a fully predictive tool. Nevertheless, this theory already represents a significant breakthrough and now provides the solution to several industrial problems.

REFERENCES

Aris, R., Reactions in continuous mixtures, *AIChE J.* **35**(4), 539–548 (1989).

Aris, R., and Gavalas, G. R., On the theory of reactions in continuous mixtures, *Philos. Trans. R. Soc. London, A* **260**, 351 (1966).

Astarita, G., and Ocone, R., Lumping nonlinear kinetics, *AIChE J.* **34**(8), 1299–1309 (1988).

Astarita, G., and Sandler, S. I. (Eds.), "Kinetic and Thermodynamic Lumping of Multicomponent Mixtures", Elsevier, Amsterdam (1991).

Baltanas, M. A., and Froment, G. F., Computer generation networks and calculation of product distributions in the hydroisomerisation and hydrocracking of paraffins on Pt-containing bifunctional catalysts, *Comp. Chem. Eng.* **9**(1), 71 (1985).

Baltanas, M. A., Van Raemdonck, K. K., Froment, G. F., and Mohedas, S. R., Fundamental kinetic modeling of hydroisomerisation and hydrocracking on noble-metal-loaded Faujasites 1: Rate parameters for hydroisomerisation, *Ind. Eng. Chem. Res.* **28**, 899–910 (1989).

Benson, S. W., "Thermochemical Kinetics". 2nd ed. Wiley (1976).

Bishop, D. M., and Laidler, K. J., Symmetry numbers and statistical factors in rate theory, *J. Chem. Phys.* **42**(5), 1688 (1965).

Bishop, D. M., and Laidler, K. J., Statistical factors for chemical reactions, *Trans. Farad. Soc.* **66**, 1685 (1969).

Blanchin, S., Galtier, P., Kasztelan, S., Kressmann, S., Penet, H., and Pérot, G., Kinetic modeling of the effect of H_2S and NH_3 on Toluene hydrogenation in the presence of a NiMo/Al_2O_3

hydrotreating catalyst: Discrimination between homolytic and heterolytic models, *J. Phys. Chem. A* **105**(48), 10860–10866 (2001).

Bonnardot, J., "Modélisation cinétique des réactions d'hydrotraitement par regroupement en familles chimiques", Thèse Lyon I (1998).

Brouwer, D. M., Reactions of alkylcarbenium ions in relation to isomerization and cracking of hydrocarbons, NATO-ASI Ser. E39, *in* "Chemistry and Chemical Engineering of Catalytic Processes" (R. Prins, and G. C. A. Schuit Eds.), pp. 137–160. Sijthoff & Noordhoff, Alphen-aan-den-Rijn (1980).

Charpentier, J. C., Did you say: Chemical, process and product-oriented engineering, *Oil Gas Sci. Technol.* **55**(4), 457–462 (2000).

Clymans, P. J., and Froment, G. F., Computer generation of reaction paths and rate equations in the thermal cracking of normal and branched paraffins, *Comp. Chem. Eng.* **8**(2), 137–142 (1984).

Cochegrue, H., "Modélisation cinétique du réformage catalytique sur catalyseur Pt-Sn/Al$_2$O$_3$", Thèse Poitiers (2001).

Dumesic, J. A., Rudd, D. F., Aparicio, L. M., Rekoske, J. E., and Treviño, A. A., "The Microkinetics of Heterogeneous Catalysis". American Chemical Society, Washington DC (1993).

Fafet, A., and Magné-Drisch, J., Analyse quantitative détaillée des Distillats moyens par couplage CG/SM, *Rev. Inst. Fr. Petrol.* **50**(3), 391–404 (1995).

Froment, G. F., Kinetic modeling of complex catalytic reactions, *Rev. Inst. Fr. Petrol.* **46**(4), 491 (1991).

Froment, G. F., and Bischoff, K. B., "Chemical Reactor Analysis and Design". Wiley (1990).

Hillewaert, L., "De thermische kraking van gasolien. Experimentale studie en modellering", PhD Thesis, Ghent University (1986).

Kodera, Y., Kondo, T., Isaito, S. Y., and Ukegawa, K., Continuous-distribution kinetic analysis for asphaltene hydrocracking, *Energy Fuels* **16**, 291–296 (2000).

Lenoir, D., and Siehl, H. U., Carbokationen, Carbokation-Radikale, *in*: "Methoden der Organische Chemie, Vierte Auflage" (M. Hamack Ed.), Georg Thieme Verlag, Stuttgart (1990).

Magné-Drisch, J., "Cinétique des réactions d'hydrotraitement de distillats par décomposition en familles et coupes étroites", Thèse Paris VI (1995).

Marin, G. B., Kapteijn, F., van Diepen, A. E., and Moulijn, J. A., Catalytic reaction and reactor engineering, *in*: "Combinatorial Catalysis and High Throughput Catalyst Design and Testing" (E. G. Derouane Ed.), pp. 239–281. Kluwer Academic Publishers (2000).

Martens, G. G. "Hydrocracking on Pt/US-Y zeolites: Fundamental kinetic modeling and industrial reactor simulation", PhD Thesis, Ghent University (2000).

Martens, G. G., and Marin, G. B., Kinetics for hydrocracking based on structural classes: Model development and application, *AIChE J.* **47**, 1607–1622 (2001).

Neurock, M., Libatani, C., and Klein, M. T., Modeling asphaltene reaction pathways : Intinsic chemistry, *AIChE Symp. Ser., Fundamentals & Resid. Upgrading* **85**(273), 7–14 (1989).

Neurock, M., Libatani, C., Nigam, A., and Klein, M. T., Monte Carlo simulation of complex reaction systems: Molecular structure and reactivity in modelling heavy oils, *Chem. Eng. Sci.* **45**(8), 2083–2088 (1990).

Neurock, M., Nigam, A., Trauth, D., and Klein, M. T., Molecular representation of complex hydrocarbon feedstocks through efficient characterization and stochastic algorithms, *Chem. Eng. Sci.* **49**(24A), 4153–4177 (1994).

Nigam, A., Neurock, M., and Klein, M., Reconciliation of molecular detail and lumping: An asphaltene thermolysis example, *in*: "Kinetic and Thermodynamic Lumping of Multi-component Mixtures" (G. Astarita, and S. I. Sandler Eds.), Elsevier, Amsterdam (1991).

Prigogine, I., "La fin des certitudes". Editions Odile Jacob, Paris (1996).

Quann, R. J., and Jaffe, S. B., Structure oriented lumping: Describing the chemistry of complex hydrocarbon mixtures, *Ind. Eng. Chem. Res.* **31**, 2483 (1992).

Read, R. C., "The enumeration of acyclic chemical compounds", *in* "Chemical Application of Graph Theory" (A. J. Balaban Ed.), (1976).

Sapre, A. V., and Krambeck, F. J. (Eds.), "Chemical reactions in complex mixtures; the MOBIL workshop", Van Nostrand Reinhold, New York (1991).

Schweitzer, J. M., "Modélisation cinétique des réactions catalytiques d'hydrocraquage par la théorie des évènements constitutifs", Thèse Lyon I (1998).

Svoboda, G. D., Vynckier, E., Debrabandere, B., and Froment, G. F., Single-event rate parameters for paraffin hydrocracking on a Pt/US-Y zeolite, *Ind. Eng. Chem. Res.* **34**, 3793 (1995).

Thybaut, J., "Production of low aromate fuels: Kinetics and industrial application of Hydrocracking", PhD Thesis, Ghent University (2002).

Thybaut, J. W., Marin, G. B., Baron, G. V., Jacobs, P. A., and Martens, J. A., Alkene protonation enthalpy determination from fundamental kinetic modeling of alkane hydroconversion on Pt/H-(US)Y-zeolite, *J. Catal.* **202**, 324–339 (2001).

Trambouze P., Van Landeghem, H., and Wauquier, J.-P., "Les Réacteurs chimiques; Conception, Calcul, Mise en œuvre" Edition Technip, Paris (1984).

Valery, E. "Application de la théorie des évènements constitutifs à l'hydrocraquage de paraffines lourdes", Thèse Lyon I, Mai (2002).

Van Raemdonck, K. K., and Froment, G. F., "Fundamental kinetic modeling of hydroisomerisation and hydrocracking on noble-metal-loaded Faujasites II: The elementary cracking steps", AIChE Meetings, San Francisco, November 5–10 (1989).

Verma, R. P., Laxminarasimhan, C. S., and Ramachandran, P. A., Continuous lumping model for simulation of hydrocracking, *AIChE J.* **42**(9), 2645 (1996).

Verstraete, J. "Kinetische studie van de katalytische reforming van nafta over een Pt-Sn/Al$_2$O$_3$ katalysator", PhD Thesis, Ghent(Belgium) (1997).

Vynckier, E. Séminaire interne IFP, September 16 (1994).

Vynckier, E., and Froment, G. F., Modeling of the kinetic of complex processes based upon elementary steps, *in*: "Kinetic and Thermodynamic Lumping of Multicomponent Mixtures" (G. Astarita, and S. I. Sandler Eds.), Elsevier, Amsterdam (1991).

Wang, M., Smith, J. M., and McCoy, B. J., Continuous kinetics for thermal degradation of polymer in solution, *AIChE J.* **41**(6), 1521–1533 (1995).

Weitkamp, J., Isomerization of long chain n-alkanes on a Pt/CaY Zeolite catalyst, *Ind. Engng. Chem., Proc. Des. Dev.* **21**, 550 (1982).

Further Reading

Schweitzer, J. M., Galtier, P., and Schweich, D., A single events kinetic model for the hydrocracking of paraffins in a three-phase reactor, *Chem. Eng. Sci.* **54**(13–14), 2441–2452 (1999).

INDEX

A

Activated complex component
  activated zones, 292
  calculation, 290–294
  cracking (s,s) C16,3→C5,1+C11,1,
    homologous activated complexes, 292
  description, 291
  homologous activated complexes, 293
  side chains, 293
Additive models, for C_1–C_4 alkane oxidation, 189–193
  applicability, 192
  GRI-Mech approach, 192
  phases, 189
    auto-accelerating chain-branched reaction, 189
    quasi-stationary chain-branched process, 190
    self-acceleration of reaction, 190
Alkylperoxy radicals, transformations of, 243–246
Alkyl radicals and oxygen, reactions between, 243–246
Aromatics, pyrolysis of, 89–90
Arrhenius equation, 206
  Arrhenius A-factor, 17
Asphaltenes, 132
Atmospheric equivalent initial boiling point (AEBPI), 96
Auto-accelerating chain-branched reaction, 189
Automated model-construction software, 12–13
Automatic generation of pyrolysis mechanism, 64–72
  Boolean algebra, 65
  combinatorial algorithm, 64
  EXGAS software, 65
  GAPP program, 66
  Graph theory, 65
  LISP programming language, 65
  lumping of reactions, 69–71
  lumping of species, 71–72
  MAMA program, 66
  NetGen program, 65
  primary propagation reactions of n-decyl radicals, 67
  reference kinetic parameters, 68
  specific reaction class based, 64
  substitution matrices, 65
  THERGAS software, 66
  THERM program, 66
  XMG (Exxon Mobil Mechanism Generation), 65

B

Benson group values, 7, 10
Bifunctional catalysis mechanisms, 270
Boltzmann constant, 99
Boolean algebra, for pyrolysis mechanism, 65
Branched alkanes, pyrolysis of, 74–78
'Burning out' of two hydrocarbons, 238

C

C_1–C_4 alkane oxidation, kinetic models of, 167–258
  additive models, 189–193, *see also separate entry*
  alkyl radicals and oxygen, reactions between, 243–246
  alkylperoxy radicals, transformations of, 243–246
  C_1–C_2 joint description, 237–239
    'burning out' of two hydrocarbons, 238

capabilities of process influencing, governing and design, 246–250
circumscription of subject and area of parameters, 176–179
combinatorial models, 193–194
complex reacting systems, modeling purposes and expectations, 172–176
comprehensive modeling, ruling principles for, 194–199, *see also under* Ruling principles
elemental base, 203–231, *see also separate entry*
elevated (moderate) temperatures, 177
expansion on higher hydrocarbons, 239–243
  propyl radicals, 241
heterogeneous processes, 179–183
heterogeneous–homogeneous catalytic reactions, modeling of, 201–203
high-temperature oxidation, 177–179
low-temperature oxidation, 176–177
macro-kinetic parameters, 183–187
modeling and experimentation, 231–237, *see also separate entry*
reduction of models, 200–201, *see also separate entry*
Carbonaceous deposits and soot particles, formation, 100–124, *see also under* Fouling processes
Catalytic processes, in petroleum process engineering, 261
CCR (Conradson carbon residue), 96
Chemical activation and fall-off, 22–23
Chemical reactions and properties, new data model for, 13–26
  functional group trees
    for reaction rate estimation, 17–23, *see also under* Reaction rate estimation
    reaction types represented using, 24–25
  hierarchical tree structure
    for functional group parameters, 14–17
CHEMKIN® software, 3–4, 9–10, 53
Combinatorial models, 193–194

Complex reacting systems, modeling
  kinetic scheme and kinetic parameters determination, 174
  purposes and expectations, 172–176
Computational fluid dynamics (CFD) simulation, 34
Computer-assisted molecular structure construction (CAMSC), 93
Continuously varying steady state (CVSS), 144–145
Crude distillation residues, 96
Crude oil fractionation, 264
Cyclo-alkanes and alkenes, pyrolysis of, 78–88
  β-decomposition reactions, 82
  1-hexene reactions, 86
  1-methyl-4-alkyl-*cyclo*-hexanes, 86
  cyclo-alkyl radicals, 79
  cyclo-hexane components, 79
  cyclo-hexanes, 79, 87
  cyclo-hexyl radical
    decomposition mechanism, 84
    primary propagation path of, 80
    radical isomerization, 84
  cyclo-pentanes, 79, 87
  decomposition reactions, 82
  dehydrogenation reactions, 82
  isomerization reactions, 82
  methyl-*cyclo*-hexane, 80, 84
  reference kinetic parameters for, 85
Cyclopentadienyl ($C_5H_5\cdot$) radical, 118

D

DAEPACK software, 31, 36
Data models for chemical kinetics 9–11
  list-of-reactions
    automated construction of, 11
    data model, deficiencies, 9
Delayed coking processes, 129–136
Density functional theory (DFT), 203
Diels–Alder reactions, 105–106
Differential-algebraic equations (DAEs), 31
Direct oxidation of methane-to-methanol (DOMM), 189–190, 246–247
Discrete sectional method (DSM), 122, 145

E

Elemental base, 203–231
  elementary reactions, 206–209
  high-pressure regime, 212
  limitations of species taken into account, 203–205
  low-pressure limit approximation, 212
  on metal catalysts
    heterogeneous–homogeneous catalytic reactions on, 227–231, see also under Heterogeneous–homogeneous catalytic reactions
  oxide catalysts
    heterogeneous–homogeneous catalytic reactions on, 213–227, see also under Heterogeneous–homogeneous catalytic reactions
  pressure-dependent reactions, 210–213
  selection of rate constants, 210
Eley–Rideal models, 202
Equilibrium constants, 209
Error bars on model predictions, estimation, 42–46
  correlated uncertainties in model input parameters, 44–45
  first-order sensitivity analysis, beyond, 44
  model truncation error, 45
Evans–Polanyi parameters, 44
  Evans–Polanyi modified Arrhenius form, 20
EXGAS software, for pyrolysis mechanism, 65
Eyring theory, 99

F

Feedstocks and effluents analysis, in petroleum processes, 262, 268
Fouling processes
  addition reactions, 105–106
  carbonaceous deposits and soot particles, formation, 100–124
    pyrolysis coils and TLE's, fouling and coking mechanisms in, 101–106
  catalytic mechanism and initial growth, 103–104
  mechanisms and features of, 102
    initial catalytic mechanism, 102
    radicalic mechanism, 102–103
  radical and concerted path growth, 105–106
  soot formation, 114–124, see also separate entry
  structure and properties of the deposit, evolution of, 107–113
    comparison with "rapid pyrolysis" data, 113
    comparisons with experimental data, 112–113
    kinetic modelling, 108–112
    poly-aromatic structure of, 107
    radical degradation mechanism, 109
Functional group trees/parameters
  for addition reactions, 19–20
  hierarchical tree structure for, 14–17

G

GAPP program, for pyrolysis mechanism, 66
Gas-phase pyrolysis processes, 51–166, see also under Pyrolysis processes
  characterization, 54
Global dynamic optimization code (GDOC) software, 41
Graph theory, for pyrolysis mechanism, 65
Graphical user interface (GUI), 14, 16
GRI-Mech approach, 192–193

H

H-abstraction reactions, 57–61
Heavy gasoils, pyrolysis of, 93–95
  model components selected, 94–95
  n-d-M method (refractive index, density and molecular weight), 93
Heavy paraffinic waxes, 294–299
  carbon atoms, hydrocracking of, 294–299
    experimental results, 296–297
    extrapolation vs. experiment comparisons, 298–299
    feeds and analysis, 296
    single-events modelling, 295

extrapolation capacities in, 294–299
n-paraffin mixtures composed of, 20–30
Heterogeneous processes, 179–183
Heterogeneous–homogeneous catalytic reactions, modeling of, 201–203
 light alkanes, characteristics of, 204
 on metal catalysts, 227–231
  complications in, 227–230
  progress in, 230
 oxide catalysts, 213–227
  limiting selectivity (S_{LIM}), 215–216
  OCM window, 214
  Polanyi–Semenov-correlation, 220
Hierarchical tree structure, for functional group parameters, 14–17
High-temperature C_1–C_4 alkane oxidation, 177–179
Hydrocarbons
 hydrocarbon gases processing
  C_1–C_4 alkane oxidation as applied to, 167–258, *see also under* C_1–C_4 alkane oxidation
 structural isomers of, 265
Hydrogen bond increments (HBI), 17

I

Internal isomerization reactions, 62–63
Intramolecular reactions, 23–26
Isomerization reactions, internal, 62–63

J

Jacobian sparse, 31

K

Kerosene feed, pyrolysis of, 90–93
Kinetic models, chemical
 construction, fundamental inputs for, 12–13
 estimations and generalizations, 13
 for predicting reaction systems, 3
 goal of, 5–7
 historical applications of, 2–5
 in petroleum process engineering, 259–304, *see also under* Petroleum process engineering
 model-prediction-data loop, 6
 of pyrolysis processes in gas and condensed phase, 51–166, *see also under* Pyrolysis processes
 predictive vs. postdictive models, 2–3
 steps involved, 6
Krestenin mechanism, 121

L

Langmuir–Hinshelwood expressions, 202, 282
Large hydrocarbons, pyrolysis of, 72–90
 aromatics, 89–90
 branched alkanes, 74–78
 cyclo-alkanes and alkenes, 78–88, *see also separate entry*
 normal alkanes, 72–74
Large kinetic simulations, solving, 29–38
 sparsity, 31–32
 using reduced chemistry models, 32–34, *see also separate entry*
Light gasoil fractions feed, pyrolysis of, 90–93
Lindemann approach, 213
Liquid feeds, characterization, 90–96
 heavy gasoils, 93–95, *see also separate entry*
 naphtha, kerosene and light gasoil fractions, 90–93
 PINA analysis, 92
Liquid-phase pyrolysis, modelling, 96–100
 Arabian residues, 97
 aromatic carbon atoms and H/C in residues, 97
 C–C bond cleavage, 98
LISP programming language, for pyrolysis mechanism, 65
Lists-of-reactions, automated construction of, 11–13
Low-temperature C_1–C_4 alkane oxidation, 176–177
Lumping approaches, in petroleum process engineering, 267
 early lumping, 267
 late lumping kinetic model, 279–286
  catalytic act, 280–281
  composition of the reaction intermediates, 281

lumped kinetics, 282–285
lumped reforming network, 285
lumping by families in equilibrium, 284
lumping by families, 281–282
regrouped kinetics, 285
summing up, 285–286
three-phase model, 280
lumping coefficients, simplified equation of, 289–290

M

MAMA program, 152–161
database structure, 157–159
end-product composition, evaluation, 152–155
for pyrolysis mechanism, 66, 73, 76, 84–88
kinetic generator, functionality of, 160–161
lumping of components, 155–157
molecule homomorphism algorithm, 159–160
Mars–van Krevelen models, 202
Maxwell–Boltzman distribution for gas species, 207
MERT (Mixing Element Radiant Tube), 125
CFD representation of MERT, 125
Method of moments, 145–147
Microkinetic concept, single-events, 273–275
Model predictions and experimental data, 38–46
error bars estimation, 42–46, *see also under* Error bars
inconsistency, proving, 39–41
standard operating practice, need for change in, 41–42
Model truncation error, 45
Modeling and experimentation, C_1–C_4 alkane oxidation, 231–237
comparison with experiment, 231–233
adequacy, problem of, 232
requirements of experiment, 233–237
'wall-less' reactor approach, 234
'yield (or selectivity) vs. conversion', 234

Molecular weight distribution (MWD), 140
Multidimensional simulations, reduced chemistry models in, 32–34

N

Naphtha feed, pyrolysis of, 90–93
n-Butane pyrolysis, 56–60
n-Decane, H-abstraction reactions of, 61
n-Decyl radicals
primary propagation reactions of, 67
β-decomposition reactions, 67
dehydrogenation reactions, 67
isomerization reactions, 67
n-d-M method (refractive index, density and molecular weight), 93
NetGen program, for pyrolysis mechanism, 65
n-Hexadecane, hydrocracking of, 286–287
model vs. experiment comparison, 287–288
cracking products, 288
yields and C16 fraction, 287–288
Normal alkanes, pyrolysis of, 72–74

O

Ordinary differential equation (ODE) systems, 30, 53
Oxidative coupling of methane (OCM), 182, 192, 202, 217–221
OCM window, 214

P

Petroleum process engineering, kinetic methods in, 259–304
analytical set, 265
catalytic processes, 261
crude oil fractionation, 264
extension to heavy paraffins, 289–294
activated complex component, calculation, 290–294, *see also under* Activated complex component
extrapolation capacities in number of carbon atoms, 294–299, *see also under* Heavy paraffinic waxes

extrapolation to heavy cuts, 286–299
  estimation of kinetic parameters, 286–288
  n-hexadecane, hydrocracking of, 286–288
feedstocks and effluents, analysis, 262
hydrocarbons, structural isomers of, 265
  paraffins and naphthalenes, composition, 266
kinetic modelling by single-events, 269–271, *see also under* Single-events theory
kinetics modeling, 267
  lumping approaches, 267
perspectives, 299–302
  extensions, 300
petroleum cuts, 261
reaction networks, generation of, 271–273
  late lumping kinetic model, 279–286, *see also under* Lumping approaches
  nC8 to nC15, computer generation of networks from, 273
  n-hexane cracking, elementary steps network for, 272
  species and chemical reactions, computer representation, 273
refining operations, 261
PINA analysis, of liquid feeds, 92
Planck's constant, 207
Polanyi–Semenov equation, 184, 220
Poly-aromatic hydrocarbons (PAH), 54
Polymers, thermal degradation of, 136–150, *see also under* Thermal degradation
Predictive kinetics, 1–50
  chemical kinetic models, 2, *see also separate entry*
  computation of reaction rates, 5
  construction, 7–29
    chemical reactions and properties, new data model for, 13–26, *see also separate entry*
    computer-aided model-construction, 7–8
    data models for chemical kinetics, 9–11, *see also separate entry*
    documenting large simulations, challenge of, 8–9
    reaction mechanism generator (RMG), 26–27
  large kinetic simulations, efficiently and accurately solving, 29–38, *see also separate entry*
  model predictions and experimental data, 38–46
  new data model for, 13–26, *see also under* Chemical reactions and properties
  rate parameters estimation, 4
  technical hurdles in the 20th century, 3–5, *see also* Technical hurdles
Pressure-dependent reactions, 210–213
Pyrolysis mechanism, using RMG, 27–29
Pyrolysis processes in gas and condensed phase, kinetic modelling, 51–166
  applications, 124–150
  automatic generation of mechanism, 64–72, *see also under* Automatic generation
  Chemkin package, 53
  crude distillation residues, 96
  fouling processes, 100–124, *see also separate entry*
  high-temperature pyrolysis, 54
  liquid feeds, characterization, 90–96, *see also separate entry*
  liquid-phase pyrolysis, modelling, 96–100, *see also separate entry*
  n-butane pyrolysis, 56
  pyrolysis of large hydrocarbons, 72–90, *see also under* Large hydrocarbons
  reaction classes during pyrolysis, 55
    ethyl radical H-abstraction reactions, 59
    H-abstraction reactions, 57
    initiation and termination, 55–56
    internal isomerization reactions, 62
    propagation reactions, 55
  steam cracking process and pyrolysis coils, 124–129

visbreaking and delayed coking
processes, 129–136, *see also separate entry* thermal degradation of polymers, 136–150, *see also separate entry*

Q

Quantum chemistry techniques, 5
Quantum Rice-Rampsberger-Kassel (QRRK), 23
Quasi-stationary chain-branched process, 190
Quasi-steady-state approximation (QSSA), 30

R

Rate constants, 210
Rate parameters estimation, 4
Reaction mechanism generator (RMG) program, 14, 26–27
  applications of, 27–29
    laser-initiated oxidation of neopentane, 27
    pyrolysis (steamcracking) of n-hexane, 27
    supercritical water oxidation of methane, 27
Reaction rate estimation
  functional group rate estimation, difficulties with, 21–26
    chemical activation and fall-off, 22–23
    intramolecular reactions, 23–26
    paucity of reliable data, 21–22
  functional group trees for, 17–23
  for addition reactions, 19
'Reaction Recipe' for radical addition to a double bond, 18
Reduced chemistry models, 32–34
  construction, 34–38
    reaction elimination via optimization, 36–37
    with user-specified valid ranges, 37–38
Reduction of models, in C_1–C_4 alkane oxidation, 200–201
  importance criterion, 200
  minimization (even neglect) of chemical differences criterion, 200–201
  partial equilibrium criterion, 200
  similarity criterion, 201
  utility criterion, 201
Refining operations, in petroleum processes, 261–263
Ruling principles for comprehensive modeling, 194–199
  independence of kinetic parameters, 197
  model fullness, 196–197
  openness of the description, 199
  parameter optimization algorithms, 195
  thermodynamic consistency, 196

S

Single-events theory, of petroleum process kinetic modeling, 269–271
  bifunctional catalysis mechanisms, 270
    14/14 hydrogenations/dehydrogenations, 270
    18/18 protonations/deprotonations, 270
    acid phase, 270
    metallic phase, 270
  kinetics by single-events, 273–279
    free enthalpies of reactants and activated complexes, 276–278
    intrinsic free enthalpy and ion type, 277–278
    intrinsic free enthalpy and number of carbon atoms, 278
    separation of chemical and structural contributions, 275–276
    thermochemical restrictions and constraints, 278–279
  single-events microkinetic concept, 273–275
Soot formation, in fouling processes, 114–124
  benzene flame, 115
  ethylene flame, 115
  formation of the first aromatic rings and aromatic growth, 116–121
    reactions involved, rate constants of, 119 PAH
    structures, growth of, 121
  soot inception and growth, 122–124
  steps involved, 114, 116

Sparsity
  fast solution of large systems of chemistry equations using, 31–32
SSITKA (Steady-State Isotopic Transient Kinetic Analysis), 216
Stabilized combustion, 182
Steam cracking process and pyrolysis coils, 124–129
  coil model and model validation, 126–129
  reference compounds involved in, 127
Stein–Rabinovitch method, 23
Stokes–Einstein theory, 99
Substitution matrices, for pyrolysis mechanism, 65
Substitutive addition reaction, 131

T

Taylor models, 35, 37
Technical hurdles, in chemical kinetic modeling, 3–5
  numerical problems, 3
THERGAS software, for pyrolysis mechanism, 66
THERM program, for pyrolysis mechanism, 66
Thermal cracking, 124
Thermal degradation of polymers, 136–150
  balance equations, 144–147
    continuously varying steady state (CVSS), 144–145
    discrete section method, 145
    method of moments, 145–147
  mechanism considerations, 137–144
    initiation reactions, 137
    propagation reactions, 138
    termination, 139
  model validation and comparisons with experimental data, 147–150
  polyethylene, pyrolysis of, 148
  polypropylene, thermal decomposition of, 148
  polystyrene, pyrolysis of, 148
  PVC pyrolysis, 141–142
    initiation reactions, 141
    molecular reactions, 142

  propagation reactions, 141
  reference components in the kinetic model of, 143
  termination or radical recombination reactions, 142
Thermochemical restrictions and constraints, of single-events theory, 278–279
Thermodynamic consistency, 196
Three-body processes, 211, 213
Transfer line heat exchangers (TLE), 101–106, 128

U

Unity Bond Index-Quadratic Exponential Potential (UBI-QEP) method, 203

V

Visbreaking and delayed coking processes, 129–136
  kinetic mechanism, 130–132
    β-scission, 130
    H-abstraction, 130
  model validation and comparisons with experimental data, 133–136
  poly-aromatic radicals, internal dehydrogenation and demethylation of, 132
  poly-aromatic species, 131
  substitutive addition reaction, 131
  visbroken residue and pyrolysis severity, stability of, 132–133

W

'Wall-less' reactor approach, 234

X

XMG (Exxon Mobil Mechanism Generation), for pyrolysis mechanism, 65

Z

Ziegler–Natta catalysts, 103

CONTENTS OF VOLUMES IN THIS SERIAL

Volume 1

J. W. Westwater, *Boiling of Liquids*
A. B. Metzner, *Non-Newtonian Technology: Fluid Mechanics, Mixing, and Heat Transfer*
R. Byron Bird, *Theory of Diffusion*
J. B. Opfell and B. H. Sage, *Turbulence in Thermal and Material Transport*
Robert E. Treybal, *Mechanically Aided Liquid Extraction*
Robert W. Schrage, *The Automatic Computer in the Control and Planning of Manufacturing Operations*
Ernest J. Henley and Nathaniel F. Barr, *Ionizing Radiation Applied to Chemical Processes and to Food and Drug Processing*

Volume 2

J. W. Westwater, *Boiling of Liquids*
Ernest F. Johnson, *Automatic Process Control*
Bernard Manowitz, *Treatment and Disposal of Wastes in Nuclear Chemical Technology*
George A. Sofer and Harold C. Weingartner, *High Vacuum Technology*
Theodore Vermeulen, *Separation by Adsorption Methods*
Sherman S. Weidenbaum, *Mixing of Solids*

Volume 3

C. S. Grove, Jr., Robert V. Jelinek, and Herbert M. Schoen, *Crystallization from Solution*
F. Alan Ferguson and Russell C. Phillips, *High Temperature Technology*
Daniel Hyman, *Mixing and Agitation*
John Beck, *Design of Packed Catalytic Reactors*
Douglass J. Wilde, *Optimization Methods*

Volume 4

J. T. Davies, *Mass-Transfer and Inierfacial Phenomena*
R. C. Kintner, *Drop Phenomena Affecting Liquid Extraction*
Octave Levenspiel and Kenneth B. Bischoff, *Patterns of Flow in Chemical Process Vessels*
Donald S. Scott, *Properties of Concurrent Gas–Liquid Flow*
D. N. Hanson and G. F. Somerville, *A General Program for Computing Multistage Vapor–Liquid Processes*

Volume 5

J. F. Wehner, *Flame Processes—Theoretical and Experimental*
J. H. Sinfelt, *Bifunctional Catalysts*
S. G. Bankoff, *Heat Conduction or Diffusion with Change of Phase*
George D. Fulford, *The Flow of Lktuids in Thin Films*
K. Rietema, *Segregation in Liquid–Liquid Dispersions and its Effects on Chemical Reactions*

Volume 6

S. G. Bankoff, *Diffusion-Controlled Bubble Growth*
John C. Berg, Andreas Acrivos, and Michel Boudart, *Evaporation Convection*
H. M. Tsuchiya, A. G. Fredrickson, and R. Aris, *Dynamics of Microbial Cell Populations*
Samuel Sideman, *Direct Contact Heat Transfer between Immiscible Liquids*
Howard Brenner, *Hydrodynamic Resistance of Particles at Small Reynolds Numbers*

Volume 7

Robert S. Brown, Ralph Anderson, and Larry J. Shannon, *Ignition and Combustion of Solid Rocket Propellants*
Knud Østergaard, *Gas–Liquid–Particle Operations in Chemical Reaction Engineering*
J. M. Prausnilz, *Thermodynamics of Fluid–Phase Equilibria at High Pressures*
Robert V. Macbeth, *The Burn-Out Phenomenon in Forced-Convection Boiling*
William Resnick and Benjamin Gal-Or, *Gas–Liquid Dispersions*

Volume 8

C. E. Lapple, *Electrostatic Phenomena with Particulates*
J. R. Kittrell, *Mathematical Modeling of Chemical Reactions*
W. P. Ledet and D. M. Himmelblau, *Decomposition Procedures foe the Solving of Large Scale Systems*
R. Kumar and N. R. Kuloor, *The Formation of Bubbles and Drops*

Volume 9

Renato G. Bautista, *Hydrometallurgy*
Kishan B. Mathur and Norman Epstein, *Dynamics of Spouted Beds*
W. C. Reynolds, *Recent Advances in the Computation of Turbulent Flows*
R. E. Peck and D. T. Wasan, *Drying of Solid Particles and Sheets*

Volume 10

G. E. O'Connor and T. W. F. Russell, *Heat Transfer in Tubular Fluid–Fluid Systems*
P. C. Kapur, *Balling and Granulation*
Richard S. H. Mah and Mordechai Shacham, *Pipeline Network Design and Synthesis*
J. Robert Selman and Charles W. Tobias, *Mass-Transfer Measurements by the Limiting-Current Technique*

Volume 11

Jean-Claude Charpentier, *Mass-Transfer Rates in Gas–Liquid Absorbers and Reactors*
Dee H. Barker and C. R. Mitra, *The Indian Chemical Industry—Its Development and Needs*
Lawrence L. Tavlarides and Michael Stamatoudis, *The Analysis of Interphase Reactions and Mass Transfer in Liquid–Liquid Dispersions*
Terukatsu Miyauchi, Shintaro Furusaki, Shigeharu Morooka, and Yoneichi Ikeda, *Transport Phenomena and Reaction in Fluidized Catalyst Beds*

Volume 12

C. D. Prater, J, Wei, V. W. Weekman, Jr., and B. Gross, *A Reaction Engineering Case History: Coke Burning in Thermofor Catalytic Cracking Regenerators*
Costel D. Denson, *Stripping Operations in Polymer Processing*
Robert C. Reid, *Rapid Phase Transitions from Liquid to Vapor*
John H. Seinfeld, *Atmospheric Diffusion Theory*

Volume 13

Edward G. Jefferson, *Future Opportunities in Chemical Engineering*
Eli Ruckenstein, *Analysis of Transport Phenomena Using Scaling and Physical Models*
Rohit Khanna and John H. Seinfeld, *Mathematical Modeling of Packed Bed Reactors: Numerical Solutions and Control Model Development*
Michael P. Ramage, Kenneth R. Graziano, Paul H. Schipper, Frederick J. Krambeck, and Byung C. Choi, *KINPTR (Mobil's Kinetic Reforming Model): A Review of Mobil's Industrial Process Modeling Philosophy*

Volume 14

Richard D. Colberg and Manfred Morari, *Analysis and Synthesis of Resilient Heat Exchange Networks*
Richard J. Quann, Robert A. Ware, Chi-Wen Hung, and James Wei, *Catalytic Hydrometallation of Petroleum*
Kent David, *The Safety Matrix: People Applying Technology to Yield Safe Chemical Plants and Products*

Volume 15

Pierre M. Adler, Ali Nadim, and Howard Brenner, *Rheological Models of Suspensions*
Stanley M. Englund, *Opportunities in the Design of Inherently Safer Chemical Plants*
H. J. Ploehn and W. B. Russel, *Interations between Colloidal Particles and Soluble Polymers*

Volume 16

Perspectives in Chemical Engineering: Research and Education

Clark K. Colton, *Editor*

Historical Perspective and Overview

L. E. Scriven, *On the Emergence and Evolution of Chemical Engineering*
Ralph Landau, *Academic—industrial Interaction in the Early Development of Chemical Engineering*
James Wei, *Future Directions of Chemical Engineering*

Fluid Mechanics and Transport

L. G. Leal, *Challenges and Opportunities in Fluid Mechanics and Transport Phenomena*
William B. Russel, *Fluid Mechanics and Transport Research in Chemical Engineering*
J. R. A. Pearson, *Fluid Mechanics and Transport Phenomena*

Thermodynamics

Keith E. Gubbins, *Thermodynamics*

J. M. Prausnitz, *Chemical Engineering Thermodynamics: Continuity and Expanding Frontiers*
H. Ted Davis, *Future Opportunities in Thermodynamics*

Kinetics, Catalysis, and Reactor Engineering

Alexis T. Bell, *Reflections on the Current Status and Future Directions of Chemical Reaction Engineering*
James R. Katzer and S. S. Wong, *Frontiers in Chemical Reaction Engineering*
L. Louis Hegedus, *Catalyst Design*

Environmental Protection and Energy

John H. Seinfeld, *Environmental Chemical Engineering*
T. W. F. Russell, *Energy and Environmental Concerns*
Janos M. Beer, Jack B. Howard, John P. Longwell, and Adel F. Sarofim, *The Role of Chemical Engineering in Fuel Manufacture and Use of Fuels*

Polymers

Matthew Tirrell, *Polymer Science in Chemical Engineering*
Richard A. Register and Stuart L. Cooper, *Chemical Engineers in Polymer Science: The Need for an Interdisciplinary Approach*

Microelectronic and Optical Material

Larry F. Thompson, *Chemical Engineering Research Opportunities in Electronic and Optical Materials Research*
Klavs F. Jensen, *Chemical Engineering in the Processing of Electronic and Optical Materials: A Discussion*

Bioengineering

James E. Bailey, *Bioprocess Engineering*
Arthur E. Humphrey, *Some Unsolved Problems of Biotechnology*
Channing Robertson, *Chemical Engineering: Its Role in the Medical and Health Sciences*

Process Engineering

Arthur W. Westerberg, *Process Engineering*
Manfred Morari, *Process Control Theory: Reflections on the Past Decade and Goals for the Next*
James M. Douglas, *The Paradigm After Next*
George Stephanopoulos, *Symbolic Computing and Artificial Intelligence in Chemical Engineering: A New Challenge*

The Identity of Our Profession

Morton M. Denn, *The Identity of Our Profession*

Volume 17

Y. T. Shah, *Design Parameters for Mechanically Agitated Reactors*
Mooson Kwauk, *Particulate Fluidization: An Overview*

Volume 18

E. James Davis, *Microchemical Engineering: The Physics and Chemistry of the Microparticle*
Selim M. Senkan, *Detailed Chemical Kinetic Modeling: Chemical Reaction Engineering of the Future*
Lorenz T. Biegler, *Optimization Strategies for Complex Process Models*

Volume 19

Robert Langer, *Polymer Systems for Controlled Release of Macromolecules, Immobilized Enzyme Medical Bioreactors, and Tissue Engineering*
J. J. Linderman, P. A. Mahama, K. E. Forsten, and D. A. Lauffenburger, *Diffusion and Probability in Receptor Binding and Signaling*
Rakesh K. Jain, *Transport Phenomena in Tumors*
R. Krishna, *A Systems Approach to Multiphase Reactor Selection*
David T. Allen, *Pollution Prevention: Engineering Design at Macro-, Meso-, and Microscales*
John H. Seinfeld, Jean M. Andino, Frank M. Bowman, Hali J. L. Forstner, and Spyros Pandis, *Tropospheric Chemistry*

Volume 20

Arthur M. Squires, *Origins of the Fast Fluid Bed*
Yu Zhiqing, *Application Collocation*
Youchu Li, *Hydrodynamics*
Li Jinghai, *Modeling*
Yu Zhiqing and Jin Yong, *Heat and Mass Transfer*
Mooson Kwauk, *Powder Assessment*
Li Hongzhong, *Hardware Development*
Youchu Li and Xuyi Zhang, *Circulating Fluidized Bed Combustion*
Chen Junwu, Cao Hanchang, and Liu Taiji, *Catalyst Regeneration in Fluid Catalytic Cracking*

Volume 21

Christopher J. Nagel, Chonghum Han, and George Stephanopoulos, *Modeling Languages: Declarative and Imperative Descriptions of Chemical Reactions and Processing Systems*
Chonghun Han, George Stephanopoulos, and James M. Douglas, *Automation in Design: The Conceptual Synthesis of Chemical Processing Schemes*
Michael L. Mavrovouniotis, *Symbolic and Quantitative Reasoning: Design of Reaction Pathways through Recursive Satisfaction of Constraints*
Christopher Nagel and George Stephanopoulos, *Inductive and Deductive Reasoning: The Case of Identifying Potential Hazards in Chemical Processes*
Keven G. Joback and George Stephanopoulos, *Searching Spaces of Discrete Soloutions: The Design of Molecules Processing Desired Physical Properties*

Volume 22

Chonghun Han, Ramachandran Lakshmanan, Bhavik Bakshi, and George Stephanopoulos, *Nonmonotonic Reasoning: The Synthesis of Operating Procedures in Chemical Plants*
Pedro M. Saraiva, *Inductive and Analogical Learning: Data-Driven Improvement of Process Operations*
Alexandros Koulouris, Bhavik R. Bakshi and George Stephanopoulos, *Empirical Learning through Neural Networks: The Wave-Net Solution*
Bhavik R. Bakshi and George Stephanopoulos, *Reasoning in Time: Modeling, Analysis, and Pattern Recognition of Temporal Process Trends*
Matthew J. Realff, *Intelligence in Numerical Computing: Improving Batch Scheduling Algorithms through Explanation-Based Learning*

Volume 23

Jeffrey J. Siirola, *Industrial Applications of Chemical Process Synthesis*
Arthur W. Westerberg and Oliver Wahnschafft, *The Synthesis of Distillation-Based Separation Systems*
Ignacio E. Grossmann, *Mixed-Integer Optimization Techniques for Algorithmic Process Synthesis*
Subash Balakrishna and Lorenz T. Biegler, *Chemical Reactor Network Targeting and Integration: An Optimization Approach*
Steve Walsh and John Perkins, *Operability and Control inn Process Synthesis and Design*

Volume 24

Raffaella Ocone and Gianni Astarita, *Kinetics and Thermodynamics in Multicomponent Mixtures*
Arvind Varma, Alexander S. Rogachev, Alexandra S. Mukasyan, and Stephen Hwang, *Combustion Synthesis of Advanced Materials: Principles and Applications*
J. A. M. Kuipers and W. P. Mo, van Swaaij, *Computational Fluid Dynamics Applied to Chemical Reaction Engineering*
Ronald E. Schmitt, Howard Klee, Debora M. Sparks, and Mahesh K. Podar, *Using Relative Risk Analysis to Set Priorities for Pollution Prevention at a Petroleum Refinery*

Volume 25

J. F. Davis, M. J. Piovoso, K. A. Hoo, and B. R. Bakshi, *Process Data Analysis and Interpretation*
J. M. Ottino, P. DeRoussel, S., Hansen, and D. V. Khakhar, *Mixing and Dispersion of Viscous Liquids and Powdered Solids*
Peter L. Silverston, Li Chengyue, Yuan Wei-Kang, *Application of Periodic Operation to Sulfur Dioxide Oxidation*

Volume 26

J. B. Joshi, N. S. Deshpande, M. Dinkar, and D. V. Phanikumar, *Hydrodynamic Stability of Multiphase Reactors*
Michael Nikolaou, *Model Predictive Controllers: A Critical Synthesis of Theory and Industrial Needs*

Volume 27

William R. Moser, Josef Find, Sean C. Emerson, and Ivo M, Krausz, *Engineered Synthesis of Nanostructure Materials and Catalysts*
Bruce C. Gates, *Supported Nanostructured Catalysts: Metal Complexes and Metal Clusters*
Ralph T. Yang, *Nanostructured Absorbents*
Thomas J. Webster, *Nanophase Ceramics: The Future Orthopedic and Dental Implant Material*
Yu-Ming Lin, Mildred S. Dresselhaus, and Jackie Y. Ying, *Fabrication, Structure, and Transport Properties of Nanowires*

Volume 28

Qiliang Yan and Juan J. DePablo, *Hyper-Parallel Tempering Monte Carlo and Its Applications*
Pablo G. Debenedetti, Frank H. Stillinger, Thomas M. Truskett, and Catherine P. Lewis, *Theory of Supercooled Liquids and Glasses: Energy Landscape and Statistical Geometry Perspectives*

Michael W. Deem, *A Statistical Mechanical Approach to Combinatorial Chemistry*
Venkat Ganesan and Glenn H. Fredrickson, *Fluctuation Effects in Microemulsion Reaction Media*
David B. Graves and Cameron F. Abrams, *Molecular Dynamics Simulations of Ion–Surface Interactions with Applications to Plasma Processing*
Christian M. Lastoskie and Keith E. Gubbins, *Characterization of Porous Materials Using Molecular Theory and Simulation*
Dimitrios Maroudas, *Modeling of Radical-Surface Interactions in the Plasma-Enhanced Chemical Vapor Deposition of Silicon Thin Films*
Sanat Kumar, M. Antonio Floriano, and Athanassiors Z. Panagiotopoulos, *Nanostructured Formation and Phase Separation in Surfactant Solutions*
Stanley I. Sandler, Amadeu K. Sum, and Shiang-Tai Lin, *Some Chemical Engineering Applications of Quantum Chemical Calculations*
Bernhardt L. Trout, *Car-Parrinello Methods in Chemical Engineering: Their Scope and potential*
R. A. van Santen and X. Rozanska, *Theory of Zeolite Catalysis*
Zhen-Gang Wang, *Morphology, Fluctuation, Metastability and Kinetics in Ordered Block Copolymers*

Volume 29

Michael V. Sefton, *The New Biomaterials*
Kristi S. Anseth and Kristyn S. Masters, *Cell–Material Interactions*
Surya K. Mallapragada and Jennifer B. Recknor, *Polymeric Biomaterias for Nerve Regeneration*
Anthony M. Lowman, Thomas D. Dziubla, Petr Bures, and Nicholas A. Peppas, *Structural and Dynamic Response of Neutral and Intelligent Networks in Biomedical Environments*
F. Kurtis Kasper and Antonios G. Mikos, *Biomaterials and Gene Therapy*
Balaji Narasimhan and Matt J. Kipper, *Surface-Erodible Biomaterials for Drug Delivery*

Volume 30

Dionisio Vlachos, *A Review of Multiscale Analysis: Examples from System Biology, Materials Engineering, and Other Fluids-Surface Interacting Systems*
Lynn F. Gladden, M.D. Mantle and A.J. Sederman, *Quantifying Physics and Chemistry at Multiple Length-Scales using Magnetic Resonance Techniques*
Juraj Kosek, Frantisek Steěpánek, and Miloš Marek, *Modelling of Transport and Transformation Processes in Porous and Multiphase Bodies*
Vemuri Balakotaiah and Saikat Chakraborty, *Spatially Averaged Multiscale Models for Chemical Reactors*

Volume 31

Yang Ge and Liang-Shih Fan, *3-D Direct Numerical Simulation of Gas–Liquid and Gas–Liquid–Solid Flow Systems Using the Level-Set and Immersed-Boundary Methods*
M.A. van der Hoef, M. Ye, M. van Sint Annaland, A.T. Andrews IV, S. Sundaresan, and J.A.M. Kuipers, *Multiscale Modeling of Gas-Fluidized Beds*
Harry E.A. Van den Akker, *The Details of Turbulent Mixing Process and their Simulation*
Rodney O. Fox, *CFD Models for Analysis and Design of Chemical Reactors*
Anthony G. Dixon, Michiel Nijemeisland, and E. Hugh Stitt, *Packed Tubular Reactor Modeling and Catalyst Design Using Computational Fluid Dynamics*

Volume 32

William H. Green, Jr., *Predictive Kinetics: A New Approach for the 21st Century*

Mario Dente, Giulia Bozzano, Tiziano Faravelli, Alessandro Marongiu, Sauro Pierucci and Eliseo Ranzi, *Kinetic Modelling of Pyrolysis Processes in Gas and Condensed Phase*

Mikhail Sinev, Vladimir Arutyunov and Andrey Romanets, *Kinetic Models of C_1–C_4 Alkane Oxidation as Applied to Processing of Hydrocarbon Gases: Principles, Approaches and Developments*

Pierre Galtier, *Kinetic Methods in Petroleum Process Engineering*

COLOR PLATE SECTION

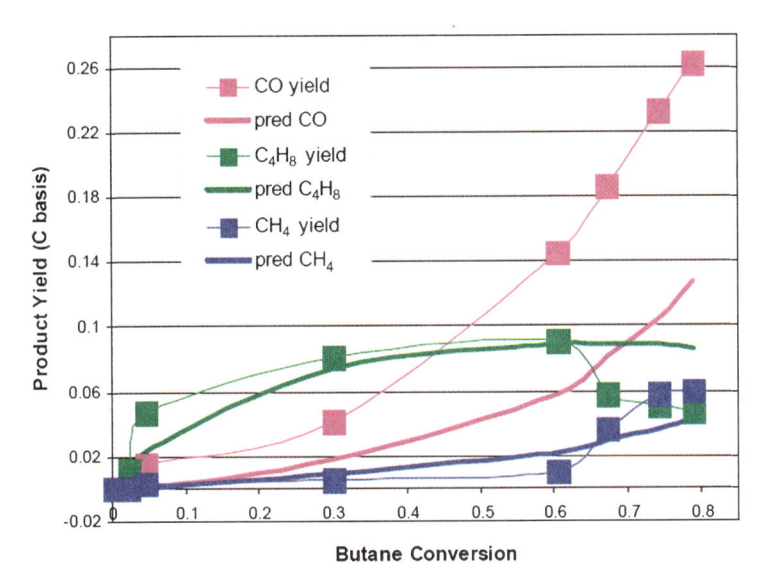

PLATE 1. Model predictions vs. experimental measurements of Wilk *et al.* (1995) from butane oxidation at 715 K. The predicted yields are generally within a factor of two of the experimental data, reflecting roughly factor of two uncertainties in rate constant estimates. As is typical, the discrepancies are largest at high conversions, both because the errors in the parameters cumulate, and because the model may be missing some reactions of the minor byproducts (For Black and White version, see page 28).

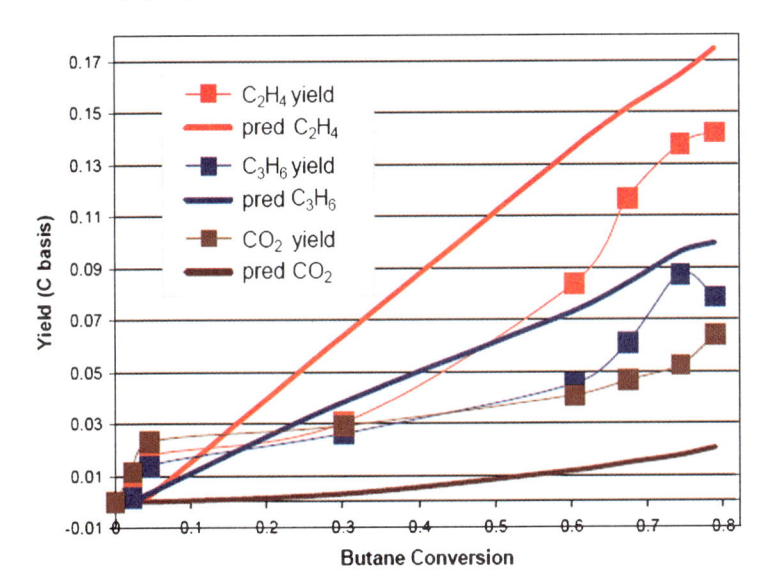

PLATE 2. Model predictions vs. experimental measurements of Wilk *et al.* (1995) for ethene, propene, and CO_2, same conditions as Fig. 12. The large discrepancy in the CO_2 predictions at the lowest conversions suggests that the model does not accurately represent the true boundary conditions at the inlet (For Black and White version, see page 29).

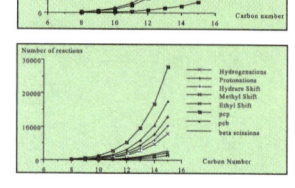

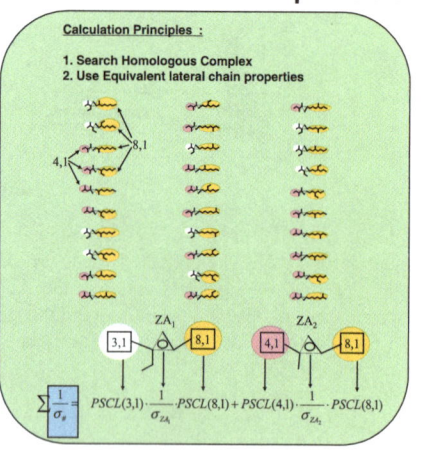

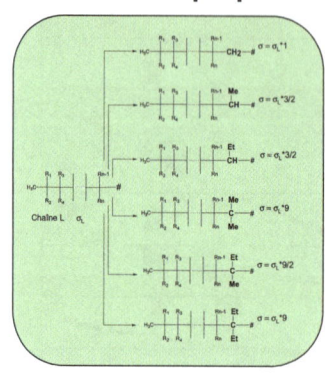

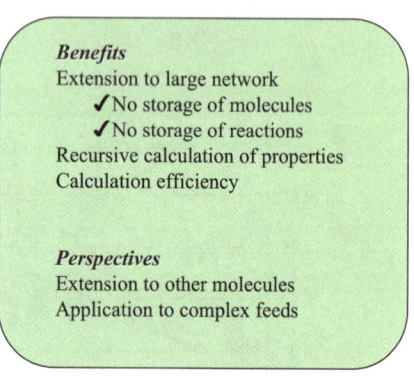

PLATE 3. Single-events modelling—extension to large networks (For Black and White version, see page 295).

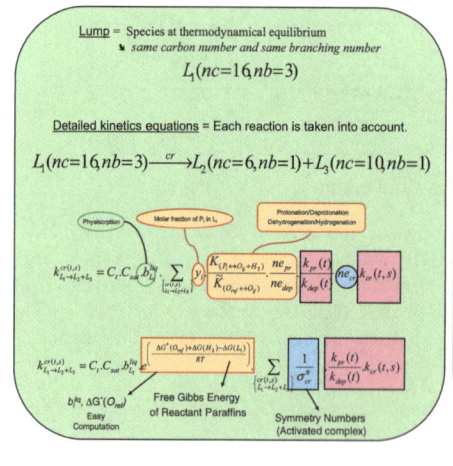

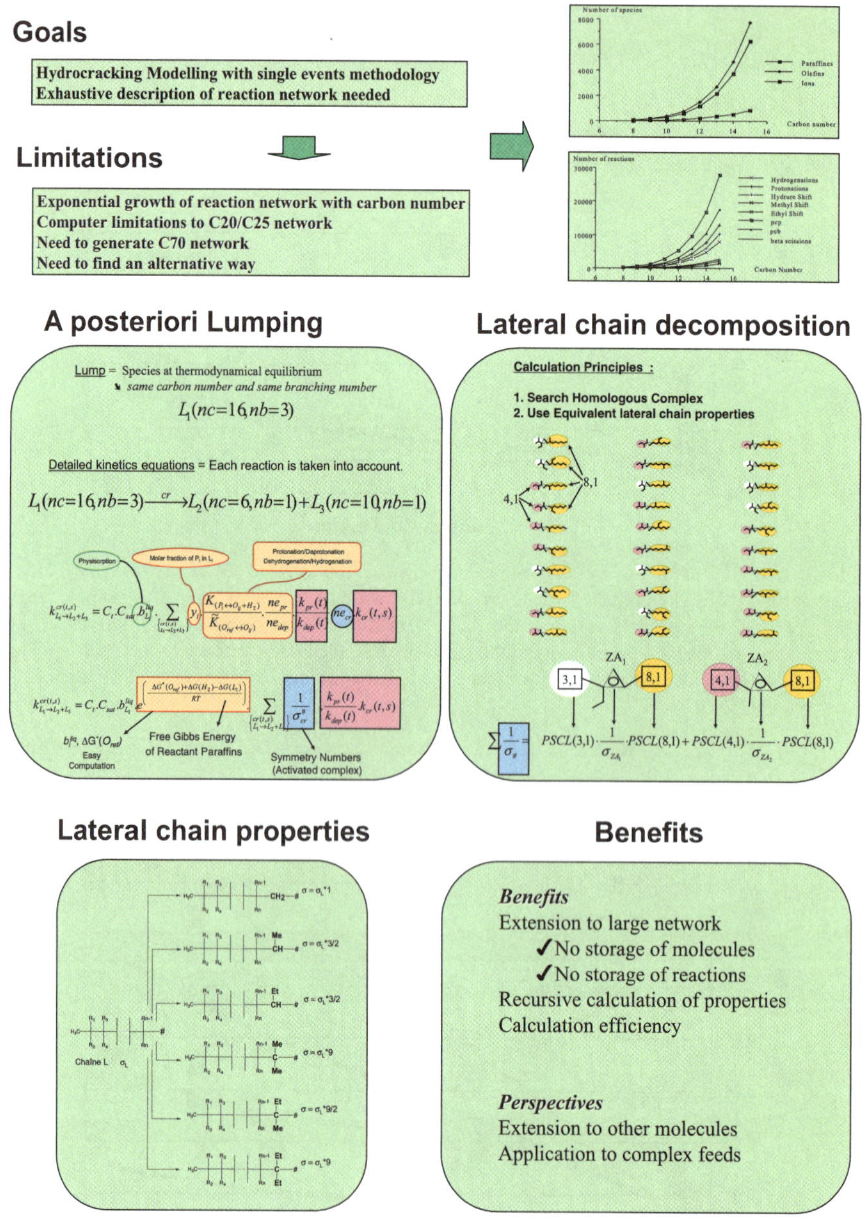